Green Chemistry and Sustainable Technology

Series Editors

Prof. Liang-Nian He
State Key Laboratory of Elemento-Organic Chemistry, Nankai University, Tianjin, China

Prof. Robin D. Rogers
Center for Green Manufacturing, Department of Chemistry, The University of Alabama, Tuscaloosa, USA

Prof. Dangsheng Su
*Shenyang National Laboratory for Materials Science, Institute of Metal Research, Chinese Academy of Sciences, Shenyang, China
and
Department of Inorganic Chemistry, Fritz Haber Institute of the Max Planck Society, Berlin, Germany*

Prof. Pietro Tundo
Department of Environmental Sciences, Informatics and Statistics, Ca' Foscari University of Venice, Venice, Italy

Prof. Z. Conrad Zhang
Dalian Institute of Chemical Physics, Chinese Academy of Sciences, Dalian, China

Green Chemistry and Sustainable Technology

Aims and Scope

The series *Green Chemistry and Sustainable Technology* aims to present cutting-edge research and important advances in green chemistry, green chemical engineering and sustainable industrial technology. The scope of coverage includes (but is not limited to):

- Environmentally benign chemical synthesis and processes (green catalysis, green solvents and reagents, atom-economy synthetic methods etc.)
- Green chemicals and energy produced from renewable resources (biomass, carbon dioxide etc.)
- Novel materials and technologies for energy production and storage (biofuels and bioenergies, hydrogen, fuel cells, solar cells, lithium-ion batteries etc.)
- Green chemical engineering processes (process integration, materials diversity, energy saving, waste minimization, efficient separation processes etc.)
- Green technologies for environmental sustainability (carbon dioxide capture, waste and harmful chemicals treatment, pollution prevention, environmental redemption etc.)

The series *Green Chemistry and Sustainable Technology* is intended to provide an accessible reference resource for postgraduate students, academic researchers and industrial professionals who are interested in green chemistry and technologies for sustainable development.

More information about this series at http://www.springer.com/series/11661

Farid Chemat • Maryline Abert Vian
Editors

Alternative Solvents for Natural Products Extraction

Springer

Editors
Farid Chemat
Maryline Abert Vian
Green Extraction Team
Université d'Avignon et des
  Pays de Vaucluse
INRA, UMR408
Avignon, France

ISSN 2196-6982 ISSN 2196-6990 (electronic)
ISBN 978-3-662-43627-1 ISBN 978-3-662-43628-8 (eBook)
DOI 10.1007/978-3-662-43628-8
Springer Heidelberg New York Dordrecht London

Library of Congress Control Number: 2014947242

© Springer-Verlag Berlin Heidelberg 2014
This work is subject to copyright. All rights are reserved by the Publisher, whether the whole or part of the material is concerned, specifically the rights of translation, reprinting, reuse of illustrations, recitation, broadcasting, reproduction on microfilms or in any other physical way, and transmission or information storage and retrieval, electronic adaptation, computer software, or by similar or dissimilar methodology now known or hereafter developed. Exempted from this legal reservation are brief excerpts in connection with reviews or scholarly analysis or material supplied specifically for the purpose of being entered and executed on a computer system, for exclusive use by the purchaser of the work. Duplication of this publication or parts thereof is permitted only under the provisions of the Copyright Law of the Publisher's location, in its current version, and permission for use must always be obtained from Springer. Permissions for use may be obtained through RightsLink at the Copyright Clearance Center. Violations are liable to prosecution under the respective Copyright Law.
The use of general descriptive names, registered names, trademarks, service marks, etc. in this publication does not imply, even in the absence of a specific statement, that such names are exempt from the relevant protective laws and regulations and therefore free for general use.
While the advice and information in this book are believed to be true and accurate at the date of publication, neither the authors nor the editors nor the publisher can accept any legal responsibility for any errors or omissions that may be made. The publisher makes no warranty, express or implied, with respect to the material contained herein.

Printed on acid-free paper

Springer is part of Springer Science+Business Media (www.springer.com)

Preface

Nowadays, we cannot find a production process in the perfume, cosmetic, pharmaceutical, food ingredients, nutraceuticals, biofuel, or fine chemicals industries which do not use solvent extraction processes, such as: maceration, percolation, steam or hydro-distillation, decoction, infusion, and Soxhlet extraction. In the food industry, besides the well-established huge extraction processes of sugar beet and sugar cane, and the preparation of decaffeinated tea and coffee, many food formulations have been developed by adding plant extracts used as ingredients (such as antioxidants, antimicrobials, colors, aromas, pH regulators, texturing agents) and nutraceutical concentrates. Bioactive compounds or their precursors (antibiotics, chemopreventive agents, alkaloids, etc.) are extracted by the pharmaceutical industry, either with conventional methods or modern technologies. Recent trends in extraction techniques have largely focused on finding solutions that minimize the use of solvents or to find alternatives for petroleum solvents. This, of course, must be achieved while also enabling process intensification and a cost-effective production of high-quality extracts.

For example, in the perfume industry, extraction of natural products was considered "clean" when compared with heavy chemical industries, but researchers and professional specialists found that its environmental impact is far greater than first appeared. The overall environmental impact of an industrial extraction cycle is not easy to estimate; however it is known that it requires at least 50 % of the energy of the whole industrial process. In spite of the high-energy consumption and the large amount of solvents, often the yield is indicated in decimals. For example, a single milliliter of rose absolute (used in famous perfumes) that weighs less than 1 g requires not only more than 1 kg of fresh roses as raw material (which became a chemical waste) but also a large quantity of solvents (n-hexane to produce the concrete and then alcohol to produce the absolute), energy (fossil) to evaporate the large quantity of solvents, and water as cooling and cleaning agent which becomes chemical wastewater.

The objective in preparing this book is to provide a complete picture of current knowledge on alternative and green solvents used at laboratory and industrial scale for extraction of natural product in terms of innovation, original methods and procedures, alternative solvents, and safe products. It will provide the necessary theoretical background and details about solvent extraction focused on solid–liquid, techniques, processes, mechanisms, protocols, industrial applications, safety precautions, and environmental impacts. This book is aimed for professionals from industry, academician's researchers and lecturers engaged into extraction engineering or natural product chemistry, and graduate-level students.

This book was prepared by a team of chemists, biochemists, chemical engineers, physicians, and food technologists who have extensive personal experience in research of innovative extraction techniques at the laboratory and industrial scales. All the collaborating authors are totally convinced that this book is the starting point for future collaborations in this new area, "alternative solvents for extraction of natural products," between research, industry, and education.

We wish to thank sincerely all of our colleagues who have collaborated in the writing of this book. We hope to express them our scientific gratitude for agreeing to devote their competence and time to ensure the success of this book.

Avignon, France
Farid Chemat
Maryline Abert Vian

Contents

1. **In Silico Search for Alternative Green Solvents** 1
   Laurianne Moity, Morgan Durand, Adrien Benazzouz,
   Valérie Molinier, and Jean-Marie Aubry

2. **Solvent-Free Extraction: Myth or Reality?** 25
   Maryline Abert Vian, Tamara Allaf, Eugene Vorobiev,
   and Farid Chemat

3. **Supercritical Fluid Extraction: A Global Perspective
   of the Fundamental Concepts of this Eco-Friendly
   Extraction Technique** 39
   Susana P. Jesus and M. Angela A. Meireles

4. **Subcritical Water as a Green Solvent for Plant Extraction** 73
   Mustafa Zafer Özel and Fahrettin Göğüş

5. **Liquefied Dimethyl Ether: An Energy-Saving, Green
   Extraction Solvent** 91
   Peng Li and Hisao Makino

6. **Ethyl Lactate Main Properties, Production Processes,
   and Applications** 107
   Carla S.M. Pereira and Alírio E. Rodrigues

7. **Ionic Liquids as Alternative Solvents for Extraction
   of Natural Products** 127
   Milen G. Bogdanov

8. **Enzymatic Aqueous Extraction (EAE)** 167
   Lionel Muniglia, Nathalie Claisse, Paul-Hubert Baudelet,
   and Guillaume Ricochon

9. **Terpenes as Green Solvents for Natural Products Extraction** 205
   Chahrazed Boutekedjiret, Maryline Abert Vian,
   and Farid Chemat

10 Emulsion Extraction of Bio-products: Influence of Bio-diluents on Extraction of Gallic Acid 221
Ka Ho Yim, Moncef Stambouli, and Dominique Pareau

11 Gluconic Acid as a New Green Solvent for Recovery of Polysaccharides by Clean Technologies 237
Juan Carlos Contreras-Esquivel, Maria-Josse Vasquez-Mejia, Adriana Sañudo-Barajas, Oscar F. Vazquez-Vuelvas, Humberto Galindo-Musico, Rosabel Velez-de-la-Rocha, Cecilia Perez-Cruz, and Nagamani Balagurusamy

12 2-Methyltetrahydrofuran: Main Properties, Production Processes, and Application in Extraction of Natural Products 253
Anne-Gaëlle Sicaire, Maryline Abert Vian, Aurore Filly, Ying Li, Antoine Bily, and Farid Chemat

13 Innovative Technologies Used at Pilot Plant and Industrial Scales in Water-Extraction Processes 269
Linghua Meng and Yves Lozano

Editors and Contributors

Editors

Farid Chemat is a Full Professor of Chemistry at Avignon University (France), Director of GREEN Extraction Team (alternative extraction techniques and solvents), Codirector of ORTESA LabCom research unit Naturex-UAPV, and scientific coordinator of "France Eco-Extraction," dealing with the dissemination of research and education on green extraction technologies. Born in 1968, he received his Ph.D. (1994) in Process Engineering from the Institut National Polytechnique de Toulouse, France. Following postdoctoral research work with Prolabo-Merck (1995–1997), he spent 2 years (1997–1999) as senior researcher at the University of Wageningen (the Netherlands). In 1999, he moved to the University of La Réunion (France DOM) to work as an assistant professor, and since 2006 he holds the position of Professor at the University of Avignon (France). His main research interests are focused on innovative and sustainable extraction techniques, protocols, and solvents (especially microwave, ultrasound, and bio-based solvents) for food, pharmaceutical, fine chemistry, biofuel, and cosmetic applications. His research activities are documented in more than 140 scientific peer-reviewed papers, 9 books and 7 patents.

Maryline Abert Vian born in 1974, she received her Ph.D. (2000) in Organic Chemistry at the University of Avignon. She spent 4 years (2000–2004) as junior researcher with industrial companies. In 2005, she moved to the University of Avignon (France) to start her independent academic career. She obtained her "Habilitation à Diriger des Recherches" in 2011 in food and natural product chemistry, and since she managed several French programs in the field of research and industrial application of alternative solvents applied for extraction of valuable compounds and biofuels from microorganisms (microalgae, yeast, etc.) with several industrial partners such as Airbus or GDF-Suez. Her research activity is documented by more than 25 scientific peer-reviewed papers and about 30 communications for scientific meetings, 9 book chapters, and 2 patents. Her research primarily focuses on the solvent extraction and analysis of natural products and has paved the way for new extraction techniques with bio-based solvents.

Contributors

Maryline Abert Vian* Green Extraction Team, Université d'Avignon et des Pays de Vaucluse, INRA, UMR 408, Avignon, France

Jean-Marie Aubry* EA 4478 Chimie Moléculaire et Formulation, Université of Lille, USTL, ENSCL, Villeneuve d'Ascq, France

Tamara Allaf ABCAR-DIC, La Rochelle, France

Nagamani Balagurusamy School of Biological Sciences, Universidad Autonoma de Coahuila, Torreon, Coahuila, Mexico

Paul-Hubert Baudelet Laboratoire Ingénierie des Biomolécules, Vandoeuvre Cedax, France

Adrien Benazzouz EA 4478 Chimie Moléculaire et Formulation, Université of Lille, USTL, ENSCL, Villeneuve d'Ascq, France

Antoine Bily R&D Director, Nutrition & Health, Naturex, Avignon, France

Milen G. Bogdanov* Faculty of Chemistry and Pharmacy, University of Sofia "St. Kl. Ohridski", Sofia, Bulgaria

*Main contact

Chahrazed Boutekedjiret* Laboratoire des Sciences et Techniques de l'Environnement (LSTE), École Nationale Polytechnique, Algiers, Algeria

Farid Chemat* Green Extraction Team, Université d'Avignon et des Pays de Vaucluse, INRA, UMR 408, Avignon, France

Nathalie Claisse Biolie SAS, Nancy Cedex, France

Juan Carlos Contreras-Esquivel* Laboratory of Applied Glycobiotechnology, Food Research Department, School of Chemistry, Universidad Autonoma de Coahuila, Saltillo, Coahuila, Mexico

Research and Development Center, Coyotefoods Biopolymer and Biotechnology Co., Saltillo, Coahuila, Mexico

Morgan Durand EA 4478 Chimie Moléculaire et Formulation, Université of Lille, USTL, ENSCL, Villeneuve d'Ascq, France

Aurore Filly Green Extraction Team, Université d'Avignon et des Pays de Vaucluse, INRA, UMR 408, Avignon, France

Humberto Galindo-Musico Laboratory of Applied Glycobiotechnology, Food Research Department, School of Chemistry, Universidad Autonoma de Coahuila, Saltillo, Coahuila, Mexico

Fahrettin Göğüş Food Engineering Department, Engineering Faculty, University of Gaziantep, Gaziantep, Turkey

Ka Ho Yim* Laboratoire Génie des Procédés et Matériaux, Ecole Centrale Paris, Châtenay-Malabry, France

Susana P. Jesus LASEFI/DEA/FEA (School of Food Engineering)/UNICAMP (University of Campinas), Campinas, SP, Brazil

Peng Li* Energy Engineering Research Laboratory, Central Research Institute of Electric Power Industry (CRIEPI), Yokosuka, Japan

Ying Li Green Extraction Team, Université d'Avignon et des Pays de Vaucluse, INRA, UMR 408, Avignon, France

Yves Lozano CIRAD, UMR CIRAD-110 INTREPID, Montpellier, France

Hisao Makino Energy Engineering Research Laboratory, Central Research Institute of Electric Power Industry (CRIEPI), Yokosuka, Japan

M. Angela A. Meireles* LASEFI/DEA/FEA (School of Food Engineering)/UNICAMP (University of Campinas), Campinas, SP, Brazil

Linghua Meng* Department of Pharmacy, School of Medicine, Shanghai Jiao Tong University, Shanghai, China

Laurianne Moity EA 4478 Chimie Moléculaire et Formulation, Université of Lille, USTL, ENSCL, Villeneuve d'Ascq, France

Valérie Molinier EA 4478 Chimie Moléculaire et Formulation, Université of Lille, USTL, ENSCL, Villeneuve d'Ascq, France

Lionel Muniglia* Laboratoire Ingénierie des Biomolécules, ENSAIA, Vandoeuvre Cedex, France

Mustafa Zafer Özel* Green Chemistry Centre of Excellence, Chemistry Department, University of York, York, UK

Dominique Pareau Laboratoire Génie des Procédés et Matériaux, Ecole Centrale Paris, Châtenay-Malabry, France

Carla S.M. Pereira LSRE – Laboratory of Separation and Reaction Engineering – Associate Laboratory LSRE/LCM, Faculdade de Engenharia, Universidade do Porto, Porto, Portugal

Cecilia Perez-Cruz Laboratory of Applied Glycobiotechnology, Food Research Department, School of Chemistry, Universidad Autonoma de Coahuila, Saltillo, Coahuila, Mexico

Guillaume Ricochon Biolie SAS, Nancy Cedax, France

Alírio E. Rodrigues* LSRE – Laboratory of Separation and Reaction Engineering – Associate Laboratory LSRE/LCM, Faculdade de Engenharia, Universidade do Porto, Porto, Portugal

Adriana Sañudo-Barajas Laboratory of Food Biochemistry, Centro de Investigación en Alimentacion y Desarrollo (CIAD)-AC, Culiacan, Sinaloa, Mexico

Anne-Gaëlle Sicaire Green Extraction Team, Université d'Avignon et des Pays de Vaucluse, INRA, UMR 408, Avignon, France

Moncef Stambouli Laboratoire Génie des Procédés et Matériaux, Ecole Centrale Paris, Châtenay-Malabry, France

Maria-Josse Vasquez-Mejia Laboratory of Applied Glycobiotechnology, Food Research Department, School of Chemistry, Universidad Autonoma de Coahuila, Saltillo, Coahuila, Mexico

Research and Development Center, Coyotefoods Biopolymer and Biotechnology Co., Saltillo, Coahuila, Mexico

Oscar F. Vazquez-Vuelvas School of Chemistry, Universidad de Colima, Coquimatlan, Colima, Mexico

Rosabel Velez-de-la-Rocha Laboratory of Food Biochemistry, Centro de Investigación en Alimentacion y Desarrollo (CIAD)-AC, Culiacan, Sinaloa, Mexico

Eugene Vorobiev Laboratoire Transformations Intégrées de la Matière Renouvelable, Équipe Technologies Agro-Industrielles, Université de Technologie de Compiègne (UTC), Compiègne, France

Chapter 1
In Silico Search for Alternative Green Solvents

Laurianne Moity, Morgan Durand, Adrien Benazzouz, Valérie Molinier, and Jean-Marie Aubry

Abstract The selection of the most appropriate alternative solvents requires efficient predictive tools that avoid resorting to time-consuming trial and error experiments. Several classifications of organic solvents exist but they most often require the knowledge of one or more experimental characteristics, which might be an obstacle in the case of emerging candidates. This chapter gives an overview of existing tools for the characterisation and classification of organic solvents and particular attention is given to purely predictive methods, such as the COnductor-like Screening MOdel for Real Solvents (COSMO-RS). A panorama of the currently available sustainable solvents is given, and these "green" alternatives are compared to the classical organic solvents, thanks to a completely in silico approach. Examples of substitutions are given to illustrate the methodology that can also be used to design new alternatives.

1.1 Tools for Solvent Selection

Solvents play an important role in a great number of unit operations in chemistry and chemical engineering. Resorting to solvents is usually required during a limited period of time during the process since they are most often expected to play a role of dissolvent, diluent, dispersant or extractant and should be removed afterwards. Nevertheless, the right choice of solvent is crucial, and through the ages, several methods for solvent selection have been developed. In former times, the choice of the most appropriate solvent was purely empirical and was often made through trial and error experiments and from empirical knowledge. This traditional approach to

---

L. Moity • M. Durand • A. Benazzouz • V. Molinier • J.-M. Aubry (✉)
EA 4478 Chimie Moléculaire et Formulation, University of Lille, USTL, ENSCL, F59652 Villeneuve d'Ascq, France
e-mail: laurianne.moity@hotmail.fr; durand_morgan@yahoo.fr; adrienbenazzouz@aol.com; Valerie.Molinier@univ-lille1.fr; Jean-Marie.Aubry@univ-lille1.fr

select a solvent usually followed the alchemist maxim *similia similibus solvuntur* that is still underlying popular contemporary approaches in use. Solvent effects were then related to the chemical structures, and several descriptors have been proposed to describe them.

1.1.1 Solvents Descriptors and Classifications

Organic chemists traditionally classify solvents as non-polar, aprotic polar and protic polar, according to their molecular structure and ability to establish hydrogen bonding. To refine this classification, solvent effects can be related to various kinds of descriptors that have evolved over time.

Initially, the only available quantitative descriptors were *physical*: enthalpy of vaporisation, dielectric constant, refractive index, boiling point, etc. However, quantifying solvent effects using physical descriptors demonstrated a moderate predictive power because these descriptors describe properly the bulk but neglect specific intermolecular interactions that are of the utmost importance whenever a second compound is added to the solvent. Therefore two types of solute-solvent interactions occur: the non-specific interactions (Van der Waals and the ion/dipole forces) and the specific interactions (hydrogen bond donor and/or hydrogen bond acceptor, electron pair donor/electron pair acceptor and solvophobic interactions) [1].

To assess these intermolecular forces, a solvent must be considered as a discontinuum, in which solvent molecules interact with each other or with the solute. For this purpose, well-chosen solutes, with a particular and quantifiable sensitivity to solvent effects, were used and allowed to access *empirical* descriptors that led to the emergence of numerous and useful empirical polarity scales, either uniparametric such as the ET(30) of Reichardt [1] or multi-parametric such as the solvatochromic parameters of Kamlet and Taft [2–4] or the Abraham parameters [5]. The last two approaches have been rationalised under the concept of linear solvation energy relationships (LSER) [6]. Recently, more than 180 polarity scales have been reviewed [7]. In the next section, the Hildebrand and Hansen solubility parameters will be emphasised since they are widely used in industry to compare and select solvents for various applications.

Over the past decade, purely *theoretical* descriptors have been introduced. They offer several advantages, the most important are being that they are easy to generate for any solvent and do not require any experiments. Different theoretical alternatives have been introduced as reviewed by Murray et al. [8]. Politzer and co-worker used electrostatic potentials computed on molecular surfaces to generate theoretical descriptors [9] that were found to be highly correlated to Kamlet and Taft solvatochromic parameters [10]. More recently, Katritzky et al. built QSPR (Quantitative Structure Property Relationship) models to predict 127 polarity scales based on theoretical descriptors. They carried out principal component analysis (PCA) of 100 solvent scales based on 703 solvents [11]. Relying on this extensive work, the authors emphasised that almost all theoretical descriptors can be related to one of the generally accepted types of intermolecular interactions.

1.1.2 Hansen Approach

The Hansen approach provides *empirical* descriptors, as presented above. At first, the solubility parameter δ_H was introduced by Hildebrand and Scott [12] and was defined as the square root of the cohesive energy density, correlated to the enthalpy of vaporisation ΔH_{vap} and to the molar volume, V (Eq. 1.1). As the difference between solute and solvent solubility parameters decreases, the tendency towards solubilisation increases.

$$\delta_H = \sqrt{\frac{\Delta H_{vap} - RT}{V}} \qquad (1.1)$$

The Hildebrand parameter was extended by Hansen by splitting it into three components called the Hansen solubility parameters (Eq. 1.2). They correspond to the three main molecular interactions, namely, dispersive (δ_d), polar (δ_p) and hydrogen-bonding contributions (δ_h).

$$\delta_H = \sqrt{\delta_d^2 + \delta_p^2 + \delta_h^2} \qquad (1.2)$$

The partial parameter for dispersive interactions, δ_d, is obtained from corresponding state principles, by considering the so-called homomorph of the molecule, while δ_p is derived from the ratio of the dipolar moment and the square root of the molar volume. The hydrogen-bonding contribution is calculated as the subtraction of δ_d and δ_p to the Hildebrand parameter. The three Hansen parameters can also be totally predicted by various group contribution methods such as the thermodynamically consistent model of Stefanis and Panayiotou [13]. Alternatively, the Hansen solubility parameters can be experimentally determined by individually mixing the solute in a ratio 1:10 to a proper set of solvents having a wide range of solubility parameters [14]. After 24 h of stirring at room temperature, the solubility is visually evaluated by a score ranging from 1 (soluble) to 6 (non-soluble). These scores are computed with a quality-to-fit function in order to build a solubility domain [15].

The three Hansen solubility parameters define a three-dimensional space, known as the Hansen space, in which all solvents and solutes can be located. A solute can be visualised as a point surrounded by its solubility sphere. All solvents and mixtures located inside this volume are likely to solubilise the solute. The closer the solute and solvent parameters are, the better the solubility is [16]. The solute-solvent distance, D, is defined according to Eq. 1.3:

$$D = \sqrt{4(\delta_{d_{solvent}} - \delta_{d_{solute}})^2 + (\delta_{p_{solvent}} - \delta_{p_{solute}})^2 + (\delta_{h_{solvent}} - \delta_{h_{solute}})^2} \qquad (1.3)$$

The ratio between the distance D and the radius R of the solubility sphere is called the "Relative Energy Difference" (*RED*) – see Eq. 1.4 – and allows a fast screening of molecules. *RED* < 1 indicates that a molecule is inside the sphere and is likely to have a high affinity with the solute while higher values of *RED* indicate a poor affinity:

$$\text{RED} = \frac{D}{R} \qquad (1.4)$$

The semi-empirical Hansen approach demonstrated its ability to correlate and predict the behaviour of solvents. It provides reasonable results for the description of molecular and macromolecular solubility and is thus a useful tool for various industrial applications ranging from polymer processing [14] to coatings [17] and cosmetics [16].

1.1.3 COSMO-RS Approach

The significant improvement in computational power and the sophistication of recent algorithms led to the possibility to use extensively *quantum* descriptors of the solvent effect. Cartier et al. showed that quantum chemistry provides a more accurate and more detailed description of electronic effects than empirical methods [18]. Thanks to a combination of a dielectric continuum solvation model and a thermodynamic treatment of the molecular interactions, Klamt developed a general approach in which a solvent can be treated in the liquid state. In the first step, the COnductor-like Screening MOdel (COSMO) [19], the solute molecule is considered to be embedded in a cavity that is surrounded by a virtual conductor. The COnductor-like Screening MOdel for Real Solvent (COSMO-RS) then allows the transfer from the state of the molecule embedded in a virtual conductor to a real solvent [20]. COSMO-RS has already been successfully used for the prediction or the modelisation of various properties in solution, as partition coefficients (for instance, octanol-water [21] and blood brain [22]), pKa [23] or solubilisation of cosmetic ingredients [16].

In a recent work [24], we have evaluated the potentialities of the COSMO-RS approach to generate quantum descriptors for an a priori classification of solvents. The descriptors obtained from COSMO-RS were treated by principal component analysis coupled with a clustering procedure to provide a classification of solvents. This a priori classification was compared to the one of Chastrette [25], who first proposed a classification of solvents by resorting to a multi-parametric statistical approach based on a selection of six physical descriptors – boiling point, molecular dipole moment, molecular refraction, index of refraction, Hildebrand solubility parameter and Kirkwood function – in conjunction with two microscopic quantum descriptors (HOMO and LUMO energies).

1 In Silico Search for Alternative Green Solvents

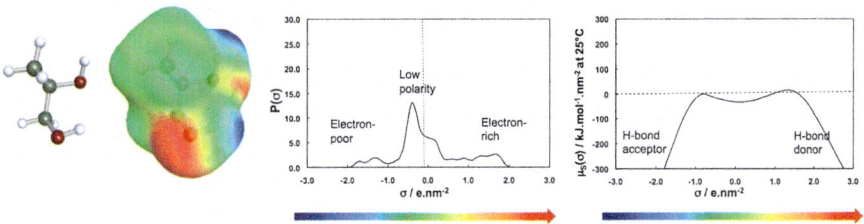

Fig. 1.1 σ-surface, σ-profile and σ-potential of 1,2-propanediol

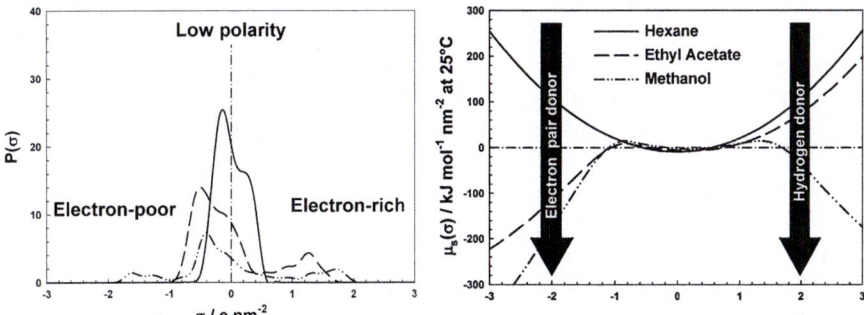

Fig. 1.2 σ-profile ($P(\sigma)$, *left*) and σ-potential ($\mu_S(\sigma)$, *right*) of three typical solvents: *n*-hexane (apolar), ethyl acetate (hydrogen bond acceptor) and methanol (amphiprotic) (Adapted from Ref. [24])

After DFT/COSMO geometry optimisations, COSMO surfaces can be generated via COSMOtherm. An example of COSMO surface is given in Fig. 1.1 for 1,2-propanediol. In the σ-surface representation, green to yellow codes the weakly polar surfaces, blue represents electron-deficient regions (δ^+) and red codes electron-rich regions (δ^-). This 3D information on the repartition of charge density on the molecular surface can be reduced to a histogram $P(\sigma)$ that expresses the redundancy of a surface density in a polarity interval. Such histograms have been defined as σ-profiles [21]. In the framework of COSMO-RS, it is also possible to generate the so-called σ-potential plots. This plot represents the chemical potential $\mu_S(\sigma)$ of a molecular surface fragment in a solvent S as a function of the polarisation charge density of this surface fragment (ranging from -3 to 3 e.nm^{-2}). This representation is of particular interest as it underlines the affinity of solvent S for a polarity of kind σ. The σ-surface, σ-profile and σ-potential of 1,2-propanediol are presented in Fig. 1.1.

Figure 1.2 shows the σ-profile and σ-potential of typical solvents. In the case of apolar solvents, exemplified by *n*-hexane, the σ-profile exhibits only one large shouldered peak centred close to 0, with a maximum at -0.1 e.nm^{-2} corresponding to the protons. Since no hydrogen bond interaction can occur with a solute surface, the corresponding σ-potential curve $\mu_S(\sigma)$ exhibits a U shape that is typical of apolar solvents: the contact between n-hexane and the molecular surface of a solute with a positive or negative charge density distant from 0 e.nm^{-2} will be energetically unfavourable ($\mu_S(\sigma) > 0$ kJ.nm^{-2}). Methanol is a typical example of an amphiprotic solvent. Its sp^3-oxygen induces at the same time a hydrogen acceptor and a hydrogen donor character. Consequently, on the σ-profile $P(\sigma)$, two secondary maxima can be observed beside the central peak corresponding to the carbon and hydrogens of the methyl group. One maximum, in the negative σ, corresponds to the hydrogen bond donor character, and the other one, in the positive σ, corresponds to the hydrogen bond acceptor character. Therefore, the σ-potential curve $\mu_S(\sigma)$ will be the opposite of that observed for apolar solvents in the regions distant from 0 e.nm^{-2}: negative, i.e. energetically favourable, $\mu_S(\sigma)$ are obtained for both negative and positive regions of a solute molecule, leading to a ∩-shaped curve. Ethyl acetate possesses only one hydrogen acceptor group due to its sp^2-oxygen. With the same analysis, it is easy to understand why its σ-potential curve is S-shaped, i.e. negative for $\sigma < 0$ and positive for $\sigma > 0$.

This is a purely qualitative interpretation of σ-profiles and σ-potentials. In actual fact, much more quantitative information is enclosed in these curves, and therefore we have used them to extract theoretical molecular descriptors of the solubilising properties further used as input parameters for solvent classification.

To attain a quantitative comparison of the σ-potential plots, 61 discrete values of $\mu_S(\sigma)$ given by COSMOtherm can be extracted for every 0.1 e.nm^{-2} increment within the interval -3 to 3 e.nm^{-2}. It was performed for the 153 solvents of the chosen dataset (the one of Chastrette [25]), and these 61 points were used to give a description of the solubilising properties of each solvent and thus make up a set of 61 descriptors. This set could be reduced by PCA to a smaller number of relevant descriptors since most of them contain redundant information. In our case, the vector space could be reduced to only four eigenvectors, still accounting for 96.4 % of the variance. By neglecting the fourth eigenvector, more than 85 % of the variance is still expressed and all solvents can then be positioned in a pseudo-3D space (F1, F2, F3) (see next paragraph).

A clustering procedure allowed gathering the 153 traditional solvents into ten classes for which the sigma profiles and sigma potentials are presented in Fig. 1.3. The description of these clusters will be discussed in the next paragraph. They are in good agreement with the ten classes defined by Chastrette, and they even allow a more accurate positioning of solvents that were mispositioned in this original work [24].

The σ-potentials derived from the COSMO-RS theory can thus be successfully employed to describe and classify solvents in a purely predictive manner, with a good consideration of hydrogen bond donor/acceptor interactions. This approach is of particular interest in the context of solvent design and will be addressed in Sect. 1.4.

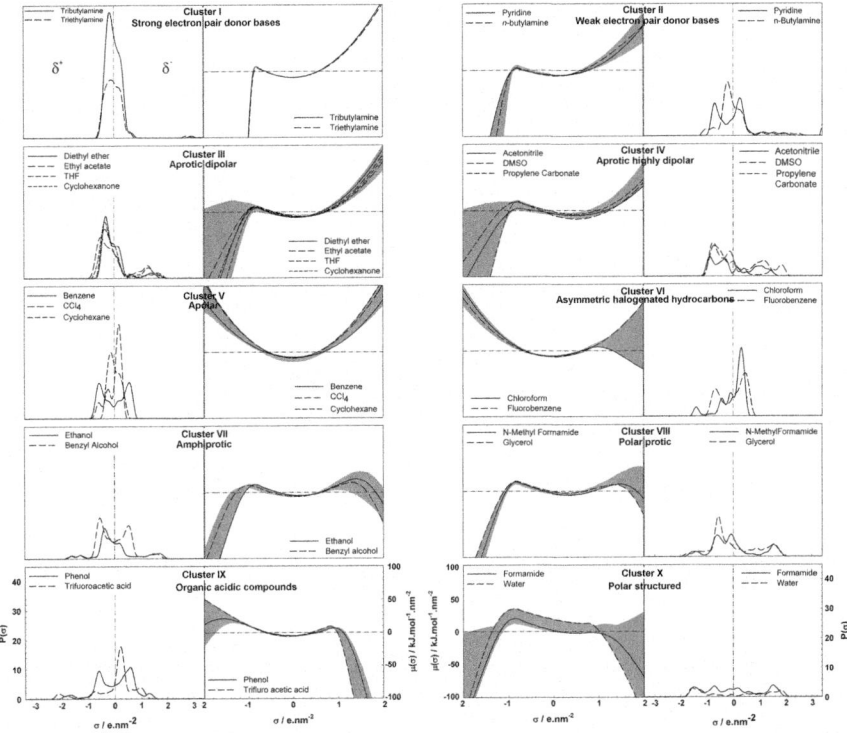

Fig. 1.3 σ-profile and σ-potential of typical solvents of the ten clusters. The *grey* regions show the dispersion of the σ-potential curves within each cluster (Adapted from Ref. [24])

1.2 Panorama of Current "Green Solvents"

1.2.1 Classes of "Green Solvents"

The adjective "sustainable" or "green" is used to describe different types of solvents including the ones that are produced from biomass feedstock and eco-friendly petrochemical-based solvents that are non-toxic and/or biodegradable. Figure 1.4 shows the different families of solvents that are generally considered as "green". It is worth noticing that the greenness of some solvents is questionable with regard to toxicity (e.g. ionic liquids) or biodegradability (fluorinated solvents and silicones).

The family of "eco-friendly" solvents is the most heteroclite one, since it gathers all kinds of solvents with a good EHS (Environment, Human, Safety) profile. These solvents may also be obtained by the valorisation of industrial by-products, as is the case for the dimethyl, diethyl and dibutyl esters of glutaric, succinic and adipic acid, and by-products of the nylon 6,6 manufacture (the so-called dibasic esters). Another example of an "eco-friendly" solvent is 3-methoxy-3-methyl-butan-1-ol (MMB)

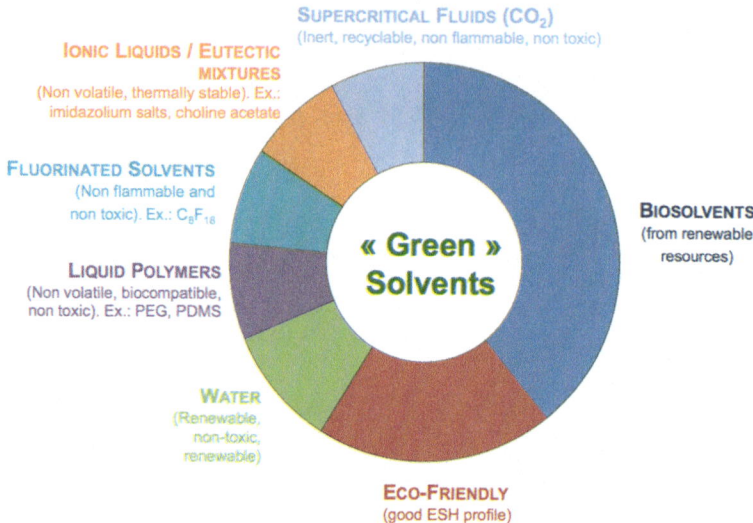

Fig. 1.4 The seven classes of solvents generally claimed as "green" solvents [26] (*EHS* Environment, Human, Safety)

which is a non-toxic and non-VOC solvent used in air freshener, household and industrial cleaner formulations. For the same reasons, the alkyl alcanolamides are considered as "eco-friendly" solvents, as well as some alkanes or dimethylsulfoxide.

Biosolvents mainly belong to three chemical families, namely, esters, alcohols and terpenes. These bio-based solvents are obtained by chemical or biochemical transformations of agro-synthons, i.e. defined molecules obtained from the biomass feedstock. Grains and oleaginous plants contain vegetable oils, which are converted to glycerol and fatty acids. They themselves are a source of solvents, giving rise to glycerol carbonate, glycerol triacetate, or vegetable oil methyl esters. The "sugar" platform (cellulose, hemicellulose, starch, sucrose) is the source of simple sugars and polyols that can be further transformed, chemically or enzymatically, to solvents. It should be stressed that, contrary to the "eco-friendly" family, all bio-based solvents do not have a good EHS profile. For instance, furfuraldehyde is a solvent readily obtained from various plant wastes, as corn stalk or sawdust, by acidic hydrolysis of hemicellulose into pentosidic units that are themselves dehydrated. This biosolvent is both toxic and carcinogenic. A first hydrogenation gives furfuryl alcohol that is also classified among the CMR (Carcinogenic Mutagenic Reprotoxic, in the European Union) substances, and a subsequent hydrogenation yields tetrahydrofurfuryl alcohol that has a good EHS profile.

Liquid polymers, such as polyethylene glycol, and silicone oils can also be considered as green solvents because of their non-volatility.

Finally, there is a growing interest for ionic liquids, i.e. salts with a melting point below 100 °C [27]. Typical ionic liquids have cations such as imidazolium, pyridinium or pyrrolidinium and anions such as hexafluorophosphate, tetrafluoroborate or triflate. They are frequently considered as green solvents because of their non-volatility. Nevertheless, the toxicity and biodegradability of such compounds are currently questioned [28], and most common ionic liquids such as [Bmim][BF4] (3-butyl-1-methylimidazolium tetrafluoroborate) do not show good biodegradation. Because the toxicity of ionic liquids is often associated to the cation part, choline has been investigated as a benign quaternary ammonium ion derived from renewable resources [29, 30].

In a recent work [31], we have listed 138 "green" solvents through the review of technical, commercial and scientific literature to provide a "panorama" of sustainable solvents through the COSMO-RS approach. Supercritical fluids have not been considered since they cannot be straightly modelled by COSMO-RS. This list is recalled in Table 1.1.

1.2.2 Positioning of Alternative Solvents

The list of "green" solvents presented in Table 1.1 has been studied via the COSMO-RS approach, as presented schematically in Fig. 1.5. One hundred fifty-three traditional solvents were also included in the analysis, as presented in Sect. 1.1, and PCA analysis of the sigma potentials followed by a clustering procedure further provided the ten classes presented in Table 1.2 and in Figs. 1.6 and 1.7.

Ionic liquids cannot be positioned in any of the ten families, and therefore, they should be considered as a full-fledged cluster. A1 and A3 coordinates of choline acetate, for example, are much higher than the ones of classical organic solvents.

The positioning of "green" solvents shows some distinctive features from the one of classical organic solvents. Cluster III (Aprotic dipolar) is the most populated family for green solvents, while cluster V (Apolar) is the largest one for classical solvents. Cluster III is made up of esters (dibasic esters or fatty acid esters) and ethers that are highly represented among sustainable solvents. If we take a closer look at fatty acid esters, we observe that they either belong to cluster III or to cluster V because of their particular chemical structure between alkane and aprotic dipolar molecules. Because of this duality, such compounds could be considered as a fully independent cluster. Cluster V contains few solvents and is mainly composed of terpenes, the main representative apolar solvents among "green" solvents. Clusters VII and VIII are also much populated by "green" solvents that are mainly alcohols coming from renewable resources. All other clusters are much less populated. Cluster II (weak electron pair donor bases) is only composed of 4 solvents, mainly amides. Since no amines were encountered among the "green" solvents listed, there is no solvent in cluster I. The same observation can be made for cluster VI since this cluster is mainly made up of halogenated compounds in the classical organic solvents. The lack of amines, aromatics and halogenated compounds among "green" solvents has already been noted using a similar approach based on Kamlet and Taft parameters [32].

Table 1.1 The 138 "green" solvents with their CAS registry numbers, positioned in the ten clusters as evidenced by the COSMO-RS analysis

Cluster I: Strong electron pair donor bases
No "green" solvents
Cluster II: Weak electron pair donor bases

Acetone	*67-64-1*
N,N-Dimethyloctanamide	1118-92-9
Methyl 5-(dimethylamino) −2-methyl-oxopentanoate	1174627-68-9
2-Pyrrolidone	*616-45-5*

Cluster III: Aprotic dipolar

Acetyltributyl citrate	*77-90-7*
Benzyl benzoate	*120-51-4*
Butyl acetate	123-86-4
Butyl laurate	106-18-3
1,4-Cineol	470-67-7
1,8-Cineol	470-82-6
Cyclopentyl methyl ether	5614-37-9
Dibutyl sebacate	*109-43-3*
Diethyl adipate	141-28-6
Diethyl glutarate	818-38-2
Diethyl phthalate	*84-66-2*
Diethyl succinate	123-25-1
Diisoamyl succinate	818-04-2
Diisobutyl adipate	141-04-8
Diisobutyl glutarate	71195-64-7
Diisobutyl succinate	925-06-4
Diisooctyl succinate	2915-57-3
Dimethyl adipate	627-93-0
Dimethyl glutarate	1119-40-0
Dimethyl phthalate	*131-11-3*
Dimethyl succinate	106-65-0
N,N-Dimethyldecanamide	14433-76-2
Dimethyl isosorbide	*5306-85-4*
Dioctyl succinate	14491-66-8
1,3-Dioxolane	646-06-0
Ethyl acetate	*141-78-6*
Ethyl laurate	106-33-2
Ethyl linoleate	544-35-4
Ethyl linolenate	1191-41-9
Ethyl myristate	124-06-1
Geranyl acetate	105-87-3
Glycerol triacetate	*102-76-1*
Glycerol-1,2,3-tributyl ether	131570-29-1
Glycerol-1,2,3-triethyl ether	162614-45-1
Glycerol-1,2,3-trimethyl ether	20637-49-4
Glycerol-1,3-dibutyl ether	2216-77-5

(continued)

Table 1.1 (continued)

Isoamyl acetate	123-92-2
Isobutyl acetate	110-19-0
Isopropylacetate	108-21-4
Isopropyl myristate	*110-27-0*
Isosorbide dioctanoate	64896-70-4
Methyl abietate	127-25-3
Methyl acetate	79-20-9
Methyl laurate	111-82-0
Methyl linoleate	112-63-0
Methyl linolenate	301-00-8
Methyl myristate	124-10-7
Methyl oleate	112-62-9
Methyl palmitate	112-39-0
Dimethyl 2-methyl glutarate	14035-94-0
2-Methyltetrahydrofuran	96-47-9
Menthanyl acetate	58985-18-5
n-Propyl acetate	109-60-4
Terpineol acetate	8007-35-0
Tributyl citrate	*77-94-1*
Triethyl citrate	*77-93-0*
Cluster IV: Aprotic highly dipolar	
Dimethylsulfoxide	*67-68-5*
2-Furfuraldehyde[a]	98-01-1
Propylene carbonate	*108-32-7*
γ-Valerolactone	108-29-2
Cluster V: Apolar	
Butyl myristate	110-36-1
Butyl palmitate	111-06-8
Butyl stearate	123-95-5
Cyclohexane	110-82-7
p-Cymene	99-87-6
β-Myrcene	123-35-3
Decamethylcyclopentasiloxane	*541-02-6*
β-Farnesene	18794-84-8
Ethyl oleate	*111-62-6*
Ethyl palmitate	628-97-7
Isopropyl palmitate	*142-91-6*
D-Limonene	5989-27-5
Methyl stearate	112-61-8
Isododecane	*31807-55-3*
Perfluorooctane	307-34-6
α-Pinene	80-56-8
β-Pinene	127-91-3
Terpinolene	586-62-9
Cluster VI: Asymmetric halogenated hydrocarbons (aprotic slightly dipolar)	
No "green" solvents	

(continued)

Table 1.1 (continued)

Cluster VII: Amphiprotic	
Benzyl alcohol	*100-51-6*
1-Butanol	71-36-3
Cyclademol	25225-09-6
1-Decanol	112-30-1
Dihydromyrcenol	18479-58-8
1,3-Dioxolane-4-methanol	5660-53-7
Ethanol	*64-17-5*
Ethylhexyl lactate	6283-86-9
Ethyl lactate	*97-64-3*
Geraniol	106-24-1
Glycerol-1,3-diethyl ether	4043-59-8
Glycerol-1,2-dibutyl ether	91337-36-9
Glycerol-1,2-diethyl ether	4756-20-1
Glycerol-1,2-dimethyl ether	40453-77-8
Glycerol-1,3-dimethyl ether	623-69-8
Glycerol-1-butyl monoether	624-52-2
Glycerol-1-ethyl monoether	1874-62-0
Glycerol-2-butyl monoether	100078-36-2
Glycerol-2-ethyl monoether	22598-16-9
Glycofurol ($n = 2$)	52814-38-7
N,N-Diethylolcapramide	136-26-5
Caprylic acid diethanolamide	3077-30-3
Isoamyl alcohol	123-51-3
Isopropyl alcohol	*67-63-0*
Methyl ricinoleate	141-24-2
Methanol	498-81-7
Nopol	128-50-7
1-Octanol	111-87-5
Oleic acid	*112-80-1*
1-Octanol	111-87-5
Oleic acid	*112-80-1*
Oleyl alcohol	*143-28-2*
Polyethylene glycol 600	25322-68-3
Solketal	100-79-8
Ricinoleic acid	*141-22-0*
α-Terpineol	98-55-5
β-Terpineol	138-87-4
Tetrahydrofurfurylic alcohol	97-99-4
Cluster VIII: Polar protic	
1,3-Dioxan-5-ol	4740-78-7
1,3-Dioxolane-4-methanol	*5464-28-8*
Ethylene glycol	107-21-1
Dipropylene glycol	*110-98-5*
Furfurylic alcohol[a]	98-00-0

(continued)

Table 1.1 (continued)

Glycerol	*56-81-5*
Glycerol carbonate	931-40-8
Glycerol-1-methyl monoether	623-39-2
Glycerol-2-methyl monoether	761-06-8
5-(Hydroxymethyl)furfural	67-47-0
3-Hydroxypropionic acid	503-66-2
3-Methoxy-3-methyl-1-butanol	56539-66-3
Polyethylene glycol 200	*112-60-7*
1,3-Propanediol	504-63-2
Propylene glycol	*57-55-6*
Cluster IX: Organic acidic compounds	
Acetic acid	64-19-7
Propionic acid	79-09-4
Cluster X: Polar structured	
Water	7732-18-5
Ionic liquids	
Choline acetate	14586-35-7
3-Butyl-1-methylimidazolium Tetrafluoroborate	174501-65-6

Adapted from Ref. [31]
Solvents coming from renewable resources (biosolvents) are indicated in italic. The solvents acceptable for pharmaceutical or cosmetic applications are in italics
[a]CMR compounds

Current sustainable solvents thus mostly belong to aprotic dipolar, amphiprotic and polar protic compounds, while strong electron pair donor bases, weak electron pair donor bases and aprotic slightly dipolar (asymmetric halogenated hydrocarbons) are scarcely or not represented at all among them.

1.3 Selection of Alternative Solvents for Extraction

Because of renewed toxicology standards and exposure guidelines, ever extending lists of volatile organic compounds (VOCs), ozone depleting substances, hazardous air pollutants (HAPs, in the USA) or CMR compounds (Carcinogenic Mutagenic Reprotoxic, in the European Union), solvent substitution is not a novel concern. When the use of a solvent becomes forbidden by new regulations, effective and quick substitution solutions have to be found. In the case of extraction, chlorinated hydrocarbons and *n*-hexane are two archetypal examples of such problematic solvents. Their replacements by biosolvents have been analysed below in light of the COSMO-RS and Hansen approaches.

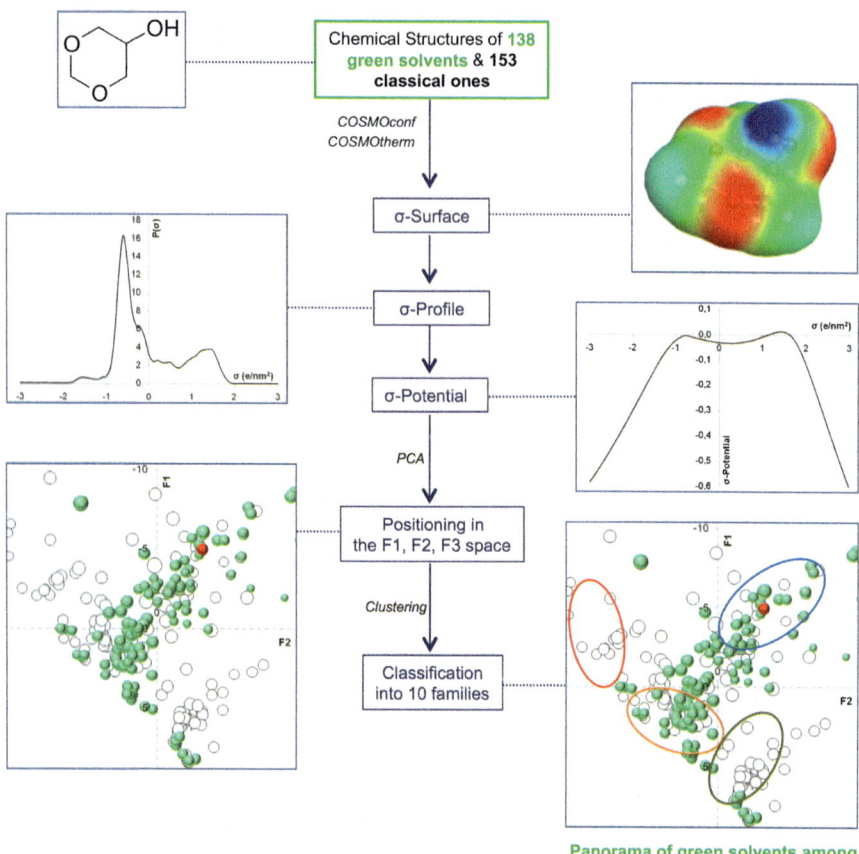

Fig. 1.5 Strategy used to position 138 green solvents (*green dots*) in the predefined 3D-space, thanks to 153 classical organic solvents (*empty dots*) using the COSMO-RS approach. The procedure is exemplified in the case of glycerol formal (1,3-Dioxolane-4-methanol) (*red dot*) (Reproduced from Ref. [31] with permission from The Royal Society of Chemistry)

1.3.1 Replacement of Chlorinated Solvents

For many years, chlorinated solvents have been in widespread use in numerous sectors, as degreasing of metallic surfaces, dry cleaning, paints (as thinner or stripper), organic synthesis and extraction among others [33]. Dichloromethane was previously used for the decaffeination of coffee which is now performed using supercritical carbon dioxide. This keen interest in chlorinated solvents is due to their outstanding physico-chemical properties, particularly their excellent solvent power, their low inflammability and high volatility. However, from a EHS point of view, chlorinated solvents exhibit a particularly bad footprint: most of them are classified among VOCs and some are blamed for stratospheric ozone depletion. Their low

1 In Silico Search for Alternative Green Solvents

Table 1.2 Clustering of "classical" and "green" solvents with typical examples in each group

Cluster	Name	Classical solvents	"Green" solvents
I	Strong electron pair donor bases	Tributylamine	–
II	Weak electron pair donor bases	Pyridine	2-Pyrrolidone
III	Aprotic dipolar	Diethyl ether	Glycerol triacetate
		Cyclohexanone	Dioxolane
IV	Aprotic highly dipolar	Sulfolane	γ-Valerolactone
		Acetonitrile	Propylene carbonate
V	Apolar	Benzene	Methyl stearate
		CCl_4	D-limonene
VI	Aprotic slightly dipolar (asymmetric halogenated hydrocarbons)	Dichloromethane	–
		Nitrobenzene	
VII	Amphiprotic	Ethanol	Isoamyl alcohol
		Benzyl alcohol	α-Terpineol
VIII	Polar protic	2-Aminoethanol	Glycerol carbonate
		Methanol	Ethylene glycol
IX	Organic acidic compounds	Phenol	Acetic acid
X	Polar structured	Water	Water
		Formamide	

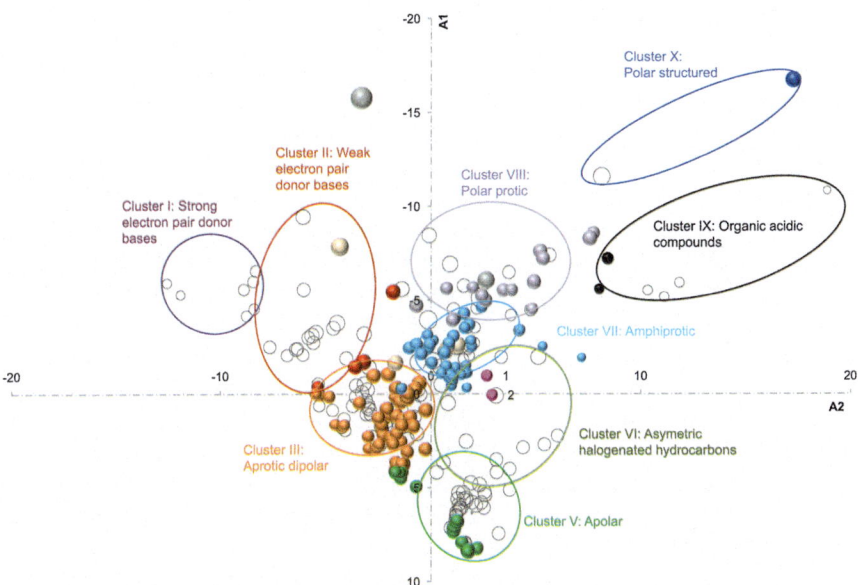

Fig. 1.6 2D – representation (A_1 vs. A_2) of green solvents (*coloured circles*) positioned within the clusters previously defined with classical solvents (*empty circles*) (Reproduced from Ref. [31] with permission from The Royal Society of Chemistry)

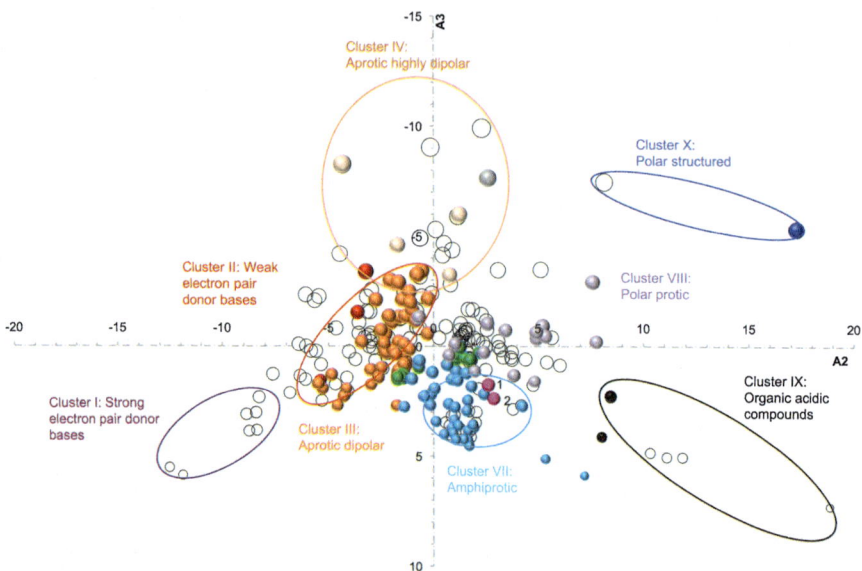

Fig. 1.7 2D – representation (A3 vs. A2) of green solvents (*coloured circles*) positioned within the clusters previously defined with classical solvents (*empty circles*) (Reproduced from Ref. [31] with permission from The Royal Society of Chemistry)

solubility in water and poor biodegradability induce long-term pollution of soil and groundwater [33]. Some of them are listed among HAP or CMR substances according to the US and/or European legislations. Therefore, for several years, substitution of chlorinated solvents has been encouraged. Among all proposed solutions, terpenes have been presented as alternatives in cleaning applications [34].

The closest neighbours of dichloromethane, tetrachloroethylene, carbon tetrachloride, 1,1,1-trichloroethane, trichloroethylene and chloroform have been looked for in the COSMO-RS classification presented before. Terpenic solvents emerge as possible substitutes for tetrachloroethylene, carbon tetrachloride and trichloroethylene (α-pinene for the first two and β-myrcene for the last one). Actually, terpenes, such as *p*-cymene, terpinolene, α-pinene or D-limonene, belong to cluster V, i.e. the cluster that contains, *inter alia*, some chlorinated solvents. The closest neighbour of 1,1,1-trichloroethane is perfluorooctane that also belongs to cluster V. This solvent has been chosen as a representative example of fluorinated solvents that are presented as "green solvents" for organic synthesis or in the electronics industry [26]. More surprisingly, the closest neighbours encountered for dichloromethane and chloroform are respectively benzyl benzoate that belongs to cluster III (aprotic dipolar) and benzyl alcohol, belonging to cluster VII (amphiprotic). Actually, chloroform and dichloromethane belong to a cluster that is not populated at all by the existing green solvents (cluster VI), which justify the current interest in the design of new bio-based solvents (see paragraph 1.4).

Fig. 1.8 σ-potential of n-hexane (*plain line*) compared to the ones of D-limonene, p-cymene, α-pinene and β-pinene (*dotted lines*)

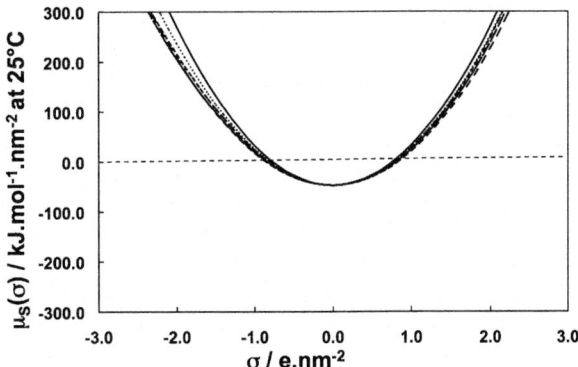

1.3.2 Replacement of n-Hexane

n-hexane is a major solvent for the extraction of natural products and particularly vegetable oils in the food industry. It has many advantages, in particular its high solubilising capacity of oily constituents and its low boiling point, which facilitates the recovery of solutes and solvent recycling. However, it is listed among VOCs and HAPs, and in Europe, it belongs to the CMR list for its reprotoxicity and its neurotoxic metabolite, 2,5-hexanedione [14]. Substitutes to n-hexane for extraction are thus wanted, and terpenes are often put forward for this application. In particular, Tanzi et al. [35] have shown that terpenes could be efficiently used for the recovery of triglycerides from the algae *Chlorella vulgaris*. This substitution solution can be investigated using the COSMO-RS and Hansen approaches.

1.3.2.1 Positioning of n-Hexane in the COSMO-RS Panorama

As already mentioned previously for chlorinated solvents, looking for solvents having the closest σ-potentials is a way to identify potential substitutes. n-hexane belongs to cluster V (apolar compounds), with a typical U-shaped σ-potential showing the lack of H-bond donor and H-bond acceptor character. D-limonene, p-cymene, α-pinene and β-pinene are common terpene solvents that also belong to this cluster, as presented in Fig. 1.8, which indicates that their solubilising properties should be close.

1.3.2.2 Combined Hansen and COSMO-RS Approaches for the Substitution of n-Hexane

The solvents highlighted by the COSMO-RS panorama can be positioned in the Hansen space. The Hansen solubility parameters of n-hexane, D-limonene, p-cymene, α-pinene and β-pinene listed in references [14, 36] are positioned in

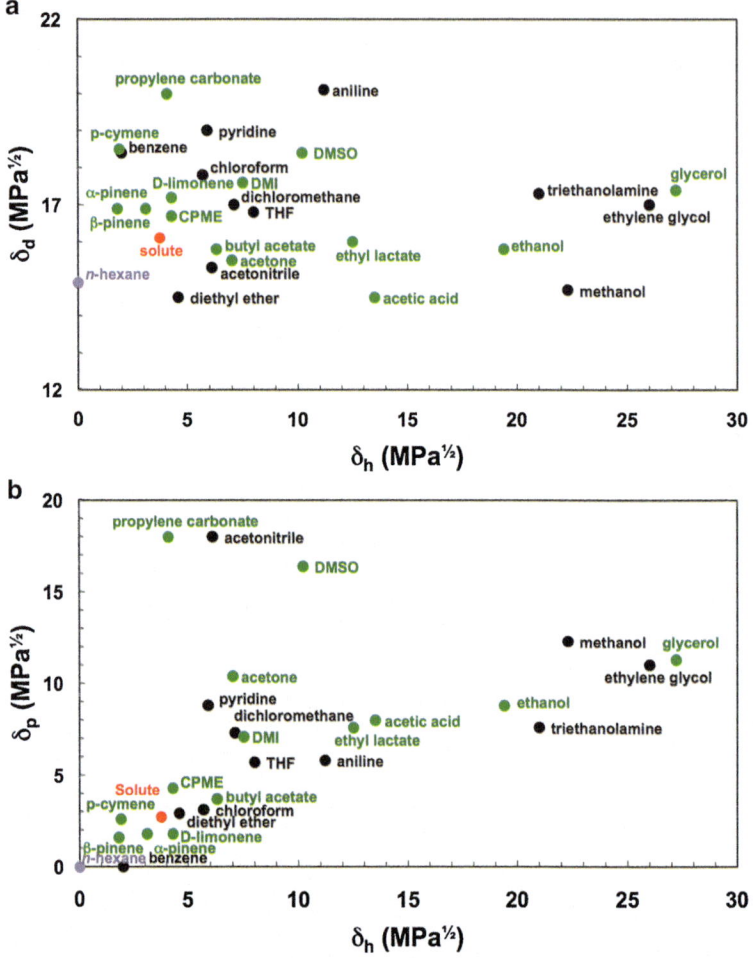

Fig. 1.9 Location of *n*-hexane, potential substitutes and fatty acid methyl esters (solute) within the (δ_p, δ_h) (*top*) and (δ_d, δ_h) (*bottom*) maps

the δ_p/δ_h and δ_d/δ_h 2-D maps in Fig. 1.9. The "solute" to extract (lipidic fraction) has been modelled by the methyl esters of palmitic, linolenic and oleic acids that are the three main fatty chains encountered in the oily fraction extracted from *Chlorella vulgaris* [35]. The position of this "solute" takes into account the relative proportions of each type of fatty acid as described in reference [35].

The close location of *n*-hexane and terpenes in the Hansen space is in good agreement with the close positioning in the COSMO-RS panorama. Other "green" solvents are found in the vicinity of *n*-hexane in the Hansen space, namely, cyclopentyl methyl ether (CPME) and *n*-butyl acetate. They both belong to cluster III (aprotic dipolar) in the COSMO-RS classification, in which solvents have an electron-donor ability (H-bond acceptor).

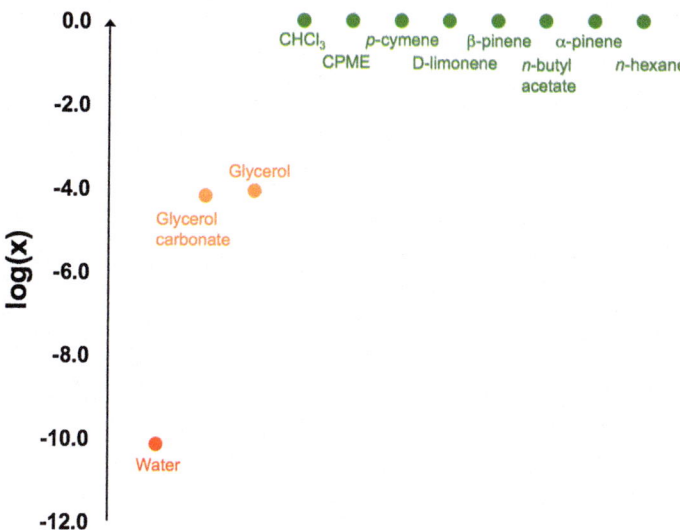

Fig. 1.10 Relative solubility of the mixture of fatty acids methyl esters considered in good (*green*), poor (*orange*) and very bad (*red*) solvents, as determined through the "solvent screening" tool of COSMOtherm

To compare the solubilising abilities of these "green" alternatives towards fatty acid methyl esters, the "solvent screening" tool implemented in COSMOtherm was used. It provides a ranking of relative solubilities from the prediction of μ^i_{solv}, the chemical potential of the compound of interest in a list of selected solvents. This solvent ranking is presented in Fig. 1.10 for the solubilisation of the mixtures of fatty acid methyl esters considered. The logarithm of solubility in mole fractions is calculated, and the logarithm of best solubility is set to 0, all other solvents being given relative to the best solvent. In the present case, *n*-hexane is the best solvent, and the "green" substitutes cannot be distinguished, they are all set at log $(x) = 0$. For comparison purposes, the relative solubilities in water (worse solvent) and in glycerol and glycerol carbonate (poor solvents) are also given.

Finally, it is also interesting to show that the Hansen approach can be used in a very simplistic manner to identify possible efficient solvent mixtures. In the present example, Fig. 1.9 shows that α-pinene can be brought closer to the position of the target solute by addition of a solvent with higher δ_p and δ_h and lower δ_d.

Figure 1.11 shows such a procedure in the case of the addition of ethanol to α-pinene. The distance D of the solvent mixture to the solute in the Hansen space is computed from Eq. 1.3 (see Sect. 1.1.2). The addition of *ca.* 6 % of ethanol to α-pinene allows reducing the distance to the solute and is expected to enhance the solubilising capacities. This effect was observed by Tanzi et al. [35] during the recovery of triglycerides from *Chlorella vulgaris* for which the addition of a small amount of methanol to chloroform proved to increase the extraction yield.

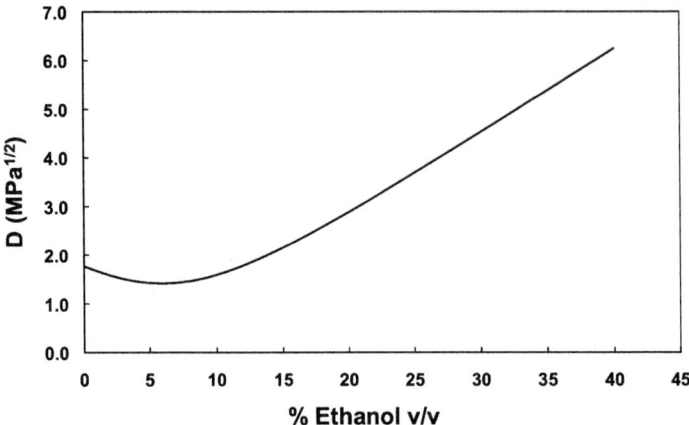

Fig. 1.11 "Synergetic effect" according to Hansen for the solubilisation of fatty acid methyl esters using an α-pinene/ethanol mixture

1.4 Design of New Solvents with Tailored Properties

1.4.1 Lack of Structures with Specific Properties

The panorama of green solvents according to the COSMO-RS approach evidences that some solvent families are little populated or even not populated at all by "green" solvents (clusters I, II, IV and VI, see Figs. 1.3, 1.6 and 1.7). In particular, the quasi-emptiness of cluster II (weak electron pair donor bases) shows that "green" amide-containing solvents should be developed. Some industrial solutions are starting to emerge, such as amide-homologues of the dibasic esters.

To enlarge the scope of substitution solutions, it is thus of utmost interest to develop new solvent structures with tailored properties for a given application. Reverse engineering is a powerful tool for such an approach, which uses a "top-bottom" strategy [37]. In a recent paper, we have presented a different approach that starts from a chosen bio-sourced building block and generates new molecules by applying chosen chemical transformations [38]. This approach has been exemplified on the generation of itaconic acid-derived solvents and is presented in the last section.

1.4.2 Automatic Generation of New Solvent Structures

The interest for using renewable resources (biomass) instead of fossil resources (coal, oil) has grown exponentially in the last years. The "biorefinery" concept,

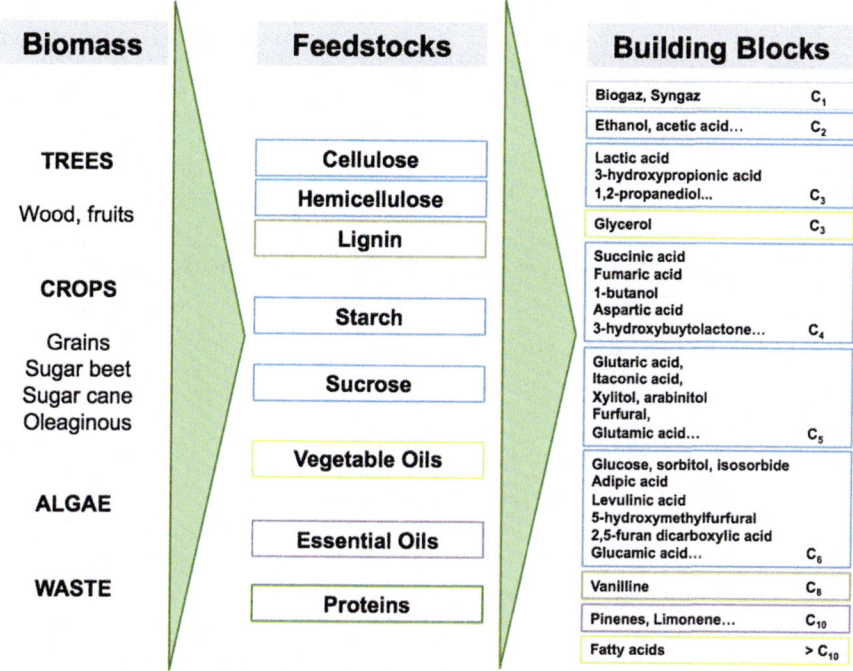

Fig. 1.12 Schematic representation of the building blocks that are currently or will soon be obtained from biorefineries, ordered by increasing number of carbon atoms. The frame colours refer to the chemical feedstocks they come from

i.e. the transposition of the petrorefinery scheme to the processing of biomass, is gaining importance, and a spectrum of products is expected to be obtained from the biomass feedstock in the forthcoming years.

Figure 1.12 shows a schematic route from biomass to defined chemicals. The main biomass sources providing incomes for biorefineries are forestry, dedicated crops and vegetable residues (cobs, straw, sugarcane bagasse, etc.). The exploitation of aquatic biomass (algae) is also a promising source of renewable carbon. The chemical or biochemical transformation of this biomass feedstock may lead to a spectrum of bio-based building blocks available for chemistry. As the main sources are polysaccharides, the available building blocks are currently sugars and sugar derivatives (polyols, organic acids obtained by fermentation).

The methodology used to generate virtual solvents is presented schematically in Fig. 1.13 [38]. The software developed has been named GRASS as the acronym of GeneratoR of Agro-based Sustainable Solvents. It requires three inputs: a bio-based building block, readily available co-reactants and a list of selected transformations that can be applied to the substrate and co-reactants. Virtual products are then automatically generated using the architecture developed by Barone et al. in the previous versions of the programme [39, 40]. This set can, in turn, be an input to GRASS and transformed again.

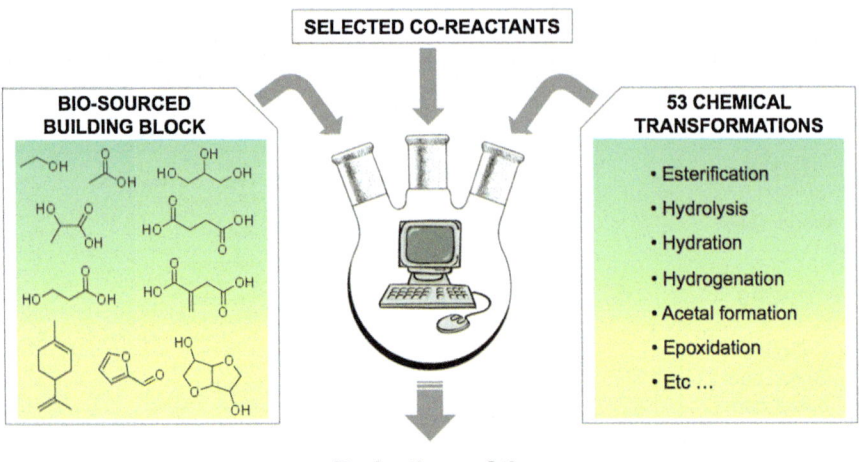

Fig. 1.13 Schematic representation of the "GRASS" programme used to generate automatically virtual solvents

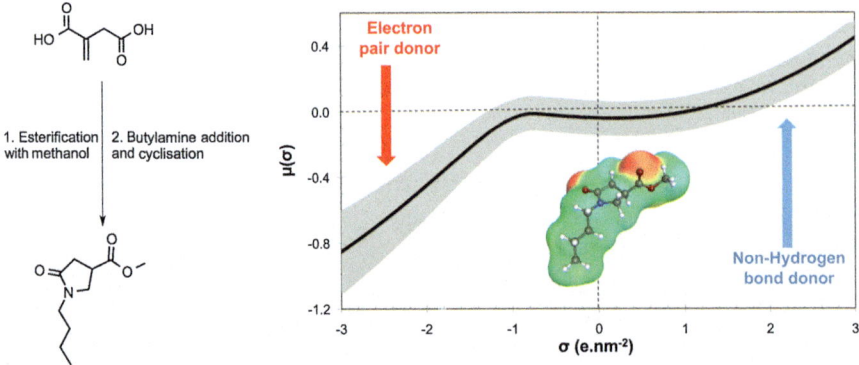

Fig. 1.14 Preparation of *N*-butyl-4-carboxypyrrolidinone ester starting from itaconic acid (*left*). σ-potential of the product $\mu(\sigma)$ compared to those of classical weak electron pair donor bases belonging to cluster II (*in grey*) (Adapted from Ref. [38])

The methodology was applied to itaconic acid, an organic acid obtained by sugar fermentation, and it highlighted a family of solvents that are readily obtained in two steps, the *N*-butyl-4-carboxypyrrolidinone esters (Fig. 1.14). This family containing a lactam function is of great interest since it is positioned in cluster II according to the COSMO-RS classification, a class of solvents that is not much populated by current "green" solvents (Fig. 1.14).

1.5 Conclusion

The search for alternative solvents is a hot topic in many industrial fields, including the extraction of natural products. A panel of so-called green solvents is already available, and several tools exist to guide the selection of the most appropriate alternative for a given application. In particular, the traditional Hansen approach coupled with more modern modelling tools such as COSMO-RS provides accurate predictions of solubilising abilities. These predictive tools can also be used to design new solvent structures with tailored properties, in order to enlarge the scope of "green" alternative solvents.

References

1. Reichardt C (1988) Solvents and solvent effects in organic chemistry, 2nd edn. Wiley, New York
2. Kamlet MJ, Taft RW (1976) The solvatochromic comparison method. 1. The β-scale of solvent hydrogen-bond acceptor (HBA) basicities. J Am Chem Soc 98:377–383
3. Taft RW, Kamlet MJ (1976) The solvatochromic comparison method. 2. The α-scale of solvent hydrogen-bond donor (HBD) acidities. J Am Chem Soc 98:2886–2894
4. Kamlet MJ, Abboud JLM, Taft RW (1977) The solvatochromic comparison method. 6. The π* scale of solvent polarities. J Am Chem Soc 99:6027–6038
5. Abraham MH (1993) Scales of solute hydrogen-bonding: their construction and application to physico-chemical and biochemical processes. Chem Soc Rev 22:73–83
6. Taft RW, Abboud JLM, Kamlet MJ, Abraham MH (1985) Linear solvation energy relations. J Solution Chem 14:153–186
7. Katritzky AR, Fara Dan C, Yang H, Tamm K, Tamm T, Karelson M (2004) Quantitative measures of solvent polarity. Chem Rev 104:175–198
8. Murray JS, Politzer P, Famini GR (1998) Theoretical alternatives to linear solvation energy relationships. Theochem-J Mol Struc 454:299–306
9. Brinck T, Murray JS, Politzer P (1993) Octanol/water partition coefficients expressed in terms of solute molecular surface areas and electrostatic potentials. J Org Chem 58:7070–7073
10. Lowrey AH, Cramer CJ, Urban JJ, Famini GR (1995) Quantum chemical descriptors for linear solvation energy relationships. Comput Chem 19:209–215
11. Katritzky AR, Fara DC, Kuanar M, Hur E, Karelson M (2005) The classification of solvents by combining classical QSPR methodology with principal component analysis. J Phys Chem A 109:10323–10341
12. Hildebrand J, Scott R (1950) The solubility of nonelectrolytes, 3rd edn. Reinhold, New York
13. Stefanis E, Panayiotou C (2008) Prediction of Hansen solubility parameters with a new group-contribution method. Int J Thermophys 29:568–585
14. Benazzouz A, Moity L, Pierlot C, Sergent M, Molinier V, Aubry JM (2013) Selection of a greener set of solvents evenly spread in the Hansen space by space-filling design. Ind Eng Chem Res 52:16585–16597
15. Gharagheizi F, Sattari M, Angaji MT (2006) Effect of calculation method on values of Hansen solubility parameters of polymers. Polym Bull 57:377–384
16. Benazzouz A, Moity L, Pierlot C, Molinier V, Aubry JM (2014) Hansen approach versus COSMO-RS for predicting the solubility of an organic UV filter in cosmetic solvents. Colloids Surf A Physicochem Eng Asp (in press). doi: 10.1016/j.colsurfa.2014.03.065
17. Hansen CM (2004) 50 years with solubility parameters – past and future. Prog Org Coat 51:77–84
18. Cartier A, Rivail JL (1987) Electronic descriptors in quantitative structure-activity relationships. Chemometr Intell Lab 1:335–347

19. Klamt A, Schueuermann G (1993) COSMO: a new approach to dielectric screening in solvents with explicit expressions for the screening energy and its gradient. J Chem Soc Perk T 2:799–805
20. Klamt A (1995) Conductor-like screening model for real solvents: a new approach to the quantitative calculation of solvation phenomena. J Phys Chem 99:2224–2235
21. Klamt A (2005) COSMO-RS: from quantum chemistry to fluid phase thermodynamics and drug design. Elsevier, Amsterdam
22. Wichmann K, Diedenhofen M, Klamt A (2007) Prediction of blood–brain partitioning and human serum albumin binding based on COSMO-RS σ-moments. J Chem Inf Model 47:228–233
23. Klamt A, Eckert F, Diedenhofen M, Beck ME (2003) First principles calculations of aqueous pKa values for organic and inorganic acids using COSMO−RS reveal an inconsistency in the slope of the pKa scale. J Phys Chem A 107:9380–9386
24. Durand M, Molinier V, Kunz W, Aubry JM (2011) Classification of organic solvents revisited by using the COSMO-RS approach. Chem-Eur J 17:5155–5164
25. Chastrette M (1979) Etude statistique des effets de solvant—I: principes et applications à l'évaluation des paramètres de solvant et à la classification. Tetrahedron 35:1441–1448
26. Kerton FM (2009) Alternative solvents for green chemistry. RSC Publishing, Cambridge
27. Plechkova NV, Seddon KR (2008) Applications of ionic liquids in the chemical industry. Chem Soc Rev 37:123–150
28. Ranke J, Stolte S, Störmann R, Arning J, Jastorff B (2007) Design of sustainable chemical products. Chem Rev 107:2183–2206
29. Klein R, Zech O, Maurer E, Kellermeier M, Kunz W (2011) Oligoether carboxylates: task-specific room-temperature ionic liquids. J Phys Chem B 115:8961–8969
30. Imperato G, König B, Chiappe C (2007) Ionic green solvents from renewable resources. Eur J Org Chem 7:1049–1058
31. Moity L, Durand M, Benazzouz A, Pierlot C, Molinier V, Aubry JM (2012) Panorama of sustainable solvents using the COSMO-RS approach. Green Chem 14:1132–1145
32. Jessop PG (2011) Searching for green solvents. Green Chem 13:1391–1398
33. Danaché B, Févotte J, Work team of Matgéné (2009) Éléments techniques sur l'exposition professionnelle à cinq solvants chlorés (trichloroéthylène, perchloroéthylène, chlorure de méthylène, tétrachlorure de carbone, chloroforme) – matrices emplois – expositions à cinq solvants chlorés. Institut de veille sanitaire, Umrestte Lyon, Saint-Maurice
34. Abel S (1990) Fate and exposure assessment of aqueous and terpene cleaning substitutes for chlorofluorocarbons and chlorinated solvents. U.S. Environmental Protection Agency Office of Toxic Substances Exposure Assessment Branch, Washington, DC
35. Tanzi C, Vian M, Ginies C, Elmaataoui M, Chemat F (2012) Terpenes as green solvents for extraction of oil from microalgae. Molecules 17:8196–8205
36. Hansen CM (2007) Hansen solubility parameters. CRC Press, Taylor & Francis Group, Boca Raton
37. Heintz J, Touche I, Teles dos Santos M, Gerbaud V (2012) An integrated framework for product formulation by computer aided mixture design. Comput Aided Chem Eng 30:702–706
38. Moity L, Molinier V, Benazzouz A, Barone R, Marion P, Aubry JM (2014) In silico design of bio-based commodity chemicals: application to itaconic acid based solvents. Green Chem 16:146–160
39. Barone R, Chanon M, Vernin G, Parkanyi C (2005) Generation of potentially new flavoring structures from thiamine by a new combinatorial chemistry program. In: Mussinan CJ, Ho CT, Tatras Contis E, Parliment TH (eds) Food flavor and chemistry: explorations into the 21st century. RSC, Cambridge, pp 175–212
40. Barone R, Chanon M, Vernin G, Parkanyi C (2010) Computer-aided organic synthesis as a tool for generation of potentially new flavoring compounds from ascorbic acid. In: Ho CT, Mussinan CJ, Shahidi F, Tatras Contis E (eds) Recent advances in food and flavor chemistry: food flavors and encapsulation, health benefits, analytical methods and molecular biology of functional foods. RSC, Cambridge, pp 81–126

Chapter 2
Solvent-Free Extraction: Myth or Reality?

Maryline Abert Vian, Tamara Allaf, Eugene Vorobiev, and Farid Chemat

Abstract One of the many environmental challenges faced by Extraction field is the widespread use of organic solvents. With a solvent based extraction the solvent necessarily has to be separated from the final extract. A large number of these solvents are toxic that pose a risk to workers and community members and virtually all of them are classified as volatile organic compounds (VOCs) that contribute to smog. In this context, the development of solvent-free extraction processes is of great interest in order to modernize classical processes making them cleaner, safer and easier to perform. This chapter presents a picture of current knowledge on innovative solvent-free methods of natural products extraction. It provides the necessary theoretical background and some details about extraction using the most innovative, rapid and green techniques such as microwaves, instant controlled pressure drop (DIC) process and Pulsed Electric Field (PEF): the technique, the mechanism and some applications.

M. Abert Vian • F. Chemat (✉)
Green Extraction Team, Université d'Avignon et des Pays de Vaucluse, INRA, UMR 408, F-84000 Avignon, France
e-mail: maryline.vian@univ-avignon.fr; Farid.chemat@univ-avignon.fr

T. Allaf
ABCAR-DIC, F-17100 La Rochelle, France
e-mail: tamara.allaf@abcar-dic.com

E. Vorobiev
Laboratoire Transformations Intégrées de la Matière Renouvelable, Équipe Technologies Agro-Industrielles, Université de Technologie de Compiègne (UTC), F-60205 Compiègne, France
e-mail: eugene.vorobiev@utc.fr

2.1 Introduction

In a typical chemical or extraction process, solvents are used extensively for dissolving reactants, solvating molecules, extracting products, separating mixtures. However the major part of the organic solvents currently found in industry, in spite of a large number of well-known advantages, are characterized by several dangerous effects for the human health and the environment. Many organic solvents are Volatile Organic Compounds (VOCs), and it means that they are highly volatile, very useful for industrial applications, contribute both to increase the risks of fire and explosion, and to facilitate the release in the atmosphere in which these solvents can act as air pollutants causing ozone depletion and global warming. Moreover, many conventional solvents are highly toxic for human beings, animals and plants, and often their toxicological properties are completely unknown.

For example, *n*-hexane, solvent of choice for extraction of oils, can be emitted during extraction and recovery; it has been identified as an air pollutant since it can react with other pollutants to produce ozone and photochemical oxidants [1, 2]. Precautions to minimize the effects of these solvents by improved recycling have limited success and cannot avoid some losses into the environment. Moreover, the risk connected to potential accidents is still present.

During the last years, a central objective in extraction field of natural products has been set to develop greener and more economically competitive processes for the efficient extraction of natural substances with potential application in the cosmetic or agrochemical industries. In this context, the development of solvent-free alternative processes is of great interest in order to modernize classical processes making them cleaner, safer and easier to perform.

Therefore the following benefits could be mentioned for solvent-free conditions: (1) Avoid large volumes of solvent which reduces emission and needs for distillation; (2) The absence of solvents which facilitates scale-up; (3) Extracts are cleaner without residues; (4) Safety is enhanced by reducing risks of overpressure and explosions.

Extraction of olive oil using mechanical pressing is recognized as a solvent-free alternative. Virgin olive oils are extracted from olive fruits by using only physical methods, which include crushing of olives, malaxation of resulting pastes and separation of the oily phase. Because of its location in mesocarp of cells and the use of purely mechanical pieces of apparatus for its extraction, virgin olive oil does not require further treatment before its consumption (Fig. 2.1).

This chapter presents a picture of current knowledge on innovative solvent-free methods of natural products extraction. It provides the necessary theoretical background and some details about extraction using the most innovative, rapid and green techniques such as microwaves, instant controlled pressure drop (DIC) process and Pulsed Electric Field (PEF): the technique, the mechanism and some applications.

Fig. 2.1 Solvent free olive oil extraction process

2.2 Solvent-Free Microwave-Assisted Extraction

The use of microwave energy was described for the first time in 1986 simultaneously by Gedbye [3] in organic synthesis and by Ganzler [4] for extraction of biological samples and analysis of organic compounds. Since then, numerous laboratories have studied the synthetic and analytical possibilities of microwaves as a non-classical source of energy. Several classes of compounds such as essential oils, aromas, pigments, antioxidants, and other organic compounds have been extracted efficiently from a variety of matrices mainly and plant materials. Advances in microwave extraction have given rise to solvent-free microwave technique namely Microwave Hydrodiffusion and Gravity.

2.2.1 Principle

Microwave hydrodiffusion and gravity (MHG) [5] is a new and green technique for the extraction of biomolecules patented in 2008. MHG was conceived for laboratory and industrial-scale applications in the extraction of pigments, aroma components, and antioxidants from different kind of plants. Based on a relatively simple principle, this method also involves placing the plant material in a microwave reactor, without adding any solvent or water. The internal heating of the *in situ* water within the plant material distends the plant cells and leads to the rupture of cells. The heating action of microwaves thus frees secondary metabolites and *in situ* water, which are transferred from the inside to the outside of the plant material. This physical phenomenon, known as hydrodiffusion, allows the extract, diffused outside the plant material, to drop by gravity out of the microwave reactor and fall through the perforated Pyrex disk.

A cooling system outside the microwave oven cooled the extract continuously. The crude extracts are collected in a receiving flask for further analysis (Fig. 2.2). MHG not only appeared as an efficient and economical technology but its chief advantage is its environmental friendly approach as it works without using any solvent just under effect of microwaves and earth gravity at atmospheric pressure.

2.2.2 Instrumentation

A Milestone NEOS-GR microwave laboratory oven (900 W maximum), as shown in (Fig. 2.2), is used to perform the microwave hydrodiffusion and gravity (MHG) extraction: this is a multimode microwave reactor of 2.45 GHz. Temperature is monitored by an external infrared (IR) sensor. MHG could also be used to produce larger quantities of extracts by using existing large-scale microwave extraction reactors called "MAC-75" (Fig. 2.3).

2.2.3 Application

The feasibility of microwave process in the preparation of samples has been investigated on different matrices, as shown in the Table 2.1. This process was applied to many kinds of plants such as aromatic plants and citrus for an essential

Fig. 2.2 Solvent free microwave extraction laboratory system (NEOS-GR)

oil extraction [6–8]. The first example is the *menthe pulegium L.* extraction [6], where 0.95 % of essential oil was obtained by the heating of 500 g of matrix at 500 W during 20 min at atmospheric pressure. For *Citrus limon L.* [7], 500 g of matrix were also treated at 500 W for 15 min and two respective yields of 0.7 and 1.6 % of essential oil were obtained at atmospheric pressure. Another example with *Rosmarinus Officinalis L.* [8] was tested by taking 500 g of plant at 500 W during 15 min, which provided 0.33 % of essential oil.

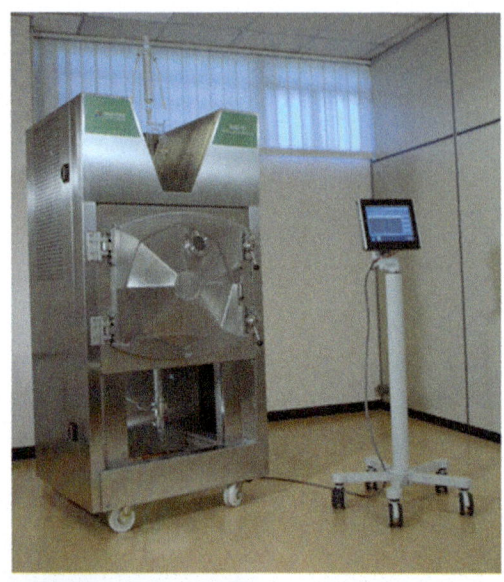

Fig. 2.3 Pilot scale Solvent free microwave extraction 'MAC 75'

Table 2.1 MHG application in extraction of natural compounds

Material	Analytes	Ref.
M. spicata L., M. pulegium L.	Essential oil	[6]
Rosemary leaves	Essential oil	[7]
Citrus peels	Essential oil	[8]
Onion (Allium cepa L.)	Phenolic compounds	[9]
Red, yellow, white and grelot onion	Phenolic compounds	[10]
Sea buckthorn (Hippophaë rhamnoides)	Phenolic compounds	[11]

Zill-e-Huma et al. reported MHG as a novel technique for extracting flavonoids from onion. The plant tissues were strongly disrupted by microwave irradiation through the microscopic observation of extracts, so that target compounds could be efficiently extracted and detected by HPLC and other analysis [9, 10]. MHG was also applied to extraction of flavonoids from sea buckthorn by-products, producing a little lower yield of flavonol in a very short time (15 min) in comparison to classic methods but a higher content of reducing compounds contained in MHG extracts [11].

2.3 Instant Controlled Pressure Drop Process (DIC, "Détente Instantanée Contrôlée")

The instant controlled pressure drop process, abbreviated DIC for 'Détente Instantanée Contrôlée', was developed by Allaf et al. in 1988 [12]. DIC extraction is based on fundamental studies with respect to the thermodynamics of instantaneity [13].

2.3.1 Principle

It consists of a thermo-mechanical process induced by subjecting the product to a fast transition from high temperature/high pressure to a vacuum. DIC extraction usually starts by establishing this high temperature/high pressure by injecting steam, microwaves, hot compressed air, etc. for some seconds, proceeding then to an abrupt pressure drop toward a vacuum (about 5 kPa with a rate higher than 0.5 $MPa.s^{-1}$). By instantly dropping the pressure, rapid autovaporization of the moisture inside the material will occur. It will swell and lead to texture change which results in higher porosity as well as a greater specific surface area and reduced mass-transfer resistance through Darcy's vapor transfer instead of Fick's similar law. The short time – high temperature operation (few seconds) and the immediate drop in temperature (to be lower than 30 °C) thanks to the pressure drop prevent further thermal deterioration and provide a final extract of great quality.

2.3.2 Instrumentation

DIC equipment is in the main part divided in three components: (1) the autoclave with heating jacket also named the processing vessel where the product is placed, (2) the vacuum tank linked to a vacuum pump, and (3) the instant valve enabling the abrupt connection between (1) and (2). The vacuum tank is cooled through a double jacket in order to condensate the extracts. Other devices are part of the DIC process such as a steam generator, an air compressor (for the electro-pneumatic actions) and the vacuum pump. A schematic diagram of the DIC apparatus and the pressure profile are presented in Figs. 2.4 and 2.5.

In order to undergo a DIC cycle, the raw material is placed within the autoclave where a vacuum is subsequently applied. This will facilitate the contact between the steam and the product enabling a homogenous heating of the product. It is afterward filled with saturated steam set and maintained at a required pressure for an optimized time. After this thermal treatment the steam is cut off and the spherical instant valve is opened in less than 0.2 s inducing an abrupt pressure drop towards a vacuum in the autoclave. After a vacuum period, the atmospheric pressure established in order to recover the solid material. The extracts are collected from the vacuum tank.

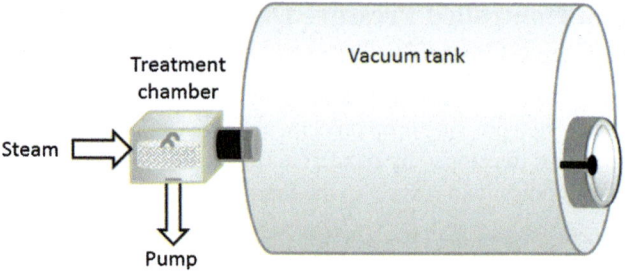

Fig. 2.4 Schematic diagram of the instantaneous controlled pressure-drop

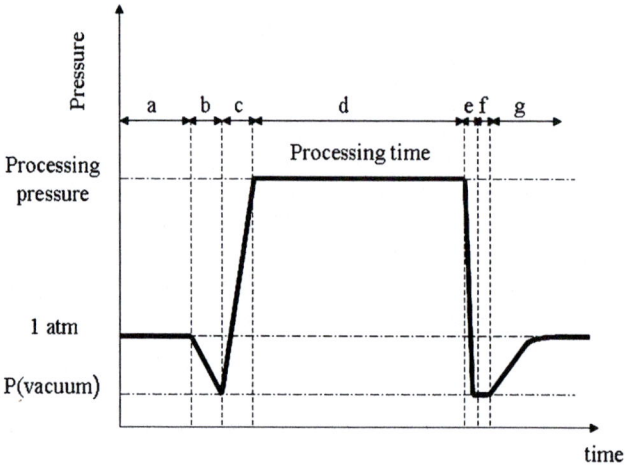

Fig. 2.5 Pressure-time profile of DIC processing cycle

Table 2.2 Instant controlled pressure drop applications

Material	Analytes	Ref.
Malaysian Roselle	Anthocyanins	[14]
Seeds of Tephrosia purpurea	Oligosaccharides	[15]
Myrtle leaves	Essential oil	[16]
Lavender	Essential oil	[17]
Ylang-ylang	Essential oil	[18]
Rapeseed	Oil	[19]

2.3.2.1 Applications

The feasibility of DIC process in the preparation of samples has been investigated on different matrices (Table 2.2).

Benamor et al. [14] have studied extraction of anthocyanins from Roselle calyces using DIC. This work has demonstrated that DIC increases kinetics and extraction yield of these compounds. The same authors have also noticed the impact of DIC treatment on the oligosaccharides (stachyose and ciceritol) extraction from the seeds of the Indian *Tephrosia purpurea* plant [15]. DIC was shown to be an effective extraction method in terms of processing time (1 h of extraction time instead of 4 h for conventional processes). DIC could be also used for the extraction of oil from various plants. One of its advantages compared to other extraction processes is the short-time contact of the oil with the apparatus heated zones to avoid the harmful thermal reactions of the different molecules, combined to an abrupt pressure variation that allows a rapid release of essential oil due to the rupture of the oil-containing glands. The DIC process is more efficient in terms of rapidity (several minutes versus several hours), essential oil yield (comparable even higher), and higher content of oxygenated compounds.

2.4 Pulsed Electric Field (PEF)

Pulsed Electric Field (PEF) technology was invented in the 1960s in order first to offer the possibility to preserve food by replacing traditional pasteurization. Electrotechnologies based on effects of PEF are currently gaining a real interest regarding food processing especially in the field of extraction [20–23].

2.4.1 Principle

Exposing a biological cell (plant, animal and microbial) to a high intensity electric field (kV/cm) in very short pulses (μs to ms) induces the formation of temporary or permanent pores on the cell membrane [24]. The cell membranes are charged and pores are formed in the membranes fostering the extraction. This phenomenon, named electroporation, causes the permeabilization of cell membrane i.e. an increase of its permeability and if the intensity of the treatment is sufficiently high, cell membrane disintegration occurs [25] (Figs. 2.6 and 2.7).

Many fields are developing cell membrane disintegration such as biotechnology, medicine and food industry [24, 26–28]. Cell membrane acts as a physical barrier in removing the intracellular substances from plant food tissues in solid–liquid extraction. The disintegration or permeabilization of the cell membrane in a plant food tissue causes the release of intracellular water and solutes (secondary metabolites) to migrate in an external medium. Thus, this method enables enhancing extraction from food plants; it enhances mass transport out of the cell.

2.4.2 Instrumentation

A PEF treatment chamber consists of at least two electrodes and insulation that forms the treatment volume (i.e. volume where the foods receive pulses). The final

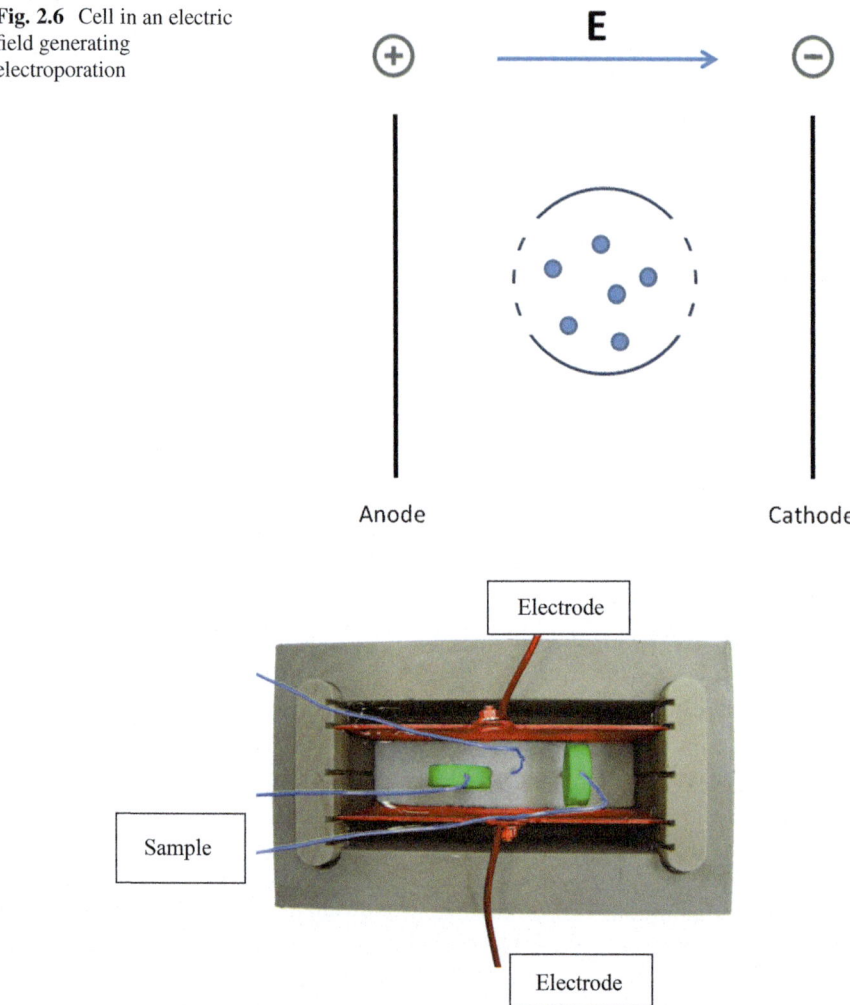

Fig. 2.6 Cell in an electric field generating electroporation

Fig. 2.7 Treatment chamber – lab scale device

distance between the electrodes can be optimized and fixed. The product exposure to a pulsed voltage can hence be done in a batch or continuous treatment chamber. Different pulse-forming networks can be used; their main components include selected voltage power supply, one or several capacitor banks, inductors or/and resistors [24]. Besides pulses of different shape can be generated, including simplest exponential decay pulses and square wave pulses should be limited for exclusion of any significant temperature elevation.

Electric field treatment is applied by a PEF generator which power is also determined depending on the needs. Shape and polarity are criterion of the generator itself. Trains of pulses are usually used for PEF treatment. An individual train consisted of n pulses with pulse duration ti and pulse repetition times Δt. A pause Δt_1 can be set after each train, N being the number of trains. The total time of electrical treatment during PEF treatments are calculated as $t_1 = N_{tot} \, t_i$ where N_{tot} is the total number of pulses $N_{tot} = Nn$.

These different parameters, whether temperature T, electric field strength E (V/cm), electric energy W (kJ/kg), number of impulsion n and duration of an impulsion t_i can be modified regarding the needs.

Some indications regarding PEF parameters:

- Permeabilisation of plant cells 0.3–1.5 kV/cm 1–10 kJ/kg
- Inactivation of microorganisms 10–30 kJ/cm 50–200 kJ/kg

2.4.3 Applications

PEF whether direct treatment or as a pre-treatment facilitate the extraction of vegetable oil, active molecules such as anthocyanins, flavonoids, etc. The application of electroporation through PEF offers a great potential for extraction purposes. It improves kinetics extraction and enhances extraction yields. PEF treatment is suitable for thermolabile fragile molecules since the extraction can be done at low temperature.

Temperature contribution to electroporation efficiency is important, which reflects the synergetic effect of the simultaneous thermal and PEF treatment and it increases at small fields [29].

PEF treatment, or pre-treatment noticeably accelerates diffusion even at low temperature (20–40 °C), which enable the "cold" soluble matter extraction [30].

When PEF is employed it is possible to regain antioxidant substances from plant processing residual material, potential of pectin recovery is enabled [31]. Regarding juice pressing with high polyphenol content, the choice of appropriate regime of pressing is required. PEF application allowed decreasing of the applied pressure and pressing time. PEF is very promising for enhancing juice and polyphenol extraction [31–33]. PEF is a promising enological technology to obtain wines with the high phenolic content necessary for the production of high quality oak aged red wines [34].

For extraction both membrane of the cell and the vacuole have to be opened. The releasing efficiency of ionic components, enzymes, proteins and other bio-products can dramatically depend on the applied method of disruption. The PEF treatment removes membrane barriers and accelerates release of the extract contents; it

Table 2.3 PEF applications

Material	Analytes	Ref.
Apple mash	Apple juice	[33]
Grape by-products	Phenolic compounds	[34]
Sugar beets	Sucrose	[35]
Inonotus Obliquus	Betulin	[36]
Soybean/Olive	Oil	[37]
Red beetroots	Betalains	[38]

however has practically no influence on the cell walls. As regards to electroporation, pores resealing is possible after the pulse application. If sufficient energy is applied, the pores are electroporated irreversibly (Table 2.3).

References

1. Wan PJ, Hron RJ, Dowd MK, Kuk MS, Conkerton EJ (1995) Alternative hydrocarbon solvents for cottonseed extraction: plant trials. J Am Oil Chem Soc 72:661–664
2. Hanmoungjai P, Pyle L, Niranjan K (2000) Extraction of rice bran oil using aqueous media. J Chem Technol Biotechnol 75:348–352
3. Gedye R, Smith F, Westaway K, Ali H, Baldisera L, Laberge L, Rousell J (1986) Tetrahedron Lett 27:279–282
4. Ganzler K, Salgó A, Valkó K (1986) J Chromatogr A 371:299–306
5. Chemat F, Abert Vian M, Visinoni F (2010) Microwave hydro-diffusion for isolation of natural products. United States Patent, US 0,062,121
6. Abert Vian M, Fernandez X, Visioni F, Chemat F (2008) Microwave hydrodiffusion and gravity, a new technique for extraction of essential oils. J Chromatogr A 1190:14–17
7. Bousbia N, AbertVian M, Ferhat MA, Meklati BY, Chemat F (2009) A new process for extraction of essential oil from citrus peels: microwave hydrodiffusion and gravity. J Food Eng 90:409–413
8. Bousbia N, AbertVian M, Ferhat MA, Peticolas E, Meklati BY, Chemat F (2009) Comparison of two isolation methods for essential oil from rosemary leaves: hydrodistillation and microwave hydrodiffusion and gravity. Food Chem 14:355–362
9. Zill-e-Huma M, Abert-Vian JF, Maingonnat FC (2009) Clean recovery of antioxidant flavonoids from onions: optimising solvent free microwave extraction method. J Chromatogr A 1216:7700–7707
10. Zill-e-Huma, Abert-Vian M, Fabiano-Tixier AS, Elmaataoui M, Dangles O, Chemat F (2011) A remarkable influence of microwave extraction: enhancement of antioxidant activity of extracted onion varieties. Food Chem 127:1472–1480
11. Périno-Issartier S, Zill-e-Huma, Abert-Vian M, Chemat F (2010) Solvent free microwave-assisted extraction of antioxidants from sea buckthorn (Hippophae rhamnoides) food by-products. Food Bioprocess Technol 4:1020–1028
12. Allaf K, Vidal P (1989) Feasibility study of a new process of drying/swelling by instantaneous decompression toward vacuum of in pieces vegetables in view of a rapid re-hydration. Gradient Activity Plotting, University of Technology of Compiegne UTC N° CR/89/103, industrial SILVA-LAON partner
13. Allaf K, Louka N, Parent F, Bouvier J, Forget M (1999) Method for processing materials to change their texture, apparatus therefor, and resulting materials. United States Patent, US 5,855,941

14. Ben Amor B, Allaf K (2009) Impact of texturing using instant pressure drop treatment prior to solvent extraction of anthocyanins from Malaysian Roselle (Hibiscus sabdariffa). Food Chem 115(3):820–825
15. Amor BB, Lamy C, Andre P, Allaf K (2008) Effect of instant controlled pressure drop treatments on the oligosaccharides extractability and microstructure of Tephrosia purpurea seeds. J Chromatogr A 1213(2):118–124
16. Berka-Zougali B, Hassani A, Besombes C, Allaf K (2010) Extraction of essential oils from Algerian myrtle leaves using instant controlled pressure drop technology. J Chromatogr A 1217(40):6134–6142
17. Besombes C, Berka-Zougali B, Allaf K (2010) Instant controlled pressure drop extraction of lavandin essential oils: fundamentals and experimental studies. J Chromatogr A 1217(44):6807–6815
18. Kristiawan M, Sobolik V, Allaf K (2008) Isolation of Indonesian cananga oil using multi-cycle pressure drop process. J Chromatogr A 1192:306–318
19. Allaf T, Allaf K (2014) Instant controlled pressure drop (D.I.C.) in food processing. Springer, New York
20. Puértolas E, López N, Condón S, Álvarez I, Raso J (2010) Potential applications of PEF to improve red wine quality. Trends Food Sci Technol 21(5):247–255
21. Grimi N, Praporscic I, Lebovka N, Vorobiev E (2007) Selective extraction from carrot slices by pressing and washing enhanced by pulsed electric fields. Sep Purif Technol 58(2):267–273
22. Toepfl S, Heinz V, Knorr D (2005) Overview of pulsed electric field processing for food. In: Emerging technologies for food processing. Academic, London, pp 69–97
23. Corrales M, Toepfl S, Butz P, Knorr D, Tauscher B (2008) Extraction of anthocyanins from grape by-products assisted by ultrasonics, high hydrostatic pressure or pulsed electric fields: a comparison. Innov Food Sci Emerg Technol 9(1):85–91
24. Lebovka N, Vorobiev E, Chemat F (eds) (2011) Enhancing extraction processes in the food industry. CRC Press, Cambridge
25. Vorobiev E, Lebovka N (eds) (2008) Electrotechnologies for extraction from food plants and biomaterials. Springer, New York
26. Jaeger H, Schulz A, Karapetkov N, Knorr D (2009) Protective effect of milk constituents and sublethal injuries limiting process effectiveness during PEF inactivation of Lb. rhamnosus. Int J Food Microbiol 134(1–2):154–161
27. Roodenburg B, Morren J, (Iekje) Berg HE, de Haan SWH (2005) Metal release in a stainless steel pulsed electric field (PEF) system: part II. The treatment of orange juice; related to legislation and treatment chamber lifetime. Innov Food Sci Emerg Technol 6(3):337–345
28. Roodenburg B et al (2010) Conductive plastic film electrodes for Pulsed Electric Field (PEF) treatment – a proof of principle. Innov Food Sci Emerg Technol 11(2):274–282
29. Grimi N, Lebovka NI, Vorobiev E, Vaxelaire J (2009) Effect of a pulsed electric field treatment on expression behavior and juice quality of chardonnay grape. Food Biophys 4(3):191–198
30. Loginova KV, Shynkaryk MV, Lebovka NI, Vorobiev E (2010) Acceleration of soluble matter extraction from chicory with pulsed electric fields. J Food Eng 96(3):374–379
31. Schilling S et al (2007) Effects of pulsed electric field treatment of apple mash on juice yield and quality attributes of apple juices. Innov Food Sci Emerg Technol 8(1):127–134
32. Boussetta N, Lebovka N, Vorobiev E, Adenier H, Bedel-Cloutour C, Lanoisellé J-L (2009) Electrically assisted extraction of soluble matter from chardonnay grape skins for polyphenol recovery. J Agric Food Chem 57(4):1491–1497
33. Turk MF, Baron A, Vorobiev E (2010) Effect of pulsed electric fields treatment and mash size on extraction and composition of apple juices. J Agric Food Chem 58(17):9611–9616
34. Puértolas E, Saldaña G, Alvarez I, Raso J (2010) Effect of pulsed electric field processing of red grapes on wine chromatic and phenolic characteristics during aging in oak barrels. J Agric Food Chem 58(4):2351–2357

35. Loginova KV, Vorobiev E, Bals O, Lebovka NI (2011) Pilot study of countercurrent cold and mild heat extraction of sugar from sugar beets, assisted by pulsed electric fields. J Food Eng 102(4):340–347
36. Yin Y, Cui Y, Ding H (2008) Optimization of betulin extraction process from Inonotus Obliquus with pulsed electric fields. Innov Food Sci Emerg Technol 9(3):306–310
37. Guderjan M, Tepfl S, Angersbach A, Knorr D (2005) Impact of pulsed electric field treatment on the recovery and quality of plant oils. J Food Eng 67(3):281–287
38. Fincan M, DeVito F, Dejmek P (2004) Pulsed electric field treatment for solid-liquid extraction of red beetroot pigment. J Food Eng 64(3):381–388

Chapter 3
Supercritical Fluid Extraction: A Global Perspective of the Fundamental Concepts of this Eco-Friendly Extraction Technique

Susana P. Jesus and M. Angela A. Meireles

Abstract Supercritical fluid extraction (SFE) is a green technology that has been applied on a commercial scale for more than three decades. SFE is a high-pressure extraction method in which a mixture of solutes is separated from a solid matrix by bringing the mixture into contact with a fluid in the supercritical state. A supercritical fluid has very particular and unique characteristics, which enable its use as an efficient extraction solvent. Carbon dioxide (CO_2) is the most commonly used supercritical fluid and has applications in food, cosmetic, pharmaceutical, and correlated industries. Many research works have already demonstrated that SFE is a technically feasible process that may also be commercially competitive in terms of economic viability. Although SFE is commercially carried out in several countries, it is nonetheless still considered as an emerging technology. This emerging status remains associated with SFE technology because the conventional low-pressure extraction methods remain the most frequently used extraction techniques, in particular due to the comparatively low cost of investment that is required for installing a low-pressure industrial plant. The physical phenomena that occur during SFE have already been extensively investigated, and there is consensus that SFE is a complex phenomenon that involves multicomponent systems. However, various simplifications can be performed to describe SFE for the purpose of process design. Presently, one of the major challenges for researchers in this area is the proposition of practical procedures (experimental and/or calculation methods) in order to simplify the determination of some process parameters which are required for the studies of economic feasibility. This chapter presents the fundamental concepts of SFE and gives special attention to the information that must be available to conduct preliminary studies of process design and cost estimation.

S.P. Jesus · M.A.A. Meireles (✉)
LASEFI/DEA/FEA (School of Food Engineering)/UNICAMP (University of Campinas),
Rua Monteiro Lobato, 80, Campinas-SP, CEP:13083-862, Brazil
e-mail: meireles@fea.unicamp.br

3.1 The Supercritical Fluid Extraction Technique

The consumers' increasing concern about environmental issues and human health has motivated the development of green technologies and the search for natural ingredients with bioactive properties. In fact, the natural products market has presented a progressive and continuous growth in the last decades. Natural matrices are complex multicomponent systems, and so the selective separation of specific substances is a difficult task that requires efficient extraction methods [1]. Rostagno and Prado [1] recently published a book that presents a global view of the state-of-the-art techniques for the extraction and processing of natural products. These authors claim that there is a need for more efficient and selective processes, which can improve the overall quality of natural products and also enable the development of innovative products [1]. Nonetheless, most of the industries still use conventional techniques that are based on outdated technologies. Considering this scenario, the supercritical fluid extraction (SFE) is a particularly interesting alternative to extract bioactive compounds from natural sources. Therefore, the SFE process has many potential applications in food, pharmaceutical, and cosmetic industries.

The SFE is a high-pressure extraction method that has been carried out on a commercial scale since the 1980s. The industrial-scale applications of SFE comprise the decaffeination of green coffee beans and black tea leaves; the production of hop extracts; the extraction of essential oils, oleoresins, and flavoring compounds from herbs and spices; the extraction of high-valued bioactive compounds from different natural matrices; the extraction and fractionation of edible oils; and the removal of pesticides from plant material [2, 3]. At the very early stages of this technology, very large vessels (up to 40 m^3) were sometimes built. Later, the extractors' capacity became smaller, and today, most extractor vessels have a volume that is equal to or smaller than 1 m^3 [3].

According to Brunner [3], the costs of SFE processes are competitive. Furthermore, in particular cases, SFE processing is the only way to satisfy the product specifications. A significant number of SFE industrial plants of various capacities have been built since the 1980s. Most of the plants are distributed within Europe, the USA, Japan, and the Southeast Asian countries. The state-of-the-art technology that is necessary to design a SFE plant is commercially available. Standard designs can be acquired from many suppliers, and special designs can be custom tailored for a particular process [3].

SFE is a unit operation that performs the separation of a mixture of solutes from a solid matrix by bringing the mixture in contact with a supercritical solvent [4]. The solid material is placed in an extraction cell, forming a fixed bed of solid particles. The supercritical fluid flows continuously through the fixed bed and dissolves the extractable components of the solid [2]. The mixture of solutes that is removed from the solid matrix is named the extract. SFE processes are usually carried out in batch and single-stage modes because solids are difficult

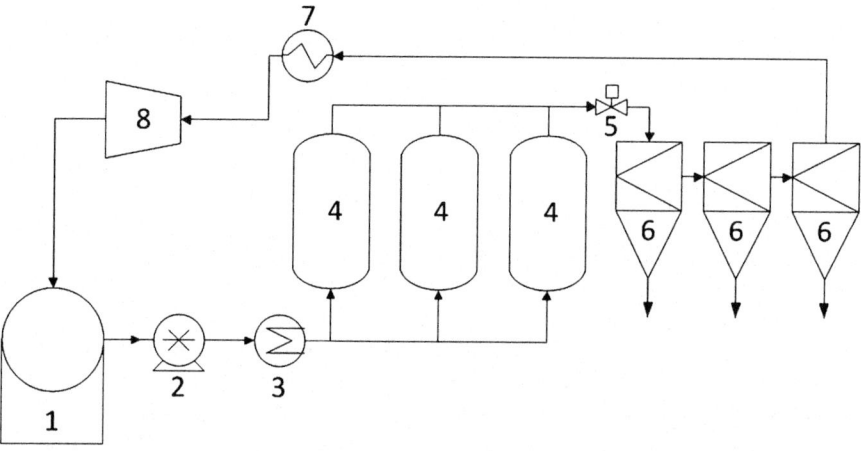

Fig. 3.1 A simplified flowchart of the SFE process (*1* CO_2 storage tank, *2* solvent pump, *3* heat exchanger, *4* extractor, *5* pre-expansion valve, *6* separator, *7* cooler, *8* compressor; the system contains several temperature and pressure controllers that are not shown) (Adapted from Pereira and Meireles [7], with kind permission from Springer Science and Business Media)

to handle continuously in pressurized vessels and separation factors are high [3]. Nonetheless, the modification of the process from batch to continuous mode can be performed by arranging two or more extractors in the process line [5, 6]. This change allows the system to operate continuously despite the occurrence of solid matrix exhaustion. Then, the arrangement of n extractors (where $n \geq 2$) operating in a parallel configuration results in the continuous production of the extract by intercalating the charge/discharge times of the n extractors in the plant. Plant operation in a continuous mode occurs according to the following format: while one extractor is in the charge/discharge step, the other $n-1$ extractors are in the extraction step [6]. This operating mode presents the advantages of reducing the process setup time and increasing productivity, which leads to a reduction of the operating costs [3, 6].

A simple SFE process comprises two major steps: extraction and separation. In the extraction step, the solvent is fed into the system and is uniformly distributed throughout the extractor. The solvent flows through the solid matrix, extracting the soluble compounds. In the separation step, the loaded solvent (the mixture formed by solvent + extract) is removed from the extraction cell and fed into the separator (flash tank), where the mixture is separated by a rapid reduction of the pressure. The extract precipitates in the separator, while the solvent is removed from the system and is delivered to a recycling step. The solvent is cooled and recompressed and then returns to a storage tank, which feeds the extraction system [2, 7]. A schematic diagram of the SFE process is shown in Fig. 3.1.

3.2 The Supercritical Fluid

A pure component is considered to be in the supercritical state when both its pressure (P) and temperature (T) are higher than their critical values (P_C and T_C, respectively) [2]. The supercritical region is illustrated in the phase diagrams presented in Figs. 3.2 and 3.3. In this region, the fluid can be considered either an expanded liquid or a compressed gas [4].

Supercritical fluids (SCFs) show very particular and unique characteristics that enable their use as efficient solvents. The densities of SCFs are relatively high (compared to gases), and consequently, SCFs have high solvation power. Furthermore, the density can be easily tuned by varying the system pressure or temperature. This particular effect provides these fluids with a certain degree of selectivity, which is useful for the extraction process and allows for easy solvent-solute separation. The separation step can be performed by either decreasing the pressure or increasing the temperature of the mixture (solvent + extract) leaving the extraction column [4]. In the supercritical state, liquid-like densities are approached, while the viscosity is near that of normal gases, and the diffusivity is approximately two orders of magnitude higher than that of the liquid forms [3]. Therefore, in comparison to a gas, a supercritical fluid (SCF) has higher density; in contrast, compared to a liquid, the SCF possesses lower viscosity and a higher diffusion coefficient. All of these characteristics result in a greater solvation power, which allows high extraction rates when SCFs are applied as solvents.

Supercritical carbon dioxide (SC-CO_2) is the most commonly used solvent for applications of SFE in the food, cosmetic, pharmaceutical, and other similar industries. According to Rosa and Meireles [4], two important justifications for the choice of CO_2 are its low critical temperature ($T_C = 304.2$ K) and mild critical pressure ($P_C = 7.38$ MPa). Additionally, CO_2 is not only cheap and readily available at high purity but is also safe to handle (nontoxic and nonflammable) and easily removed by simple expansion to common environmental pressure values [3]. Some

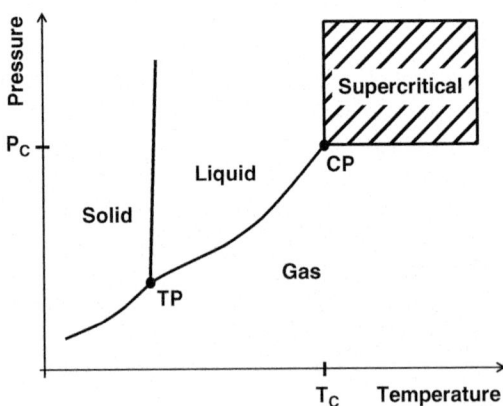

Fig. 3.2 A pure component P × T (pressure vs. temperature) diagram: the supercritical region is indicated by the hatched lines (*TP* triple point, *CP* critical point, P_C critical pressure, T_C critical temperature) (Adapted from Brunner [2], with kind permission from Springer Science and Business Media)

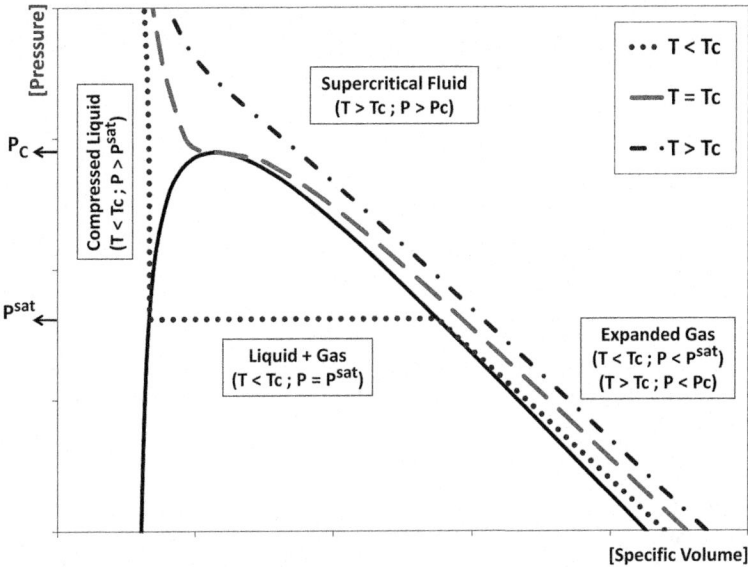

Fig. 3.3 Schematic illustration of a pure component P × V (pressure vs. volume) diagram (*T* temperature, T_C critical temperature, *P* pressure, P_C critical pressure, P^{sat} saturation pressure)

well-noted advantages of the SFE process are the solvent recycling possibility, low energy consumption, adjustable solvent selectivity, prevention of oxidation reactions, and production of high-quality extracts.

The properties of SC-CO_2 can be modified over relatively wide ranges. The solvent power of SC-CO_2 is high for hydrophobic or slightly hydrophilic components and decreases with increasing molecular weight [3]. Generally, when the operational pressure is increased, more hydrophilic compounds can also be extracted. If the goal is the extraction of more hydrophilic compounds, then the solvent polarity can be increased by the addition of a polar solvent. The added solvents are named the cosolvents or modifiers [4]. The cosolvent is generally a solvent of high polarity, such as water or ethanol. These two solvents are conveniently selected because both are classified as GRAS (generally recognized as safe). Therefore, the green concept of supercritical technology is perfectly maintained. The cosolvent takes the form of a compressed liquid (see Fig. 3.3) when held in the usual operational conditions of the SFE process.

3.3 The Solid Matrix

In natural sources, the soluble portion of the solid matrix is generally composed of several different classes of organic compounds. As a result, the extract (or solute) is a complex mixture of chemical species, such as terpenes, terpenoids, flavonoids, alkaloids, and many other compounds [8]. The soluble fraction may be located inside cellular structures and may interact very strongly with the nonsoluble components of the raw material. Therefore, vegetable raw materials often pass through a pretreatment process to facilitate solvent access to the solute and to increase the solute-solvent interactions.

3.3.1 Raw Material Pretreatment

In SFE, the raw material commonly passes through a pretreatment stage before it is fed into the fixed bed extractor. Pretreatment is performed to prepare the solid particles, allowing the best possible efficiency to be achieved in the extraction process. In most cases, the pretreatment process comprises one or more of the following steps:

- *Drying*: A drying step is often used to adjust the water content of the solid matrix. If the target compound is a nonpolar or slightly polar substance, then the water content is reduced to increase the extraction efficiency. However, if the target compound has a more polar structure, the drying process may not be necessary or adequate. In some cases, the initial water contained in the solid particles can act as a cosolvent and improve the extraction efficiency of certain polar compounds.
- *Milling*: The main purpose of the milling step is the reduction of the solid particle sizes to enlarge the interfacial solid-fluid mass transfer area. Furthermore, the milling process may also cause the destruction of some plant cellular structures and, consequently, facilitate solvent access to the solute. Nonetheless, reducing the particle size also increases the degree of compaction of the solid substrate. Excessive bed compaction must be avoided because it can result in the formation of preferential pathways of solvent access, preventing the solvent from reaching all of the extractable material [5].
- *Sieving*: A sieving step is generally applied to standardize the size of the solid particles. Some particles may be discarded according to the particle diameter range of interest.
- *Chemical reaction*: A reaction step is not commonly applied, but it can be useful in particular cases. A chemical reaction may be performed to free the target solutes and improve the extraction efficiency.

3.4 The Definition of the Pseudoternary System

In SFE from natural matrices, the obtained extracts are complex mixtures composed of different groups of chemical compounds. Therefore, the extract is always a multicomponent system. Additionally, the solid matrix is a very complex mixture that can contain intact cellular structures, as well as broken cellular structures [8, 9].

Knowledge of the system's composition and the physical phenomena that occur inside the extraction bed is essential for creating a detailed description of the SFE process. This knowledge is also fundamental to decision making with respect to simplifying the description of the phenomena that take place within the extraction cell. With respect to composition, some assumptions may be used to facilitate the description of the SFE system (solid material + solvent). According to Rodrigues et al. [8], a very simplified picture of the system is developed when it is treated as been formed by three pseudocomponents (extract + cellulosic structure + solvent), which are defined below:

- *Extract (or solute)*: The extract is a multicomponent mixture composed of the solids that are soluble in the extraction solvent. The extract interacts with both the supercritical solvent and the cellulosic structure [8].
- *Cellulosic structure (or inert material)*: The cellulosic structure is formed by a multicomponent mixture that contains all of the solids that are insoluble in the supercritical solvent. It is crucial to note that although being inert to the solvent action, the cellulosic structure interacts strongly with the extract [8].
- *Solvent*: The solvent can be either a pure component (the fluid in the supercritical state) or a mixture of the supercritical fluid and a cosolvent. In the typical operating conditions of SC-CO_2 extraction, the cosolvent (water, ethanol, among others) is a compressed liquid.

3.5 Thermodynamic Aspects

The design of an engineering project of a SFE system requires knowledge of the limitations that control the extraction process. According to Ferreira and Meireles [10], the constraints of the SFE are related to two aspects: (a) the thermodynamics (solubility and selectivity) and (b) the mass transfer phenomena. A discussion of the first is presented in this section, while the second aspect is treated in Sect. 3.6.

3.5.1 Equilibrium Solubility (Y^*)

The driving potential for mass transfer is determined by the difference relative to the equilibrium state. According to Brunner [3], the phase equilibrium provides information regarding (a) the capacity of the supercritical solvent, which is directly related to the solubility of a specific solute in the solvent (the solubility is the amount of a solute that is dissolved by the supercritical solvent at thermodynamic equilibrium); (b) the selectivity of a supercritical solvent, which can be described as the ability of a solvent to selectively dissolve one or more compounds; and (c) the dependence of these two solvent properties on the conditions of state (P and T). If the capacity and selectivity are known, a guess can be made regarding whether a separation problem can be solved using a supercritical solvent [3].

It should be noted that two different approaches can be adopted when considering the equilibrium solubility of an extract within a supercritical fluid, including (a) the solubility of the pseudobinary system ($Y_{BIN}*$), which is composed only of the extract + solvent, and (b) the solubility of the pseudoternary system ($Y_{TER}*$), as described in Sect. 3.4 (cellulosic structure + extract + solvent). It is well known that the cellulosic structure strongly interacts with the extract. Thus, the solubility of a solute as measured in the pseudobinary system differs significantly from the solubility of the same solute when measured in the pseudoternary system [9]. A good example of the influence of the cellulosic structure on the solubility value is given by Brunner [2]: the solubility of pure caffeine in SC-CO_2 (binary system) is approximately 20 times greater than the solubility of caffeine measured for the pseudoternary system (caffeine + coffee grains + SC-CO_2) at the same conditions of temperature and pressure. Brunner [2] also mentioned that the concentration of caffeine in the supercritical solvent throughout most time of the SFE process is less than 100 ppm. This value is significantly below the solubility of caffeine as measured for the pseudoternary system ($Y_{TER}* = 200$ ppm at $T = 350$ K and $P = 30$ MPa). Then, it can be said that when the solubility of the pseudoternary system is relatively high (as in the caffeine example), the mass ratio of the solute in the fluid phase (Y) will likely be significantly lower than $Y_{TER}*$ during typical SFE operational conditions.

Equilibrium solubility is only reached under specific processing conditions. A detailed discussion of the experimental determination of the pseudoternary solubility is presented by Rodrigues et al. [8]. These authors used the dynamic method to measure the pseudoternary solubility of extracts from three vegetable raw materials (clove buds, ginger, and eucalyptus). In the dynamic method, a typical SFE experiment is performed: the solvent is continuously fed into an extraction column at a given pressure and temperature using a solvent flow rate (Q_{CO2}) that assures saturation at the exit of the column [4]. Rodrigues et al. [8] demonstrated that there is a particular solvent flow rate (denoted $Q*$) at which the equilibrium is achieved and the solubility must be measured. Therefore, the use of the dynamic method requires that a certain set of experiments must be performed to determine the specific solvent flow rate at which the solvent leaves the extraction cell under the saturation condition [11]. This is necessary because, under large flow rates, there is insufficient contact time to guarantee that the solvent is saturated. However, at very low solvent flow rates, axial dispersion may interfere with the measurement of solubility. Hence, there is an optimum solvent flow rate that is a function of the raw material and the thermodynamic state (P and T) used in the SFE process [4, 11].

In the dynamic method, the equilibrium solubility is given by the slope of the linear part of the overall extraction curve (OEC) (this curve is discussed extensively in Sect. 3.6.2). The work presented by Rodrigues et al. [8] showed that the experimental determination of $Y_{TER}*$ requires a slow, tedious, and costly experimental investigation because it is necessary to determine the CO_2 flow rate that can be used safely for the measurement of the equilibrium solubility [9]. In some works, the solubility is simply calculated by using the slope of the linear part of an OEC determined under a random solvent flow rate (i.e., $Q_{CO2} \neq Q*$). Meireles

[9] states that in this case, the measured value should be referred to as $Y_{S/F}*$ and that there is a clear difference between $Y_{TER}*$ and $Y_{S/F}*$. This author also mentioned that the difference can be understood by recalling that to measure the first value (the true solubility in the pseudoternary system), it is expected that equilibrium is achieved during the extraction experiment (i.e., Q_{CO2} must be equal to $Q*$). In the second case, the "solubility" ($Y_{S/F}*$) is measured at a given solvent-to-feed (S/F) mass ratio using a random solvent flow rate. In the latter case, there is no guarantee that the saturation of the solvent is reached; thus, the value of $Y_{S/F}*$ cannot be treated as the real equilibrium solubility.

3.5.2 Global Yield Isotherms (GYI)

When studying a SFE system, one of the first fundamental steps is the selection of the temperature and pressure parameters, which must be chosen by taking into account the quality and purity of the obtained extract. The quality of an extract is determined by its chemical composition, which is directly related to the selectivity of the solvent. Thus, a set of experiments must be performed based on various combinations of temperature and pressure because both thermodynamic parameters are strongly related to selectivity and solubility. These experiments deliver information regarding the solvent density, which is directly associated with the solvent power and consequently with the adjustable selectivity of SC-CO_2. Moreover, these experiments also provide information regarding the solubility of the solute in the supercritical solvent. According to Carvalho et al. [12], the investigation of a SFE process requires some knowledge of the behavior of the system of "solid material + CO_2." The interactions of the extract with both the solvent and cellulosic structure are fundamental to understanding the extraction process. However, very little is known regarding these interactions because they involve multicomponent systems of high complexity. The extension of these "solute-solvent" and "solute-cellulosic structure" interactions can be evaluated through two types of experiments: (a) the determination of the solubility of the pseudoternary system (as previously discussed in Sect. 3.5.1) under different conditions of temperature and pressure and (b) the results of the global yield isotherms (GYI) [12]. In GYI experiments, an exhaustive extraction is conducted under different conditions of temperature and pressure.

Meireles [9] claimed that to obtain reliable results for $Y_{TER}*$, the experiments used to determine solubility must be performed in a SFE unit containing an extractor vessel with a volume of at least 50 cm^3. This requirement is because in these experiments, an overall extraction curve (OEC) (see Sect. 3.6.2) must be built; thus, the use of small amounts of feed material is generally associated with relatively high experimental errors. Moreover, the solubility measurements require difficult experimental work (as discussed in Sect. 3.5.1). However, the GYI experiments are comparatively easy to conduct because they only require an exhaustive extraction.

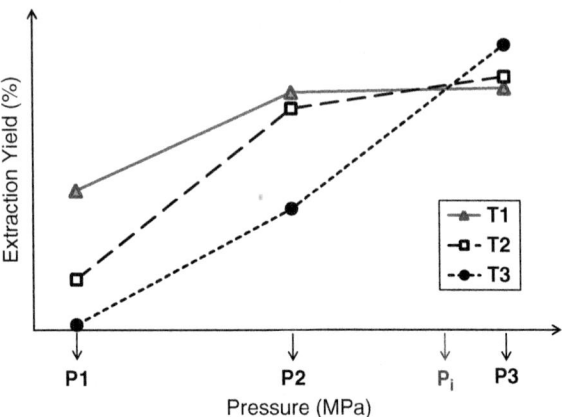

Fig. 3.4 Schematic illustration of the global yield isotherms (T1 < T2 < T3, P1 < P2 < P3, *Pi* crossover pressure) (the experimental points were connected by lines only to evidence the crossover point) (Adapted from Jesus et al. [13])

In this case, extractor vessels of small volumes (such as 5 cm^3) and, consequently, small amounts of the feed material can be safely used to perform GYI assays because there is no need to build an OEC [9]. Therefore, taking into account all the aspects cited above, it is apparent that the choices of operating temperature and pressure may be easier upon consideration of the results of GYI experiments.

In terms of the total extraction yield or the yield of a specific target compound, the results from GYI assays are generally plotted on a graph similar to the schematic illustration presented in Fig. 3.4. From this plot, it is possible to evaluate the effects of the parameters temperature and pressure on the extraction yield. Taking into account an isothermal condition, the effect of operational pressure can be understood. It is clear that a rising pressure results in an increasing extraction yield. This effect is attributed to the increase in CO_2 density and, consequently, the enhancement of its solvation power (although, a higher solvation power may be associated with lower selectivity) [13]. The effect of the operational temperature in SFE is typically more complex due to the combination of two variables, density and vapor pressure. The vapor pressure of the solute increases with temperature, causing increased solubility. However, the solvent density decreases with increasing temperature, causing reduced solubility [11]. As a result, these two variables cause inverse effects on the extraction yield. It is well known that the dominant effect depends on the magnitudes of both effects individually.

At relatively low pressures ($P < Pi$, according to Fig. 3.4), the effect of solvent density prevails; thus, increasing the temperature results in a reduction of the extraction yield. However, at relatively high pressures ($P > Pi$, according to Fig. 3.4), the effect of vapor pressure dominates; as a result, increasing the temperature enhances the extraction yield [2, 11, 13]. The pressure at which the inversion of the dominant mechanism occurs is known as either the crossover point or the crossover pressure. From the GYI graph (Fig. 3.4), it can be said that the crossover pressure (*Pi*) falls somewhere between P2 and P3. At pressures less than *Pi*, the solvent density always dominates, while at pressures higher than *Pi*, the dominant mechanism is the solute vapor pressure. The crossover point is a characteristic of each SFE system (solvent + solute + cellulosic structure) and must be experimentally determined for each distinct pseudoternary system.

Generally, when working with SFE from natural matrices, the major goal is to produce extracts that are enriched in bioactive compounds. As a result, it is important to hold in mind that the selection of the operating temperature and pressure must be made by taking into account the extract characterization in terms of its chemical composition and functional properties. To do so, the extracts obtained in the GYI experiments should be characterized using appropriate methods, such as gas chromatography with flame ionization detection (GC-FID), gas chromatography-mass spectrometry (GC-MS), high-performance liquid chromatography (HPLC), and ultraviolet spectrophotometry, among others [9]. Additionally, the bioactive properties of the material should also be investigated, particularly if the production of nutraceutical products is the purpose of the extraction process.

3.6 Mass Transfer Aspects

The mass transfer mechanisms that occur in SFE from natural solid matrices are not readily understood. The difficulties encountered in describing and modeling the SFE process arise from the fact that SFE involves multicomponent systems with a significant number of components, which can belong to many different chemical classes. Therefore, it is very difficult to establish the interactions between the solvent, the solutes, and the solid matrix [10].

3.6.1 The Mass Balance Equations in the Fixed Bed Extractor

The SFE process is generally performed in a fixed bed extractor of cylindrical shape. The solid particles are packed in the extraction cell, forming a fixed bed through which the supercritical solvent is continuously flowed. A schematic representation of the fixed bed extractor is shown in Fig. 3.5.

It is crucial to propose simplifications when carrying out calculations of the process design. Some simplifications must be assumed to reduce the problem to one that is mathematically tractable. To simplify the description of the SFE process, the extraction system is usually treated as a pseudoternary (cellulosic structure + extract + solvent) and biphasic system (fluid phase + solid phase). The fluid phase (solvent + extract) and the solid phase (cellulosic structure + extract) are both pseudobinary systems [9, 14]. A schematic diagram of the components inside the fixed bed extractor is presented in Fig. 3.6.

When evaluating the mass balance of SFE, it is typical to assume that the extraction cell is a cylindrical bed in which the solid particles are homogeneously distributed. The solvent flows in the axial direction (z), and the extractor geometry

Fig. 3.5 A typical fixed bed extractor of the SFE process (z axial coordinate, H_B bed height)

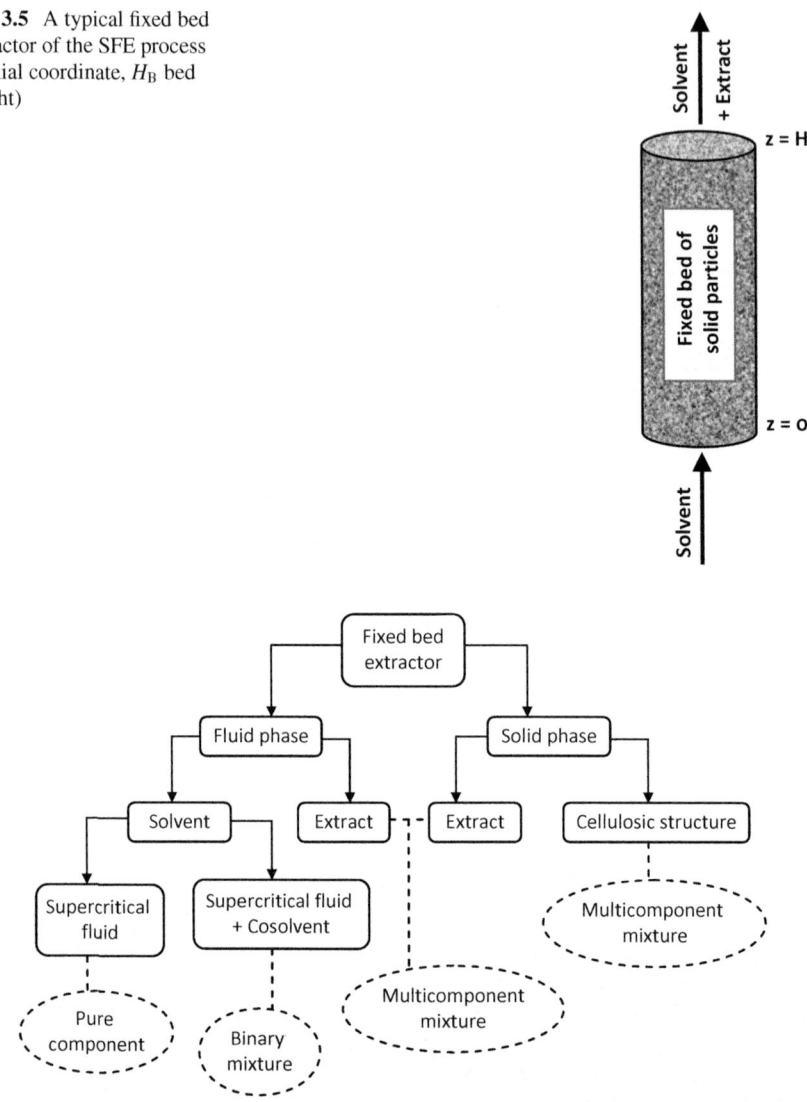

Fig. 3.6 Diagram of the fixed bed extractor composition in SFE from natural matrices

is such that the bed height can be considered infinitely larger than the bed diameter ($H_B >>> d_B$). Then, the terms of the radial (r) and tangential ($\ominus$) directions can be neglected in the mass balance equations. Moreover, the solid and fluid phases can be taken as nonreactive systems. By taking into account all of these assumptions, the mass balance in the extraction bed can be described by Eqs. 3.1 and 3.2 [14, 15]. It is interesting to note that in SFE, the fluid phase can be treated as a diluted solution; therefore, the solvent properties can replace the fluid-phase properties [10].

- *fluid phase*: [Accumulation] + [Convection] = [Dispersion] + [Interfacial Mass Transfer]

$$\frac{\partial Y}{\partial t} + u_i \frac{\partial Y}{\partial z} = \frac{\partial}{\partial z}\left(D_{aY}\frac{\partial Y}{\partial z}\right) + \frac{J(X,Y)}{\varepsilon} \quad (3.1)$$

- *solid phase*: [Accumulation] = [Diffusion] + [Interfacial Mass Transfer]

$$\frac{\partial X}{\partial t} = \frac{\partial}{\partial z}\left(D_{aX}\frac{\partial Y}{\partial z}\right) + \frac{J(X,Y)}{(1-\varepsilon)}\frac{\rho_{CO_2}}{\rho_S} \quad (3.2)$$

where Y is the mass ratio of the solute in the fluid phase (kg/kg), X is the mass ratio of the solute in the solid phase (kg/kg), t is the extraction time (s), u_i is the interstitial velocity of the solvent (m/s), z is the axial direction (m), D_{aY} is the dispersion coefficient in the fluid phase (m^2/s), D_{aX} is the diffusion coefficient in the solid phase (m^2/s), ρ_{CO2} is the solvent density (kg/m^3), ρ_S is the true density of the solid matrix (kg/m^3), $J(X,Y)$ is the interfacial mass transfer term (s^{-1}), and ε is the bed porosity (dimensionless).

The mass balance equations of the fluid and solid phases have been applied by several authors who have proposed many mathematical models based on the mass transfer phenomena that occur inside the extraction bed. One of the main differences among the proposed mathematical models is how each author describes the interfacial mass transfer term. This description depends on the personal assumptions that are made by each author when developing a different mass transfer model. Some of the mathematical models available in the literature are discussed in Sect. 3.7.

3.6.2 The Overall Extraction Curve (OEC)

According to Brunner [2], the course of SFE can be evaluated by analyzing the variables of (a) the total amount of extract, (b) the extraction rate, (c) the remaining amount of extract in the solid, and (d) the concentration of the extract in the supercritical solvent at the extractor outlet. All of the cited variables can be plotted as a function of the extraction time (or solvent consumption) to obtain curves that give important information regarding the SFE process. In most cases, variable (a) is selected such that the course of the extraction process is followed by determining the accumulated mass of the extract against the extraction time (or solvent consumption). This representation is the most commonly used and is well known as the *overall extraction curve* (OEC). The information provided by the OEC is useful for comparing the extraction results within a series of experiments when using the same solid matrix [2, 3].

The mass of the extract that accumulates during the SFE process is typically shaped as shown in the schematic curve presented in Fig. 3.7. The first part (*P–I*) of the curve is a straight line and, therefore, corresponds to a constant extraction

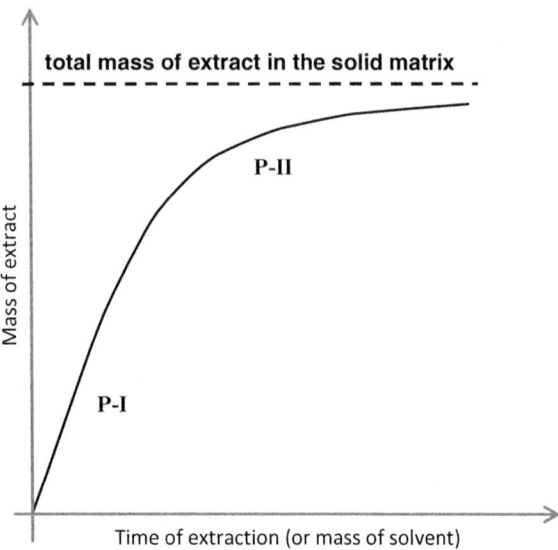

Fig. 3.7 The typical overall extraction curve (OEC) (*P-I* part 1, *P-II* part II) (Adapted from Brunner [2], with kind permission from Springer Science and Business Media)

rate period. The second part (P-II) is a nonlinear function that approaches a limiting value, that is, the total amount of extractable substances in the solid matrix [2]. Under certain processing conditions (as discussed in Sect. 3.5.1), the slope of the linear part of the graph may be given by the equilibrium solubility. However, it is fundamental to remember that the straight line generally occurs because the mass transfer resistance remains constant in the early stages of the extraction process. Therefore, the presence of the linear region is not a proof that equilibrium conditions have been attained during SFE [2, 3].

The shape of the OEC depends on the kinetics of solute extraction from the solid matrix and the solvation power of the SC-CO_2, which in turn depends on the operational conditions [3]. The course of the SFE from a solid matrix follows two types of curves for the extraction rate, as can be seen in Fig. 3.8. Curve 1 (C1) represents the extraction rate when a high initial concentration of solute in the solid substrate exists or when the solute is readily available to the solvent. Curve 2 (C2) represents the extraction rate when a low initial concentration of solute exists in the solid substrate or when the solute is not readily available to the solvent. Curve 2 also corresponds to the second part (P-II) of curve 1 because a depletion phase always comes after the first part (P-I, where a constant extract concentration is observed in the fluid phase at the outlet of the extraction cell) [2].

According to Brunner [2], the first part (P-I) of curve 1 (C1) has several main characteristics: (a) in the fluid phase, the mass transfer resistance dominates the process, (b) the solute compounds are readily available at the interface solid/fluid, and (c) a constant amount of extract is transferred to the bulk of the supercritical solvent, resulting in a constant concentration at the bed outlet. In the second part (P-II) of curve 1 (C1), as well as in curve 2 (C2), the extract concentration decreases with increasing extraction time due to the increasing mass transfer resistances and

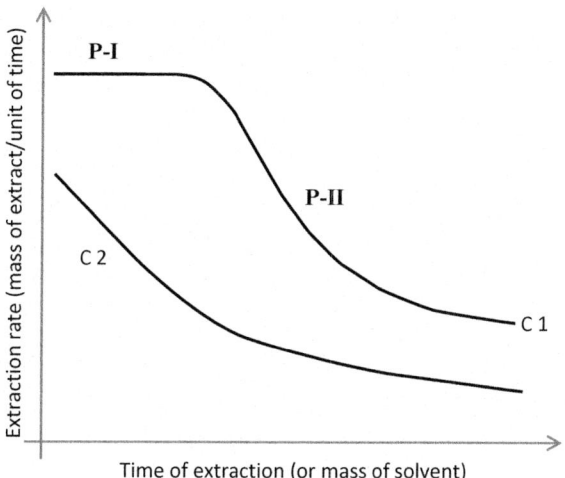

Fig. 3.8 Extraction rate curve: schematic illustration of curve 1 (C1) and curve 2 (C2) as described by Brunner [2] (the OEC from curve 1 has the shape previously presented in Fig. 3.7; *P-I* part 1, *P-II* part II) (Adapted from Brunner [2], with kind permission from Springer Science and Business Media)

the depletion of the extract in the solid phase. The solid matrix will be depleted of the extractable material in the direction of flow. The concentration of extract components increases in the direction of flow both in the SCF and in the solid material [2, 3].

3.7 Mathematical Modeling

Mathematical models based on the mass transfer phenomena, or even with merely empirical basis, are important tools in SFE investigations. The mathematical modeling of extraction curves may help develop an understanding of the kinetic behavior of SFE through the definition of extraction rates, steps, time, and/or mass transfer parameters with strong physical meaning [5]. The modeling of OECs helps the determination of the extraction time (cycle time), which is important for achieving the optimal utilization of an industrial-scale plant [2]. The main goal of using a mathematical model is the determination of parameters that may be applied to key aspects of process design, such as equipment dimensions, the solvent flow rate, particle size, and the solvent-to-feed (S/F) mass ratio, among others [16]. Thus, mathematical models can be useful tools for scale-up prediction, process design, and/or cost estimation purposes.

Knowledge of the initial distribution of a solute in the solid substrate directly affects the selection of the models that can adequately describe a given SFE system. The extractable substances may be distributed within the solid matrix in various ways. The solute can be (a) located freely on the surface of the solid material, (b) adsorbed on the outer surface of the solid material, (c) heterogeneously distributed inside the solid particle (located inside the pores or other specific cell structures), or (d) evenly distributed within the solid particles [17].

Many mathematical models have been developed to describe the OEC, ranging from simple equations to very complex equations. Some extensive reviews concerning the mathematical modeling of SFE were presented by Oliveira et al. [18] and Sovová [19], among other authors. In this chapter, it is not our intention to deliver a detailed discussion of all models available in the literature. Thus, we take a classical approach and focus on the fundamental concepts while presenting some well-known models from the SFE literature. According to Reverchon [17], the mathematical models used to describe the OEC can be divided into three main categories based on the approaches of (a) empirical evidence, (b) heat transfer analogy, or (c) differential mass balance integration.

The models developed from the first category are based on the hyperbolic shape of the typical OEC. One example is the model proposed by Esquível et al. [20] for describing the SFE of oil from olive husk. The empirical models use a hyperbolic function to fit the experimental data. The general form of the models from this category can be given by Eq. 3.3 [4]:

$$m_{\text{EXT}} = X_0 F \left(\frac{t}{C_1 + t} \right) \tag{3.3}$$

where m_{EXT} is the mass of the extract (kg), t is the time of extraction (s), F is the mass of the feed material (kg), X_0 is the initial mass ratio of the extractable solute in the solid substratum (kg/kg), and the constant C_1 is an adjustable parameter that has no physical meaning (s). The empirical model may give good fits in some particular cases, but it does not give any phenomenological information regarding the SFE process. Thus, this model has limited application in terms of scale-up and process design.

In the second category, an analogy is considered between SFE and the heat transfer by diffusion. In this case, all mass transfer is considered to happen based only on the mechanism of diffusion, allowing an apparent diffusion coefficient to be obtained [4]. The model presented by Crank [21] for the description of heat transfer in a solid particle cooling in a uniform medium was adapted by Reverchon [17] and was used to fit SFE data [15]. Reverchon [17] applied Fick's second law of diffusion to obtain a model that describes the OEC according to Eq. 3.4:

$$m_{\text{EXT}} = X_0 F \left[1 - \frac{6}{\pi^2} \sum_{n=1}^{\infty} \frac{1}{n^2} \exp\left(\frac{-n^2 \pi^2 D_{\text{ef}} t}{r^2} \right) \right] \tag{3.4}$$

where m_{EXT} is the mass of the extract (kg); t is the time of extraction (s); F is the mass of the feed material (kg); X_0 is the initial mass ratio of the extractable solute in the solid substratum (kg/kg); n is an integer number; r is the radius of the spherical particle (m); and D_{ef} is the adjustable parameter, which represents the effective diffusion coefficient of the solute within the solid matrix (m^2/s). The application of the diffusion model is restricted to very few systems because in most cases, it results in a poor fit. This behavior is expected because mass transfer in SFE may not be properly described by diffusion alone because convective mass transport dominates the beginning of the process [4].

The third category comprises the majority of the mathematical models proposed for the description of SFE processes. The starting point is the evaluation of the differential mass balance (see Eqs. 3.1 and 3.2, which were presented in Sect. 3.6.1) inside the fixed bed extractor [4]. Then, each author gives a personal interpretation of the mass transfer phenomena that happen in both the fluid and solid phases. An example from this category is the model presented by Sovová [22], which has been extensively used by various researchers of SFE. A fundamental characteristic of this model is that the solute is distinguished in two different fractions, one present in broken cells and the other in intact cells [4]. As a result, this model was developed for application when the raw material passes through a milling process before extraction (see Sect. 3.3.1). The solute fraction present in the broken cells is denoted as the easily accessible solute (X_P), which is located at the particle surface and is the first fraction extracted. The fraction contained in the intact cells is denoted as the hardly accessible solute (X_K) and is located inside the solid particle. The OEC follows the shape of the type 1 curve (C1) described by Brunner [2] (as discussed in Sect. 3.6.2).

Sovová [22] divided the OEC into three distinguishable regions [10, 11] as follows:

- *Constant extraction rate (CER)*: In the CER period, the external surfaces of the solid particles are assumed to be fully covered with the easily accessible solute. In this region, the solute is essentially removed by convection; thus, the mass transfer resistance exists in the fluid phase.
- *Falling extraction rate (FER)*: In the FER period, flaws in the superficial solute layer begin to appear, and so the hardly accessible solute starts to be extracted. As a result, the solute is extracted by both convection and diffusion mechanisms. This is a transition period that is caused by the continuous depletion of the solute layer in the external surface.
- *Diffusion-controlled (DC)*: In the DC period, the solute at the particle surface is completely exhausted, and only the hardly accessible solute is available for extraction. As a result, mass transfer is controlled by intraparticle diffusion. The mass transfer resistance exists in the solid phase due to the low diffusivity of the solute in the solid matrix.

The model developed by Sovová [22] takes into account the solute solubility (Y^*) in the fluid phase and the mass transfer coefficients in both the fluid and solid phases (k_{YA} and k_{XA}, respectively) [10]. This model neglects the terms of dispersion and accumulation in the fluid phase, as well as the diffusion in the solid phase. Accumulation in the fluid phase was disregarded because the residence time of the solvent was considered to be low enough to support this assumption. Hence, the accumulation term was considered only in the solid phase [4]. The model also assumes pseudo-steady-state and plug flow. The parameters temperature, pressure, and solvent velocity are taken as constant throughout the entire extraction process. The fixed bed is assumed to be homogeneous with respect to the particle size and the initial solute distributions [10]. The mass balance equations proposed by Sovová [22] are presented in Eqs. 3.5 and 3.6 for the fluid and solid phases, respectively.

$$u_i \frac{\partial Y}{\partial z} = \frac{J(X,Y)}{\varepsilon} \tag{3.5}$$

$$\frac{\partial X}{\partial t} = \frac{J(X,Y)}{(1-\varepsilon)} \frac{\rho_{CO_2}}{\rho_S} \tag{3.6}$$

where Y and X are the mass ratios of the solute in the fluid and solid phases, respectively (kg/kg); t is the extraction time (s); u_i is the interstitial velocity of the solvent (m/s); ρ_{CO_2} and ρ_S are the solvent and solid matrix densities, respectively (kg/m^3); ε is the bed porosity (dimensionless); z is the axial direction (m); and $J(X,Y)$ is the interfacial mass transfer term (s^{-1}) as described by Eqs. 3.7 and 3.8, which must be applied when $X > X_K$ and $X \leq X_K$, respectively. The initial and boundary conditions for the mass balance equations are presented in Eqs. 3.9 and 3.10.

$$J(X,Y) = k_{YA}(Y^* - Y) \tag{3.7}$$

$$J(X,Y) = k_{XA} X \left(1 - \frac{Y}{Y^*}\right) \tag{3.8}$$

$$X(z, t=0) = X_0 \tag{3.9}$$

$$Y(z=0, t) = 0 \tag{3.10}$$

Sovová [22] solved the model equations and developed an analytical solution that is presented in Eqs. 3.11, 3.12, and 3.13, which must be applied, respectively, to the CER ($t \leq t_{CER}$), FER ($t_{CER} < t \leq t_{FER}$), and DC regions ($t > t_{FER}$). The extraction times that identify the ends of the CER and FER periods are denoted t_{CER} and t_{FER}, respectively.

$$m_{EXT} = Q_{CO_2} Y^* [1 - \exp(-Z)] t \tag{3.11}$$

$$m_{EXT} = Q_{CO_2} Y^* [1 - t_{CER} \exp(Z_W - Z)] \tag{3.12}$$

$$m_{EXT} = m_{SI} \left\{ X_0 - \frac{Y^*}{W} \ln \left[1 + \exp\left(\frac{W X_0}{Y^*}\right) - 1 \right] \exp \left[\frac{W Q_{CO_2}}{m_{SI}} (t_{CER} - t) \right] \left(\frac{X_P}{X_0} \right) \right\} \tag{3.13}$$

Considering that

$$Z = \frac{m_{IS} k_{YA} \rho_{CO_2}}{Q_{CO_2} (1-\varepsilon) \rho_S} \tag{3.14}$$

$$W = \frac{m_{IS} k_{XA}}{Q_{CO_2}(1-\varepsilon)} \quad (3.15)$$

$$Z_W = \frac{ZY^*}{WX_0} \ln\left\{\frac{X_0 \exp\left[\frac{WQ_{CO_2}}{m_{IS}}(t_{CER}-t)\right] - X_K}{X_0 - X_K}\right\} \quad (3.16)$$

$$t_{CER} = \frac{m_{IS} X_P}{Y^* Z Q_{CO_2}} \quad (3.17)$$

$$t_{FER} = t_{CER} + \frac{m_{IS}}{Q_{CO_2} W} \ln\left[\frac{X_K + X_P \exp\left(WX_0/Y^*\right)}{X_0}\right] \quad (3.18)$$

$$m_{IS} = F - m_0 = F - (X_0 F) \quad (3.19)$$

The nomenclature used in Eqs. 3.11, 3.12, 3.13, 3.14, 3.15, 3.16, 3.17, 3.18, and 3.19 is specified as follows:

m_{EXT} = the mass of the extract (kg)
t = the extraction time (s)
F = the mass of the feed material (kg)
X_0 = the initial mass ratio of extractable solute in the solid substratum (kg/kg)
m_0 = the initial amount of extractable solute in the solid substratum (kg)
m_{IS} = the mass of the inert solid (kg)
Q_{CO2} = the solvent flow rate (kg/s)
ρ_{CO2} and ρ_S = the densities of CO_2 and the solid material, respectively (kg/m^3)
ε = the bed porosity (dimensionless)
Y^* = the solubility of the extract in the supercritical solvent (kg/kg)
X_P = the mass ratio of the easily accessible solute in the solid substratum (kg/kg)
X_K = the mass ratio of the hardly accessible solute in the solid substratum (kg/kg)
k_{YA} and k_{XA} = the mass transfer coefficients in the fluid and solid phases, respectively (s^{-1})

The application of the model developed by Sovová [22] generally results in good fits to experimental data for many different raw materials. A significant advantage of this model is that it provides a good physical description of the mass transfer phenomena in SFE processes [11]. Therefore, it is a convenient choice for the purposes of process design because the adjustable parameters (k_{YA}, k_{XA}, and X_K) can be applied in scale-up investigations. Years later, Sovová [23] presented another model that is also based on the concept of broken and intact cells. In this new model, the term for accumulation in the fluid phase was considered, and some changes were applied to the term of interfacial mass transfer. As a result, the complexity of the mathematical model increased significantly, and then the model was solved numerically because an analytical solution was no longer suitable [4, 23]. Furthermore, the number of adjustable parameters increased, and more information was required for the application of the new mathematical model, thereby limiting its practical use.

All of the models discussed thus far assume that the solute is a pseudocomponent. Martínez et al. [24] proposed a model that can be applied under two different assumptions regarding the solute composition, that is, to either a pseudocomponent or a multicomponent system. The assumption of a multicomponent system may be useful if there exists interest in knowing the kinetic behavior of specific compounds that are present in the extract. In this chapter, we present the model for a pseudocomponent solute, and we refer to it as the "logistic model." Further extension of this model to multicomponent systems is easily carried out because the same considerations and analogous equations are used.

According to Martínez et al. [24], the model begins by applying the differential mass balance inside the extraction bed for solid and fluid phases. This author neglected the terms of accumulation and dispersion in the fluid phase because he assumed that both phenomena lack significant influence relative to the convection term. The main peculiarity of this model is the definition of the term of interfacial mass transfer, which is described by one of the solutions from the logistic equation. The model equation for a pseudocomponent system is presented in Eq. 3.20. The logistic model has two adjustable parameters, named C_2 and t_m. No physical meaning is attributed to the first parameter (C_2), while the second (t_m) is defined as the time during which the extraction rate reaches its maximum value [24].

$$m_{EXT} = \frac{X_0 F}{\exp(C_2 t_m)} \left\{ \frac{1 + \exp(C_2 t_m)}{1 + \exp[C_2 (t_m - 1)]} - 1 \right\} \quad (3.20)$$

where m_{EXT} is the mass of the extract (kg), t is the time of extraction (s), F is the mass of the feed material (kg), X_0 is the initial mass ratio of the extractable solute in the solid substratum (kg/kg), and C_2 (s^{-1}) and t_m (s) are the adjustable parameters.

The logistic model generally provides a relatively good fit to experimental data gleaned from different raw materials. However, when applying this model to common OEC shapes, many authors have obtained negative values for t_m; when this happens, no physical meaning can be attributed to the parameter t_m [13]. The absence of physical meaning brings an empirical character to this model; thus, the model has limited application in terms of process design and scale-up.

3.7.1 The Spline Model

Many different mathematical models have been used to describe and understand the kinetics of SFE processes, ranging from simple equations to very complex equations. An example of a simplified approach used to model the extraction curve is the so-called spline model, as presented by Meireles [9]. This model, which has an empirical basis, is based on the assumption that the OEC can be described by a family of N straight lines. The lines from the spline model can be calculated using Eq. 3.21 [9, 25][1] written for the 1st, 2nd, 3rd, ..., and N-th lines:

$$m_{EXT} = \left(b_0 - \sum_{i=1}^{i=N-1} t_i a_{i+1}\right) + \sum_{i=1}^{i=N} a_i t \qquad (3.21)$$

where m_{EXT} is the mass of the extract (kg), t is the time of extraction (s), N is the number of straight lines, b_0 is the linear coefficient of line 1 (kg), $\sum a_i$ (for $i = 1$ to $i = N$) are the slopes of lines 1 to N (kg/s), and t_i (for $i = 1$ to $i = N-1$) is the time in which the intercept between line "i" and line "$i + 1$" occurs (s). Equation 3.21 is greatly simplified for two or three straight lines, as presented in Eqs. 3.22, 3.23, and 3.24. When considering an OEC described by 3 straight lines, the m_{EXT} for the three different periods of extraction should be calculated using the following equations [13]:

- For the first straight line ($t \leq t_1$), the m_{EXT} is obtained by Eq. 3.22:

$$m_{EXT} = b_0 + a_1 t \qquad (3.22)$$

- For the second straight line ($t_1 \leq t \leq t_2$), the m_{EXT} is obtained by Eq. 3.23:

$$m_{EXT} = (b_0 - t_1 a_2) + (a_1 + a_2) t \qquad (3.23)$$

- For the third straight line ($t \geq t_2$), the m_{EXT} is obtained by Eq. 3.24:

$$m_{EXT} = (b_0 - t_1 a_2 - t_2 a_3) + (a_1 + a_2 + a_3) t \qquad (3.24)$$

The spline model has been extensively used by our research group (LASEFI/FEA/UNICAMP) to model the kinetic data obtained from SFE studies [11, 13, 26–28]. This model has been applied based on considerations that the OEC can be described by two or three straight lines, depending on the shape of the extraction curve. Although the use of two straight lines may be adequate in some cases, the model with three lines is more versatile because it can be applied to any possible OEC shape. Moreover, when the OEC is described by three straight lines, it is possible to make a useful analogy with the three different extraction regions (the CER, FER, and DC periods, as previously discussed in Sect. 3.7) that are observed in a typical OEC. In this case, the parameters t_1 and t_2 (from Eqs. 3.23 and 3.24) correspond to t_{CER} and t_{FER}, the extraction times that mark the ends of the CER and FER periods, respectively. A schematic representation of an OEC that was fitted with three lines is presented in Fig. 3.9.

---

[1]The model presented here (Eq. 3.21) is the revised form of the equations previously published by Meireles [9, 25] because, in the original reference [9], typographical errors were present in the equation that describes the spline model.

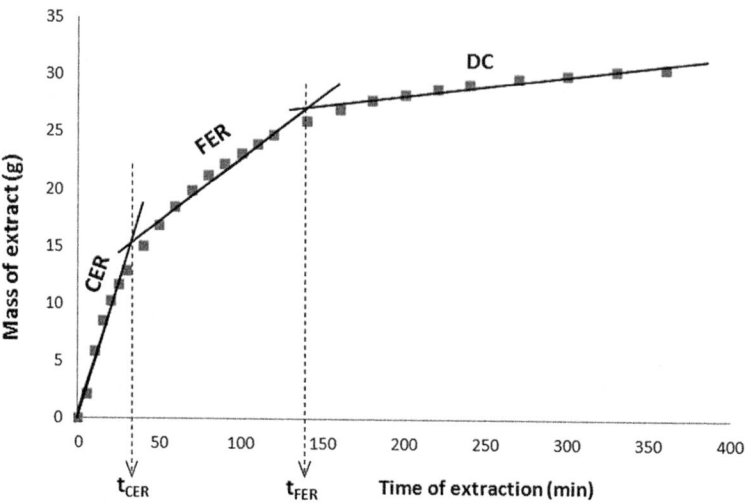

Fig. 3.9 Schematic representation of the spline model: extraction curve of SFE from clove bud (313 K/15 MPa, 226 g of feed material, solvent flow rate $= 9.6 \times 10^{-5}$ kg/s) fitted to three straight lines, which were prolonged to evidence the intercept points (t_{CER} and t_{FER}). Experimental data were obtained from Prado [27] (*CER* constant extraction rate, *FER* falling extraction rate, *DC* diffusion-controlled, t_{CER} is the time span of the CER period, t_{FER} is the time that marks the end of the FER period)

To fit the experimental OEC to a spline containing three straight lines, a nonlinear fit must be performed since the intercept points (t_{CER} and t_{FER}) are unknown. This can be carried out by using the procedures PROC REG and PROC NLIN of the SAS® software package (SAS Institute Inc., Cary, NC, USA) [13]. According to Jesus et al. [13], the fitted lines may be associated with three different mass transfer mechanisms (as illustrated in Fig. 3.9), following the classic description of the CER, FER, and DC periods [22]. Thus the first, second, and third lines can be related to the CER, FER, and DC regions, respectively. When studying SFE kinetics, it is a very common procedure to apply the spline model for the determination of various kinetic parameters that characterize the CER period. These parameters are [9, 13] the time span of the CER period (t_{CER}), the extraction rate of the CER period (M_{CER}), the mass ratio of the extract in the fluid phase at the bed outlet (Y_{CER}), the extraction yield of the CER period (R_{CER}), and the solvent-to-feed mass ratio of the CER period (S/F$_{CER}$). Both t_{CER} (s) and M_{CER} (kg extract/s) are adjustable parameters from the spline model (t_1 and a_1, respectively, as presented in Eq. 3.23). Y_{CER} (kg extract/kg CO_2) is obtained by dividing M_{CER} by the mean solvent flow rate (Q_{CO2}, kg CO_2/s). The parameters R_{CER} (%, kg extract/kg feed material) and S/F$_{CER}$ (kg CO_2/kg feed material) should be calculated using modeled data (the values obtained for t_{CER} and m_{EXT} at the end of the CER period) [13].

The spline model generally presents a good fit to experimental data; thus, it is capable of delivering a good description of the OEC quantitative behavior [13]. Furthermore, although the model possesses an empirical basis and is comparatively simple in terms of its mathematical complexity, it nonetheless delivers helpful information regarding the SFE process. The association that can be made between the first line and the CER period is particularly useful because the CER region is the most important in terms of process design. According to Pereira and Meireles [7], between 50 and 90 % (w/w) of the total amount of extract can be recovered before the end of the CER period. Therefore, for many industrial applications, the extraction process may be ended shortly after t_{CER} because the best operational conditions are likely those in which a significant amount of extract is produced within a relatively short process time [7]. Therefore, the values of t_{CER} and R_{CER} approximately represent the minimum time that a SFE cycle should last and the minimum extraction yield expected under the given process conditions [9].

Some works on scale-up (see Sect. 3.8 for details) have demonstrated that the extraction yields and kinetic behaviors observed in laboratory assays can be reproduced on a pilot scale [16, 28–31]. Hence, it is possible that the same extraction yields may be achievable in an industrial plant. In this case, the parameters t_{CER}, S/F_{CER}, and R_{CER} can be used in preliminary studies of economic feasibility (aspects concerned with cost estimation are discussed in Sect. 3.9). According to Leal [32], when using the spline model, the intersection between lines 1 and 3 (CER and DC, respectively, as illustrated in Fig. 3.9) defines an additional parameter of time, which is named t_{CER2}. This parameter can also be used as a good estimation of the process time in preliminary studies of COM predictions [13, 26].

In the literature on SFE, several additional complex mathematical models are presented for the description of the OEC. These models, which have a phenomenological basis, may provide reliable descriptions of the mass transfer mechanisms involved in the extraction process. This means that the adjustable parameters can have significant physical meanings and, as a result, may be used for scale-up purposes. Nonetheless, to apply phenomenological models, additional specific data are required. The model proposed by Sovová [22], for example, requires information concerning the extract solubility (Y^*) in supercritical CO_2, representing data that are not always available; in many cases, such data may not be available in the literature, and the associated experimental determination would be a difficult task (as discussed in Sect. 3.5.1). Thus, considering the difficulties encountered in finding specific data for many natural extracts, it is clear that one advantage of the spline model is that only kinetic data are necessary to carry out OEC mathematical modeling. Moreover, even with an empirical basis, this model provides useful and practical information concerning the SFE process, particularly with respect to the CER period.

3.8 Scale-Up

Scale-up is the task of achieving on a larger scale the same process behavior that was previously obtained in laboratory assays by considering the differences that are inherent to the processes conducted on equipment of significantly different sizes [5, 30]. By scaling up a process, a product with the same characteristics can ideally be obtained at a larger production rate with no or minimal modifications required. The prediction of a process behavior at the industrial scale is one of the most challenging tasks for food and chemical engineers [5].

After many decades of intensive research, the theoretical basis of SFE is now well established. Hundreds of publications concerning the optimization of process parameters in SFE from different raw materials are reported in published books, articles, and patents based on the results obtained on the laboratory scale. However, few data can be found for pilot-plant scales, and less data are available at the industrial scale [5]. Open and accessible knowledge regarding commercial-scale processes and equipment is very scarce. Information regarding industrial processes depends on the policies of the companies that use and sell SFE units [33].

The available scale-up data in the open literature are inconclusive, so there is no consensus regarding a general scale-up criterion that may be applicable to SFE from solid matrices [30]. To validate scale-up criteria, it is necessary to assess their applicability to different types of raw materials [34] because the mass transfer mechanisms depend on the specific characteristics of the solid substrates and respective solutes. The works that explore scale-up methods are usually limited to specific raw materials and process conditions; as a result, significant care is necessary when proposing a generalization. The process of defining universal scale-up criteria is very complex. However, when considering the main process parameters of SFE and how they affect extraction yield and kinetics, it may be possible to find ways of achieving some effective scale-up procedures [5].

In SFE, the scale-up objective is the reproduction of the same extraction curve at a larger scale by preserving some of the extraction parameters used at the laboratory scale. Therefore, the biggest challenge is the discovery of which parameters, when conserved, will lead to the same results (extraction rates, yields, and chemical compositions of the products) when performing the scale-up procedure. The solution to this type of problem is tricky, and the challenge involves deep knowledge of the limiting factors of the SFE process, which may be based on either thermodynamics or mass transfer [5]. Del Valle et al. [35] suggested that caution is required when working with simple scale-up procedures because in SFE, the relationships between extraction rates and extraction conditions depend on several parameters and may be very complex. Moreover, differences between the mass transfer phenomena may occur when significantly increasing the process scale [35]. However, Prado et al. [30] emphasize that the use of some simple criteria could help the development of easily applicable scale-up methods, which would decrease the time and cost utilized in the design of a SFE process.

According to Clavier and Perrut [36], a simple scale-up procedure for SFE processes can be conducted by following two main steps: (a) perform small-scale experiments to define the optimal extraction conditions by scanning over the operational parameters (different pressures, temperatures, solvent-to-feed ratios, and others) and then (b) select the scale-up method based on the factors that limit mass transfer during extraction. Depending on the complexity and kinetic limitations of the process, different strategies may be applied to the design of the production unit. The easiest scale-up method consists of holding one or both of the ratios of Q_{CO2}/F and S/F constant, where Q_{CO2} is the solvent flow rate, F is the feed mass in the extractor, and S is the solvent mass required for the extraction [36]. Then, three scale-up criteria can be proposed [36]: (a) in the case of an extraction limited by solubility, the S/F ratio should be held constant between the small and large scales; (b) for a process limited by internal diffusion, the Q_{CO2}/F ratio should be conserved from the small to the large scale; and (c) when both diffusion and solubility are limiting mechanisms, both ratios (S/F and Q_{CO2}/F) should be held constant in the scale-up process.

The Q_{CO2}/F ratio is inversely proportional to the residence time of the solvent inside the extractor, as can be seen in Eq. 3.25. It is important to emphasize that the solvent density (ρ_{CO2}), the bed porosity (ε), and the bed apparent density (ρ_B) should be preserved when studying the abovementioned scale-up criteria. Therefore, it is clear that the residence time (t_{RES}) will be conserved if the ratio between the solvent flow rate (Q_{CO2}) and the feed mass in the extractor (F) is held constant. Clavier and Perrut [36] note that the contact time between the solvent and solid matrix is a determining factor for processes limited by internal diffusion; as a result, the residence time should be conserved from the small to the large scale.

$$t_{RES} = \frac{\varepsilon \rho_{CO2} F}{\rho_B Q_{CO2}} \qquad (3.25)$$

where t_{RES} is the residence time of the solvent (s), ε is the bed porosity (dimensionless), ρ_{CO2} is the solvent density (kg/m^3), ρ_B is the bed apparent density (kg/m^3), F is the feed mass in the extractor (kg), and Q_{CO2} is the solvent flow rate (kg/s).

The criterion that necessitates maintaining the Q_{CO2}/F ratio as a constant (and consequently preserving the residence time) has been effective when applied to the scale-up of SFE from clove [16], peach almond [31], and striped weakfish wastes [37]. However, it is considered unsatisfactory for the scale-up of SFE data from vetiver roots [16]. This may have resulted from the physical properties of vetiver oil (particularly, its high viscosity), which could have affected the mass transport properties in small-scale experiments and may have contributed to a significant loss of the extract at some locations within the equipment [16]. In the just-mentioned works [16, 31, 37], the large-scale experiments were conducted on SFE equipment with capacities no larger than 300 cm^3; hence, no assays were performed on pilot-scale units. The same criterion (constant Q_{CO2}/F) was used to investigate the scale-up of SFE from red pepper by performing large-scale experiments in a pilot-plant unit [5]. The authors observed that the extraction curves obtained at the laboratory (300 cm^3 capacity) and pilot (5,150 cm^3 capacity) scales exhibited significantly different kinetic behaviors, so the applied scale-up criterion could not be used to accomplish the authors' goal [5]. According to Martínez and Silva [5], the divergences observed between applications at small and large scales may have occurred as a result of bed compaction, variations in the efficiencies during the separation step, distinct bed geometries, and mechanical dragging. Martínez et al. [16] also investigated another scale-up proposal that consisted of holding constant the superficial velocity of the solvent; however, this criterion was ineffective because the results obtained for large-scale experiments were far from those achieved for small-scale experiments.

In recent works from our research group, the criterion based on holding constant both the S/F and Q_{CO2}/F ratios has been successfully applied to the scale-up of SFE from different raw materials [27–30]. In these works, the small-scale extraction curves were obtained on laboratory-scale equipment (an extraction vessel measuring 290 cm^3 in volume) and were then used as references for scaling up the SFE process. The large-scale experiments were performed in a pilot-plant unit (an extraction vessel measuring 5,150 cm^3 in volume), containing three separators that were arranged in series. The proposed criterion was effective for the scale-up data of SFE from clove [30], sugarcane residue [30], grape seeds [28], ginger [27], and annatto seeds [29]. Taking into account the feed mass in the extractor, a 15-fold scale-up was achieved for clove and sugarcane residue [30], a 17-fold scale-up was performed for grape seeds [28] and ginger [27], and a 12-fold scale-up was accomplished for annatto seeds [29]. The extraction curves obtained in small- and large-scale experiments had similar shapes, but in all cases, the authors found that the pilot-scale yields were higher (ranging from 5 to 20 % higher, depending on the raw material used) than those achieved in the small-scale assays [27–30]. According to Prado et al. [30], the manufacturers of SFE equipment claim that the extraction process is more efficient at larger scales, so the higher yields achieved in pilot-scale experiments are in agreement with the information delivered by manufacturers.

The scale-up procedure suggested by Clavier and Perrut [36] (holding one or both of the S/F and Q_{CO2}/F ratios constant) provides the significant advantage of simplicity. Nonetheless, this approach does not take into account several important factors that may affect the extraction process (radial diffusion, axial mixing, bed compaction, etc.) and is incapable of predicting the effects of using a series of extractors. A refined scale-up method that integrates all of the relevant factors in SFE processes requires a numerical simulation that may estimate any possible plant configuration and may lead to the optimization of industrial units [36].

3.9 Economic Analysis

It is apparent that industries must earn profits, so even the most brilliant technology will never be accepted unless it can provide a product with a price tag that is at least compatible to that of similar products that are already available in the market [38]. This means that demonstrating the economic feasibility of an emerging technology is the only way to attract potential investors. Therefore, researchers should blend their scientific enthusiasm with economic awareness [38] because the cost aspects are fundamental to the process design.

According to Meireles [9], SFE from solid matrices was shown to be a technically feasible process. However, despite the increasing number of industrial plants in operation all over the world, in many regions (e.g., Latin America), SFE is not applied on a commercial scale [9]. Thus, although SFE has been used as an industrial operation since the 1980s [2], it can still be considered an emerging technology because the conventional techniques continue to be the most commonly used approaches in various applications of solid-fluid extraction. One reason for this is the restraints imposed by the high investment costs, which are usually associated with the high-pressure aspect of the processes [9, 39]. Therefore, to spread SFE technology, it is critical to find ways of demonstrating that this technique can be profitable. Indeed, this is a task of major importance with respect to preventing the elimination of SFE at the very early stages of the process design. Hence, efforts must be undertaken to develop simple and reliable methods for estimating the cost of manufacturing (COM) of SFE products because cost information is a determinant factor in the initial stages of business plan analyses [9, 40]. Moreover, it is also important to emphasize that a preliminary analysis of the COM must be performed with minimal experimental information [40].

The COM of various SFE extracts has been systematically studied by our research group for more than a decade. Based on the knowledge acquired from this systematic investigation, we can state that the following information must be available to perform cost estimations:

- The *operating conditions of temperature and pressure* should be selected by taking into account the results from GYI experiments (see Sect. 3.5.2). Both parameters are strongly related to equipment specifications and the utilities demand.
- The *extraction yield for a given extraction time and solvent-to-feed ratio*, which are process parameters that should be obtained from the OEC (see Sect. 3.6.2). These parameters are necessary to determine the rates of solvent consumption and extract production, as well as the cycle time.
- The *description of the raw material pretreatment*, which are the process steps that must be conducted prior to the extraction process (as discussed in Sect. 3.3.1). The pretreatment requirements are important for estimating the preprocessing costs.
- The *bed apparent density* is required to calculate the mass of feed material that must be packed into a certain bed volume. If a given production rate is desired, then the plant capacity and the raw material demand can be determined using the bed apparent density.
- The *extract composition* is valuable information, although it is not necessary when calculating the COM. Nonetheless, characterization of the extract, in terms of its chemical compounds and functional properties, is essential information for defining the selling price of SFE products. If a reliable estimation of the selling price can be made, then it is possible to also make a good prediction of the payback period, which is a cost parameter that may attract investors and aid decision makers.

Rosa and Meireles [39] presented a simple procedure for estimating the COM of extracts obtained by SFE. These authors applied the methodology described by Turton et al. [41], in which the COM is calculated as a sum of the direct costs, fixed costs, and general expenses [9, 39]. The direct costs are directly dependent on the production rate, that is, they are composed of the costs of raw materials, operating labor, and utilities, among others. The fixed costs are independent of the production rate and involve taxes, insurance, depreciation, etc. The general expenses are associated with business maintenance, such as administrative costs, research and development, and sales expenses, among others [39]. The three components of the COM (direct costs + fixed costs + general expenses) are then estimated in terms of five main costs, as expressed in the model (Eq. 3.26) proposed by Turton et al. [39, 41]:

$$\text{COM} = 0.304 F_{\text{CI}} + 2.73 C_{\text{OL}} + 1.23 \left(C_{\text{UT}} + C_{\text{WT}} + C_{\text{RM}} \right) \quad (3.26)$$

where COM is the cost of manufacturing, which is expressed in US\$/kg; F_{CI} is the fixed cost of investment; C_{OL} is the cost of the operating labor; C_{UT} is the cost of the utilities; C_{WT} is the cost of waste treatment; and C_{RM} is the cost of the raw materials.

The fixed cost of investment (F_{CI}) can be calculated on a yearly basis as the product of the total investment by the annual depreciation rate (normally, a 10 % rate is considered). In addition to the expenses associated with equipment and installations, the investment cost should also include the initial amount of CO_2 that is required to fill the solvent reservoir [39]. The cost of operational labor (C_{OL}) is related to the number of workers that are needed to operate the process equipment (extractors, separators, heat exchangers, compressors, pumps, storage tanks, etc.). The cost of the utilities (C_{UT}) is calculated by considering the demand for heating steam, cooling water, and electric power [26, 39].

In the SFE of natural products, the raw material is a plant or animal substrate, which may require one or more pretreatment steps (cleaning, selection, drying, milling, etc.) before extraction can be performed. The cost of the raw materials (C_{RM}) is composed of expenses that include the solid substrate (both the solid matrix and all of the pretreatment costs) and the loss of CO_2 during the process. The solvent lost is associated with the leaking of CO_2 from the system, either as a result of dissolution in the extract after the separation process or entrapment in the solid substrate that is removed from the extractor [39]. Rosa and Meireles [39] considered that a factor of 2 % (taking into account the total amount of solvent used in a cycle of extraction) was adequate for estimating the CO_2 lost. Regarding the generation of waste, the only waste accumulated is the exhausted solid, which is harmless and can be reused in other industrial applications or is simply disposed of as an ordinary organic waste [26]. In particular cases, the exhausted solid is the main desired product, as in the removal of caffeine from coffee, the reduction of nicotine in tobacco, and the removal of cholesterol from foods, among others. Therefore, the cost of waste treatment (C_{WT}) can be completely neglected and is assumed to be zero [26, 39].

As long as the production requirements of a particular SFE process are known, the optimal configuration of the industrial plant can be determined [36]. A typical SFE unit (see Fig. 3.1) is composed of two or more extraction columns; two or more separators (flash tanks), which are arranged in series to allow a certain degree of extract fractionation; a CO_2 reservoir; a solvent pump; heat exchangers; a compressor for CO_2 recycling; several valves; and temperature and pressure controllers [7, 39].

To determine the input and output mass rates and the energy demands of the industrial process, the mass and energy balance equations must be solved. This can be achieved by using software (either homemade or commercial packages) that addresses process engineering calculations. In recent years, our research group has adopted the commercial software SuperPro Designer® (Intelligent Inc., Scotch Plains, NJ, USA) as a useful tool for studying the economic feasibility of SFE [26, 28, 34, 42–46]. This software allows calculations of the process and economic parameters, so it can be used to perform simulations of industrial-scale processes. The COM and the payback period are some of the output data obtained from simulations performed in SuperPro Designer®. According to the Association for the Advancement of Cost Engineering International, cost estimations can be divided into five classes (1–5), which are defined by taking into account the degree of accuracy between the predicted value and the real COM. The class 5 estimation is based on the lowest level of project definition, while the class 1 estimation is closer to the final definition of the industrial project. The SuperPro Designer® software is capable of estimating COMs that may be classified as classes 2–3 [26].

It is well known that the COM of a SFE product is significantly influenced by extraction time (t_{EXT}), which is the time required for one cycle of extraction. Therefore, it is very important to know the extraction curve because kinetic data can be used to estimate the time in which the COM reaches its minimum value. Prado et al. [28] studied the economic viability of the production of grape seed oil by SFE. These authors investigated different times of extraction (from 60 to 300 min) and plant capacities (0.005, 0.05, and 0.5 m^3). The minimum COM (12 US$/kg) was found for a plant size of 0.5 m^3 by considering an extraction time equal to 240 min [28]. Taking into account the selling price (40–80 US$/kg) of a similar product (grape seed oil obtained by cold pressing), the SFE process was considered to be economically viable [28]. Other examples of recent works in which similar cost analyses were performed are summarized in Table 3.1.

The economic feasibility of a SFE product depends on a comparative analysis between COM and the product's selling price [9, 28, 39]. If the preliminary COM estimated for a certain extract is lower than the market price of a similar product, then there is a very strong indication that the process under investigation can be economically viable. However, defining a selling price may not be a trivial task because SFE extracts are still innovative products. Therefore, in many situations, an equivalent product is not yet available in the market, preventing a selling price from being accurately determined. Moreover, in the natural products market, the prices are directly dependent on the extract quality, which can be evaluated in terms of its chemical composition and functional properties. Then, depending on the composition and properties of the extract, different selling prices are possible. It is well established that, in most cases, SFE extracts tend to possess quality advantages compared to extracts obtained by other techniques, particularly in comparison to extracts produced with low-pressure solvent techniques. This happens because

Table 3.1 Cost of manufacturing (COM) of extracts obtained by supercritical fluid extraction (SFE)

Raw material	Botanic name	Target compounds	T (K)	P (MPa)	t_{EXT} (min)	S/F (kg/kg)	Yield (%)	COM (US$/kg)	Selling price (US$/kg)	Ref.
Clove buds	Eugenia caryophyllus	Volatile oil	313	15	52	3.65	14.2	31	100	[45]
Grape seeds	Vitis vinifera	Unsaturated fatty acids and antioxidants	313	35	240	6.6	9.9	12	40–80	[28]
Annatto seeds	Bixa orellana	Tocotrienols	313	20	105	8.7	2.75	115	NI	[29]
Cashew leaves	Anacardium occidentale	Volatile oil, flavonoids, alkaloids, and antioxidant compounds	318	20[a]	47	11.5	1	24	NI	[42]
Lemon verbena leaves	Aloysia triphylla	Volatile oil and flavonoids	333	35	180	9.1	1.8	1070	1375	[34]
Mango leaves	Mangifera indica	Variety of bioactive compounds (flavonoids, alkaloids, terpenes, terpenoids, and antioxidant compounds)	323	30	90	4.2	1.8[b]	92	10–500	[46]

T temperature, P pressure, t_{EXT} extraction time, S/F solvent-to-feed ratio, NI not informed, Ref. reference
[a]SFE was performed using ethanol (5 %) as a cosolvent
[b]Approximated value (obtained by visual observation of the overall extraction curve)

SFE is a green, selective, and mild extraction method, resulting in an extract that is enriched in desirable compounds, free of toxic solvents, and without the loss of compounds due to thermal degradation or oxidative reactions [7]. Thus, SFE products may be given higher prices in comparison to extracts obtained using other extraction methods. Prado and Meireles [45] reported that the selling price of clove oil extracted by SFE is 110 US$/kg, whereas the price of clove volatile oil obtained by steam distillation varies between 26 and 86 US$/kg. Generally, when the SFE product is still not available in the market, the selling prices of oils produced by steam distillation or cold pressing may be used as initial references in the cost analyses of oils obtained by SFE [28, 39, 45].

In some cases, a preliminary cost analysis can indicate that the COM of a SFE extract is too close to or even higher than the market price of a similar product [39, 44]. Even so, it is important to bear in mind that certain considerations must be made before disregarding SFE as a viable process [39]. Some of the important factors that should be considered when evaluating the results obtained in a preliminary cost analysis are listed below [39, 44, 46]:

- *Optimization of the process parameters*: Generally, further and detailed studies of process parameters can result in significant cost reductions. If the extraction rates are increased, then the extraction time and the COM will be reduced [45]. Additionally, the evaluation of different plant configurations and operating modes (by varying the number and arrangement of extractors) may lead to increasing productivity, which can lead to a decrease in the operating costs and COM [3, 6].
- *Different selling prices*: The prices of natural products can vary significantly according to the concentration of one or more target compounds. Extracts obtained by SFE are generally recognized as nutraceutical products; as a result, they may possess special uses and distinct prices. Therefore, the amount and availability of specific bioactive compounds should be carefully evaluated to verify the quality of the product and to specify the market price of the extract [39].
- *Scale increase*: Many authors have demonstrated that the COM of a SFE product tends to be reduced when the plant capacity is increased [26, 28, 34, 42, 45, 46]. Albuquerque and Meireles [26] reported that the COM (SFE extract obtained from annatto seeds) decreased from 125 to 109 US$/kg as the extraction vessels' capacities were increased from 0.1 to 0.5 m^3.
- *Advancements in project detailing*: In a preliminary analysis, the COM tends to be overestimated because the worst-case scenarios are normally assumed to avoid cost underestimations. Uncertainties in the process design are diminished as the project advances, allowing more accurate cost calculations to be performed.

It is common knowledge that high-pressure plants are associated with high investment costs. However, the cost of SFE units has decreased in recent years due to competition between suppliers, which has motivated significant technical improvements and cost reductions [44]. Furthermore, it is important to emphasize that the COM is calculated as a sum of five main costs (as previously presented in Eq. 3.26) [39]; hence, several other cost aspects (not only the investment costs) must be considered to evaluate the economic feasibility of SFE processes. Many recent works have reported that SFE can be an economically viable method for obtaining bioactive extracts [28, 29, 34, 39, 45]. Thus, it is clear that a promising business opportunity is available [9] because SFE has shown true potential as a profitable alternative for the production of high-quality and high value-added products.

References

1. Rostagno MA, Prado JM (eds) (2013) Natural product extraction: principles and applications. The Royal Society of Chemistry, Cambridge
2. Brunner G (1994) Gas extraction: an introduction to fundamentals of supercritical fluids and the application to separation processes. Steinkopff, Darmstadt
3. Brunner G (2005) Supercritical fluids: technology and application to food processing. J Food Eng 67(1–2):21–33. http://dx.doi.org/10.1016/j.jfoodeng.2004.05.060
4. Rosa PTV, Meireles MAA (2009) Fundamentals of supercritical extraction from solid matrices. In: Meireles MAA (ed) Extracting bioactive compounds for food products: theory and application. CRC Press/Taylor & Francis Group, Boca Raton, pp 272–288
5. Martínez J, Silva LPS (2013) Scale-up of extraction processes. In: Rostagno MA, Prado JM (eds) Natural product extraction: principles and applications. The Royal Society of Chemistry, Cambridge, pp 363–398
6. Moraes MN, Zabot GL, Meireles MAA (2013) Assembling of a supercritical fluid extraction equipment to operate in continuous mode. In: Proceedings of the III Iberoamerican conference on supercritical fluids, Cartagena de Indias, 1–5 April 2013
7. Pereira CG, Meireles MAA (2010) Supercritical fluid extraction of bioactive compounds: fundamentals, applications and economic perspectives. Food Bioprocess Technol 3(3):340–372. http://dx.doi.org/10.1007/s11947-009-0263-2
8. Rodrigues VM, Sousa EMBD, Monteiro AR, Chiavone-Filho O, Marques MOM, Meireles MAA (2002) Determination of the solubility of extracts from vegetable raw material in pressurized CO_2: a pseudo-ternary mixture formed by cellulosic structure + solute + solvent. J Supercrit Fluids 22(1):21–36. http://dx.doi.org/10.1016/S0896-8446(01)00108-5
9. Meireles MAA (2008) Extraction of bioactive compounds from Latin American plants. In: Martinez JL (ed) Supercritical fluid extraction of nutraceuticals and bioactive compounds. CRC Press/Taylor & Francis Group, Boca Raton, pp 243–274
10. Ferreira SRS, Meireles MAA (2002) Modeling the supercritical fluid extraction of black pepper (*Piper nigrum* L.) essential oil. J Food Eng 54(4):263–269. http://dx.doi.org/10.1016/S0260-8774(01)00212-6
11. Sousa EM, Chiavone-Filho O, Moreno MT, Silva DN, Marques MOM, Meireles MAA (2002) Experimental results for the extraction of essential oil from *Lippia sidoides* Cham. using pressurized carbon dioxide. Braz J Chem Eng 19(2):229–241. http://dx.doi.org/10.1590/S0104-66322002000200003
12. Carvalho RN Jr, Moura LS, Rosa PTV, Meireles MAA (2005) Supercritical fluid extraction from rosemary (*Rosmarinus officinalis*): kinetic data, extract's global yield, composition, and antioxidant activity. J Supercrit Fluids 35(3):197–204. http://dx.doi.org/10.1016/j.supflu.2005.01.009
13. Jesus SP, Calheiros MN, Hense H, Meireles MAA (2013) A simplified model to describe the kinetic behavior of supercritical fluid extraction from a rice bran oil byproduct. Food Public Health 3(4):215–222. http://dx.doi.org/10.5923/j.fph.20130304.05
14. Vasconcellos CMC (2007) Extração supercrítica dos óleos voláteis de *Achyrocline satureioides* (Macela) e *Vetiveria zizaioides* (Vetiver): determinação da cinética de extração e estimativa de custos de manufatura. UNICAMP (University of Campinas), Brazil

15. Martínez J (2005) Extração de óleos voláteis e outros compostos com CO_2 supercrítico: desenvolvimento de uma metodologia de aumento de escala a partir da modelagem matemática do processo e avaliação dos extratos obtidos. UNICAMP (University of Campinas), Brazil
16. Martínez J, Rosa PTV, Meireles MAA (2007) Extraction of clove and vetiver oils with supercritical carbon dioxide: modeling and simulation. Open Chem Eng J 1(1):1–7
17. Reverchon E, Marrone C (1997) Supercritical extraction of clove bud essential oil: isolation and mathematical modeling. Chem Eng Sci 52(20):3421–3428. http://dx.doi.org/10.1016/S0009-2509(97)00172-3
18. Oliveira ELG, Silvestre AJD, Silva CM (2011) Review of kinetic models for supercritical fluid extraction. Chem Eng Res Des 89(7):1104–1117. http://dx.doi.org/10.1016/j.cherd.2010.10.025
19. Sovová H (2012) Modeling the supercritical fluid extraction of essential oils from plant materials. J Chromatogr A 1250:27–33. http://dx.doi.org/10.1016/j.chroma.2012.05.014
20. Esquível MM, Bernardo-Gil MG, King MB (1999) Mathematical models for supercritical extraction of olive husk oil. J Supercrit Fluids 16(1):43–58. http://dx.doi.org/10.1016/S0896-8446(99)00014-5
21. Crank J (1975) The mathematics of diffusion, 2nd edn. Oxford University Press, Oxford
22. Sovová H (1994) Rate of the vegetable oil extraction with supercritical CO_2—I. Modelling of extraction curves. Chem Eng Sci 49(3):409–414. http://dx.doi.org/10.1016/0009-2509(94)87012-8
23. Sovová H (2005) Mathematical model for supercritical fluid extraction of natural products and extraction curve evaluation. J Supercrit Fluids 33(1):35–52. http://dx.doi.org/10.1016/j.supflu.2004.03.005
24. Martínez J, Monteiro AR, Rosa PT, Marques MO, Meireles MAA (2003) Multicomponent model to describe extraction of ginger oleoresin with supercritical carbon dioxide. Ind Eng Chem Res 42(5):1057–1063. http://dx.doi.org/10.1021/ie020694f
25. Meireles MAA (2013) Supercritical CO_2 extraction of bioactive components from algae. In: Domínguez H (ed) Functional ingredients from algae for foods and nutraceuticals. Woodhead Publishing, Cambridge, pp 561–584
26. Albuquerque CLC, Meireles MAA (2012) Defatting of annatto seeds using supercritical carbon dioxide as a pretreatment for the production of bixin: experimental, modeling and economic evaluation of the process. J Supercrit Fluids 66:86–95. http://dx.doi.org/10.1016/j.supflu.2012.01.004
27. Prado JM (2010) Estudo do aumento de escala do processo de extração supercrítica em leito fixo. UNICAMP (University of Campinas), Brazil
28. Prado JM, Dalmolin I, Carareto NDD, Basso RC, Meirelles AJA, Vladimir Oliveira J, Batista EAC, Meireles MAA (2012) Supercritical fluid extraction of grape seed: process scale-up, extract chemical composition and economic evaluation. J Food Eng 109(2):249–257. http://dx.doi.org/10.1016/j.jfoodeng.2011.10.007
29. Albuquerque CLC (2013) Obtenção de sementes desengorduradas e de óleo rico em tocotrienóis de urucum por extração supercrítica: estudo dos parâmetros de processo, do aumento de escala e da viabilidade econômica. UNICAMP (University of Campinas), Brazil
30. Prado JM, Prado GHC, Meireles MAA (2011) Scale-up study of supercritical fluid extraction process for clove and sugarcane residue. J Supercrit Fluids 56(3):231–237. http://dx.doi.org/10.1016/j.supflu.2010.10.036
31. Mezzomo N, Martínez J, Ferreira SRS (2009) Supercritical fluid extraction of peach (*Prunus persica*) almond oil: kinetics, mathematical modeling and scale-up. J Supercrit Fluids 51(1):10–16. http://dx.doi.org/10.1016/j.supflu.2009.07.008
32. Leal PF (2008) Estudo comparativo entre os custos de manufatura e as propriedades funcionais de óleos voláteis obtidos por extração supercrítica e destilação por arraste a vapor. UNICAMP (University of Campinas), Brazil
33. Brunner G (2010) Applications of supercritical fluids. Annu Rev Chem Biomol Eng 1:321–342. http://dx.doi.org/10.1146/annurev-chembioeng-073009-101311

34. Prado JM, Veggi PC, Meireles MAA (2014) Supercritical fluid extraction of lemon verbena (*Aloysia triphylla*): process kinetics and scale-up, extract chemical composition and antioxidant activity, and economic evaluation. Sep Sci Technol 49(4):569–579. http://dx.doi.org/10.1080/01496395.2013.862278
35. del Valle JM, Rivera O, Mattea M, Ruetsch L, Daghero J, Flores A (2004) Supercritical CO_2 processing of pretreated rosehip seeds: effect of process scale on oil extraction kinetics. J Supercrit Fluids 31(2):159–174. http://dx.doi.org/10.1016/j.supflu.2003.11.005
36. Clavier J-Y, Perrut M (2004) Scale-up issues for supercritical fluid processing in compliance with GMP. In: York P, Kompella UB, Shekunov BY (eds) Supercritical fluid technology for drug product development. Marcel Dekker, New York, pp 565–601
37. Aguiar AC, Visentainer JV, Martínez J (2012) Extraction from striped weakfish (*Cynoscion striatus*) wastes with pressurized CO_2: global yield, composition, kinetics and cost estimation. J Supercrit Fluids 71:1–10. http://dx.doi.org/10.1016/j.supflu.2012.07.005
38. Prasad NK (2012) Downstream process technology: a new horizon in biotechnology. PHI Learning Private Limited, New Delhi
39. Rosa PTV, Meireles MAA (2005) Rapid estimation of the manufacturing cost of extracts obtained by supercritical fluid extraction. J Food Eng 67(1–2):235–240. http://dx.doi.org/10.1016/j.jfoodeng.2004.05.064
40. Meireles MAA (2003) Supercritical extraction from solid: process design data (2001–2003). Curr Opinion Solid State Mater Sci 7(4–5):321–330. http://dx.doi.org/10.1016/j.cossms.2003.10.008
41. Turton R, Ballie RC, Whiting WB, Shaeiwitz JA (1998) Analysis, synthesis, and design of chemical process. Prentice Hall PTR, Upper Saddle River
42. Leitão NCMCS, Prado GHC, Veggi PC, Meireles MAA, Pereira CG (2013) *Anacardium occidentale* L. leaves extraction via SFE: global yields, extraction kinetics, mathematical modeling and economic evaluation. J Supercrit Fluids 78(0):114–123
43. Prado IM, Albuquerque CLC, Cavalcanti RN, Meireles MAA (2009) Use of a commercial process simulator to estimate the cost of manufacturing (COM) of carotenoids obtained via supercritical technology from palm and buriti trees. In: Proceedings of the 9th international symposium on supercritical fluids, Arcachon, 18–20 May 2009
44. Prado JM, Assis AR, Maróstica-Júnior MR, Meireles MAA (2010) Manufacturing cost of supercritical-extracted oils and carotenoids from Amazonian plants. J Food Process Eng 33(2):348–369. http://dx.doi.org/10.1111/j.1745-4530.2008.00279.x
45. Prado JM, Meireles MAA (2011) Estimation of manufacturing cost of clove (*Eugenia caryophyllus*) extracts obtained by supercritical fluid extraction using a commercial simulator. In: Proceedings of the 11th international congress on engineering and food, Atenas, 22–26 May 2011
46. Prado IM, Prado GHC, Prado JM, Meireles MAA (2013) Supercritical CO_2 and low-pressure solvent extraction of mango (*Mangifera indica*) leaves: global yield, extraction kinetics, chemical composition and cost of manufacturing. Food Bioprod Process 91(4):656–664. http://dx.doi.org/10.1016/j.fbp.2013.05.007

Chapter 4
Subcritical Water as a Green Solvent for Plant Extraction

Mustafa Zafer Özel and Fahrettin Göğüş

Abstract Subcritical water extraction (SWE), as a method, is non-toxic, non-flammable, cheap, readily available, safe, environmental friendly and uses a green solvent. Chemicals with different functional groups such as flavonoids, vitamins, antioxidants and antimicrobials, can be extracted selectively using SWE. SWE has become a popular green extraction method for different classes of compounds present in numerous kinds of matrices and samples, such as those from environmental, food or botanical sources. Plant oils normally contain a complex mixture of organic compounds. They are largely composed of a range of saturated or partially saturated cyclic and linear molecules of relatively low molecular mass and within this range a variety of hydrocarbons and oxygenated compounds occur. SWE is a technique based on the use of water as an extractant, at temperatures between 100 and 374 °C and at a pressure high enough to maintain the liquid state. SWE of plant materials is a powerful alternative to traditional methods because it enables a rapid extraction, and the use of moderate temperatures. This avoids the loss and degradation of volatile and thermo labile compounds. Additional positive aspects of the use of SWE are its simplicity, low cost, and a more favourable environmental impact than traditional solvents. The extraction of phenolic compounds, essential oil, carotenoids, flavonoids, flavour and fragrance compounds has been carried out using SWE. SWE is also selective in that the operator is able to extract various polar and non-polar organic compounds by choice, through varying the temperature.

M.Z. Özel (✉)
Green Chemistry Centre of Excellence, Chemistry Department, University of York, York YO10 5DD, UK
e-mail: mustafa.ozel@york.ac.uk

F. Göğüş
Food Engineering Department, Engineering Faculty, University of Gaziantep, 27310 Gaziantep, Turkey

When doing this, the water must be kept in a liquid state using minor adjustments in pressure. In the extraction of essential oils from herbs, SWE has been seen to give recoveries comparable to those of steam distillation and solvent extraction.

4.1 Introduction

Subcritical water extraction (SWE) is a promising "green" technique based on the use of water as the sole extraction solvent [1], at temperatures between 100 and 374 °C and at a pressure high enough to maintain the liquid state. Under subcritical conditions, liquid water is less polar than at ambient temperature and has an increased capacity for dissolving organic molecules [2]. There have been many reports on SWE applications especially in the flavour and fragrance industries [2–4].

Subcritical water extraction is also known under the terms of superheated water extraction, hot water extraction, pressurized hot water extraction, pressurized low polarity water extraction, high-temperature water extraction and hot liquid water extraction. Its popularity as a technique is increasing. Table 4.1 shows the results of an internet search carried out on 15 January 2014 using the Science Citation Index (SCI) [5]. The search was performed to see how many times subcritical water extraction (under its various terms) had been used between 1990 and 2013 both in the titles of papers and as topics of papers. It can be seen that subcritical water, superheated water and pressurized hot water were the most commonly used terms to describe the technique in the literature. Subcritical water, superheated water and pressurized hot water, as terms, were used in the titles of papers 436, 156 and 94 times, respectively, between the years 1990–2013. Subcritical water extraction seems to have been the most preferred term with others being less popular. In this chapter we have used the term subcritical water (SW) or subcritical water extraction (SWE). Figure 4.1 demonstrates the increasing use of SW and SWE as a studied technique. The first paper to mention SW in its title was in 1991. It was then used again only once in 1992, in 1993 and in 1995. From 1997, it started to appear

Table 4.1 Search using the SCI (on 15/01/2014) showing number of times subcritical water under its various terms had been used in titles and topics of papers published 1990–2013

Term	Used in *titles* of papers (1990–2013)	Used as *topics* of papers (1990–2013)
Subcritical water	436	1,185
Superheated water	156	504
Pressurized hot water	94	223
Subcritical water extraction	103	268
Superheated water extraction	26	51
Pressurized hot water extraction	53	88

Source: Web of Science [5]

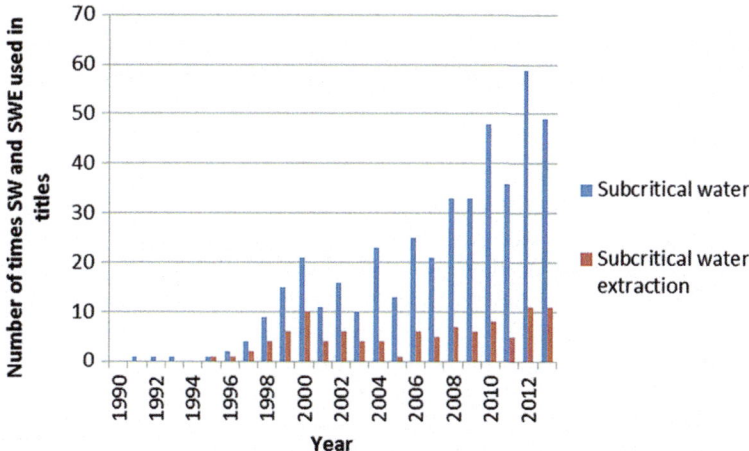

Fig. 4.1 Result of search using SCI (on 15/01/2014) showing increase in the use of SW and SWE in titles of papers published 1990–2013 (Source: Web of Science [5])

more often. In 2012, SW was used 59 times in the titles of papers. Trends are going up gradually. It has become an increasingly popular extraction technique especially for use with plant materials.

In the mid-1990s, SW began to be used in the extraction of non-polar to mid-polar groups of compounds such as PAHs [6] and essential oils [7]. Its use gained popularity and now has more uses such as the extraction of flavour and fragrance compounds, essential oils and phenolic compounds from matrices such as fruit, vegetables, plants, herbs and flowers. SW can be classed as a type of green technology because one of the ways to adapt to the principles of green chemistry is to reduce the use of toxic organic solvents and instead encourage the use of green extraction techniques [8]. Traditional and commercial extraction techniques such as steam distillation, solvent extraction and solid-liquid extractions require long extraction times and usually use a large amount of toxic solvents. Thus the environmental impact can be negative [8].

The literature on toxic solvents has been increasing from year to year. SW uses an alternative green solvent to these toxic organic solvents. The use of toxic solvents in chemical laboratories and the chemical industry is considered a very important problem for the health and safety of workers and in environmental pollution. The majority of solvents are organic molecules with hazardous and toxic properties and can be costly in themselves and/or in their disposal. Their use, storage and disposal also cause environmental problems. Many non-green organic solvents are still used in the food industry such as in the extraction of food additives.

Bio-based solvents, ionic liquids, supercritical fluids and subcritical water are becoming important alternative green solvents to be used in the future. As an example, a green pilot process using SWE obtained potent antioxidants from rosemary leaves [9]. They extracted dry antioxidant powder directly using SWE followed by drying using SC-CO$_2$ (80 bar) and N$_2$ flow (0.6 mL/min). SC-CO$_2$ is one of the most promising ways to dry compounds from organics and showed better recoveries than vacuum drying and freeze drying techniques.

The solubility of a compound can be an indicator of how well it can be extracted. In the literature, there are many reports on the solubilities of many organic compounds for temperatures near 25 °C [10]. However, there are only a few articles about solubilities under SWE conditions. Miller and Hawthorne [10] looked at solubilities of hydrophobic organic compounds such as benzo[a]pyrene and chlorothalonil in subcritical water at temperatures from 25 to 250 °C and pressures (30–70 bar). Increasing temperature elevated both compounds' solubilities rapidly. When the temperature of chlorothalonil was raised from 25 to 250 °C, the solubility increased by a factor of 130,000. Solubilities of some flavour and fragrance compounds such as limonene, carvone, eugenol, 1,8-cineole and nerol were determined in subcritical water between 25 and 200 °C [11]. Solubilities increased with temperature by a factor of 25–60. Other scientists have also found temperature is very effective in increasing the solubilities of terephthalic acid, fatty acids, gallic acids, catechin and protocatechuic acid under subcritical water conditions.

4.2 Properties of Subcritical Water

Water is highly polar and a weak solvent for most organic compounds under ambient conditions [12] but raising the temperature significantly above ambient has a dramatic effect. Pressure is applied to keep water in a liquid state. Optimum conditions for SWE of a given substance depend on temperature, pressure, pH, flow rate, extraction kinetics, water properties and analyte chemical structure. Some scientists found altering the pH of water to 3.5 using buffer solution to be better for extracting from anti-cancer drugs [13, 14]. However many people use water satisfactorily without pH adjustment [15–17]. Flow rate can also affect the extraction rate. Mixing the water with solvent in some cases can enhance the extraction yield of target compounds, although of course this can make the method less green depending on which solvent is used. Using a mixture of water and methanol improved the recoveries of catechin from tea leaves and grape seeds and also of phenolic compounds from grapes [18, 19].

Increasing temperature means water becomes less polar (Fig. 4.2 and Table 4.2). The polarity of SW is measured by the value of the dielectric constant. When water is heated above 100 °C its dielectric constant decreases and water becomes similar to organic solvents [13]. At 214 °C the dielectric constant of water is the same as that of methanol at room temperature. At 295 °C water becomes similar to acetone. The ability to tune the dielectric constant of water to mimic the dissolving power of non-polar organic solvents has been used to selectively extract large organic molecules from plants and foods [20, 21]. SWE yields are comparable to organic solvent extraction techniques. Above 200 °C water may be an acid or base catalyst

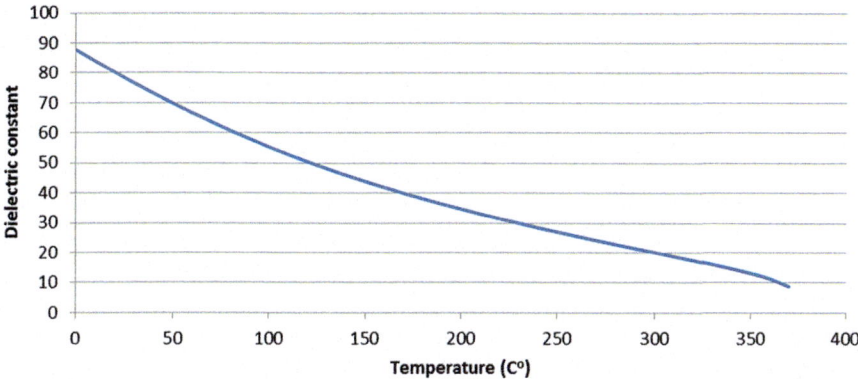

Fig. 4.2 Dielectric constant of liquid water with temperature changes (Data taken from the International Association for the Properties of Water and Steam (IAPWS) [23]. Source: IAPWS [23])

Table 4.2 Dielectric constants of selected solvents at 25 °C showing their equivalent dielectric constants in terms of SW temperature

Solvent	Dielectric constant at 25 °C	SW temperature (°C)
Acetonitrile	37.5	181.45
Methanol	32.7	213.78
Ethanol	24.5	269.00
Acetone	20.7	294.59

Source: IAPWS [23]

because its H_3O^+ and OH^- ion concentrations are perhaps orders of magnitude higher than in ambient water. Subcritical water is therefore a much better solvent for hydrophobic organics than ambient water. It can itself be a catalyst for reactions which normally require an added acid or base.

The basic SWE system set up is similar to the accelerated solvent extraction and supercritical fluid extraction system [15]. In a laboratory built system, degassing of water is needed to prevent potential oxidative corrosion of the extraction line and cell [22]. Typically, a SWE system consists of a syringe or HPLC pump, oven, valves, water reservoir, extraction cell, collection vial, pre-heating and pre-cooling coils. A 5–10 m long pre-heated coil can equilibrate the water to the desired temperature and, after the extraction, a 1–5 m cooling loop (in iced water) outside of the oven cools the hot water containing extract back to room temperature. A pressure control valve is placed between the cooling loop and the collection vial.

4.2.1 Temperature

Temperature is one of the critical factors that affect the extraction efficiency. The use of high temperatures improves the extraction efficiency as it helps the disruption of analyte-sample matrix interactions caused by van der Waals forces, hydrogen bonding and dipole-dipole interaction [24]. The thermodynamic properties of water are typically described in terms of hydrogen bond strength and structure [25]. Changes in hydrogen bonding strength are reflected in the heat of vaporization values and in the dielectric constant [26]. When SW temperature is lower, hydrogen bonds are stronger and the dielectric constant value is higher. With increasing SW temperature, the increased thermal agitation reduces the strength of each hydrogen bond and leads to an amplified reduction in the dielectric constant value [27].

The reduction of the polarity of water generally leads to an increase in the solubility of organic molecules in it [13]. Hawthorne and Miller [28] showed that by applying high temperature during extraction, the polarity (dielectric constant) of water decreases substantially. Increasing temperature decreases the surface tension of any solvent, solutes and matrix and therefore enhances the solvent wetting of the sample [29]. Higher temperatures will also decrease the viscosity of the liquid used, thus allowing better penetration of matrix particles and enhancing extraction [22]. Likewise with water, increasing the temperature also decreases its viscosity and its surface tension. This fact makes water a suitable solvent to extract polar, moderately polar and non-polar organic compounds.

SWE temperatures are generally between 100 and 374 °C. However, many thermolabile compounds such as those in essential oil can be degraded at high temperatures (more than 200 °C). Some compounds can even degrade at 175 °C [15]. The use of high temperatures in the extraction process has been shown to result in the generation of new bioactive compounds during the extraction process via Maillard caramelization reactions. In general, more components are extracted when the temperature is elevated, brought about by their increasing solubility. Jimenez-Carmona et al. [30] for marjoram extraction and Gamiz-Gracia and Luque de Castro [31] for fennel extraction by SWE found that the yields reached their maximum at 150 °C over a temperature range of 50–175 °C. At 175 °C, the SW extracts were seen to be dark brown with a strong burning smell especially in the case of the flowers [32]. Some components appearing at 175 °C from the flowers may have been browning reaction products (furfural, acetylfuran, 5-methylfurfural). The appearance of these components in the case of the flowers might be explained due to the high sugar content of the flowers compared to the leaves [32]. Kubatova et al. [33] for savory and peppermint, Gamiz-Gracia and Luque de Castro [31] for fennel and Ozel et al. [15] for *Thymbra spicata* extraction using SWE, also found degradation at 175 °C [32]. In addition, Rovio et al. [16] for clove extraction and Basile et al. [17] for rosemary extraction selected 150 °C as an optimum water extraction temperature because of processing difficulties in further stages at higher temperatures.

At higher temperatures, where extraction yields often increase, the risk of degradation to the extract also increases. The tendency of a molecule to undergo hydrolysis, oxidation, methylation, isomerization and other reactions depends on the molecule [13]. For example, the bioactive and marker compounds in some medicinal plants may be non-polar to polar and thermally labile. In order to extract non-polar compounds effectively, it may be necessary to increase the temperature from 150

Table 4.3 The minimum pressure range required to keep water in a liquid state at selected temperatures

Temperature (°C)	Pressure (bar)
100	1.0
120	2.0
140	3.6
160	6.1
180	10.0
200	15.5
300	85.8
374	219.1

Calculated from the Critical Process website [48]

to 250 °C. An increase from 100 to 180 °C usually results in higher recoveries. Further increases in the extraction temperature however, would cause the production of degradation products. Such examples include beberine, strychnine, aristoloctic acids, baicalein, glcyrrhizin, tanshinone from medicinal plants [34–36] phenolics from origanum and thymbra species, catechin and epicatechin from tea leaves and grape seeds [15, 17, 20, 33, 34, 37–41] The temperatures that can be used safely depend on the type of material being extracted from. A high amount of bio-oil (37 %) was obtained at a temperature of 360 °C from marine microalgae using SWE [42].

By changing the temperature, solvent properties of water can be better tuned to match the polarity of the target analytes and there is no need for a large selection of solvents to be kept for different polarities. Figure 4.2 shows dielectric constants of water at different temperatures. The dielectric constant is the key parameter in interpreting solute-solvent interactions. The dielectric constant of water is high at room temperature (80.26 at 20 °C), which favours the solubility of ionic and very polar compounds [16]. The dielectric constant decreases with increasing temperature which favours the solubility of non-ionic and non-polar compounds [6].

4.2.2 Pressure

Pressure has a lesser effect on extraction efficiency [15, 28]. A sufficient pressure is required to maintain the liquid state above boiling point for efficient extraction. The pressure forces the hot liquid water into areas of the matrixes that would not normally be contacted by fluid under atmospheric pressure conditions [43]. The necessary pressure range therefore is at least 16 bars at 200 °C and up to 219 bars at 374 °C (Table 4.3). Scientists have mostly used temperatures between 100 and 200 °C for plant extraction [11, 15]. The effect of pressure has no significant difference in the amounts of extract obtained [15]. Varying the pressure has also not changed extraction efficiencies for the extraction of essential oils from medicinal plants and ginsenoisides [15, 44–47].

4.2.3 Extraction Kinetics

The influence of the extraction time on the extraction kinetics is important. Four kinetic steps control the extraction efficiency of SW. They are desorption of solutes from the matrix, diffusion of SW into the organic matrix, the dissolution of the analyte into the SW and, finally, the elution of the extract from the sample matrix. It is clear when considering these four kinetic steps that higher temperatures will lead to improvement of extraction efficiency [13].

SWE is very fast extraction technique over conventional essential oil production techniques. For example, the efficient extraction of essential oil of *Thymbra spicata, Origanum micranthum and Achillea monocephala* using SWE has been achieved within 30 min of extraction time [15, 32, 49]. Essential oil extraction is very slow using traditional steam distillation and Soxhlet extraction (4–24 h) [8]. The essential oil composition has changed with changing temperature [15]. The browning reaction products have observed at a temperature of 175 °C [15, 32, 49]. Extraction kinetic studies have showed on *Thymbra spicata* that the extraction is very fast at high temperatures (150–175 °C) [15].

4.3 Applications

The extraction of phenolic compounds, carotenoids, flavonoids, fragrances and essential oils have been carried out using SWE and both qualitative and quantitative results obtained.

4.3.1 Essential Oil

Essential oils are volatile aroma compounds found mainly in herbal plant materials. They are well known for their use in food, medicinal and cosmetic products. Some essential oils are known for their antioxidant, antimicrobial and antifungal activities [8]. Previous workers [16, 31, 50] reported that SW for the extraction of essential oils is a powerful alternative, because it enables a rapid extraction, and the use of low working temperatures. This avoids the loss and degradation of volatile and thermo labile compounds. Additional positive aspects of the use of SWE are its simplicity, low cost, and more favourable environmental impact.

Essential oils normally contain a complex mixture of organic compounds. They are largely composed of a range of saturated or partly unsaturated cyclic and linear molecules of relatively low molecular mass and within this range a variety of terpenes, sesquiterpenes and oxygenated compounds occur. The essential oils of Origanum species have been proven to have antibacterial, antifungal and antioxidant activities [7]. Oregano, thymus and thymbra are accepted as essential oil yielding plants, their essential oils consisting of mainly carvacrol and thymol (phenolic type compounds). Carvacrol and thymols have been extracted from *Origanum onites*, *Origanum migrathum* and *Thymbra spicata* using SWE [7, 15, 41, 49]. Fragrance compounds have been extracted from *Rosa canina*, *Rosa damascana* using SWE [3, 4, 11]. SWE has been used for the extraction of essential oils from rosemary, marjoram, savory, peppermint, clove, salvia, sideritis, *Teucrium chamaedrys* and *Ziziphora teurica* [16, 20, 30–33, 51, 52].

Ibanez et al. [20] extracted the most active antioxidant compounds from rosemary, such as carnosol, rosmanol, carnosic acid, methyl carnosate and flavonoids such as cirsimaritin and genkwanin, using SWE. The data indicated high selectivity for this method, and the antioxidant activity of the fractions obtained by extraction at different water temperatures was very high. Kim and Mazza [53] reported that SWE of phenolic compounds, including phydroxybenzaldehyde, vanillic acid, vanillin, acetovanillone and ferulic acid, from flax shive was maximized at combined conditions of high temperature and high NaOH concentration.

4.3.2 Phenolic Compounds

Phenolic compounds are one of the four major secondary metabolites found in plants. The main source of phenolic compounds in the human diet comes from plant based foods, namely vegetables, fruits, cereals, legumes and nuts [8]. The type of extraction process is an important parameter for determination of phenolic compounds from plant matrices.

A number of studies have been carried out using SWE and some are outlined below.

SWE was found to be an appropriate extraction technique for obtaining a greater quantity of polyphenolic compounds from winery by-products, and compared well with conventional methods [54–57]. SWE has been used to extract ginsenosides from American ginseng [58]; from tea leaves and grape seeds [37]; anthraquinones (antibacterial, antiviral and anticancer compounds) from roots of *Morinda citrifolia* [59]; and flavones, anilines and phenols from orange peels [60]. Catechin and epicatechin were extracted using magnetic stirring-assisted extraction, ultrasound-assisted extraction, static extraction and also SWE [37]. They found that SWE was a better method in terms of recovery.

Lignans were also extracted from whole flaxseed by SWE [45]. Maximum amounts of lignans and other flaxseed bioactives, including proteins, were extracted at 160 °C. However, these authors reported that on a dry weight basis, the most concentrated extracts in terms of lignans and other phenolic compounds were extracted at 140 °C. Herrero et al. [46] extracted lignans, carbohydrates and proteins from flaxseed meal. The maximum yield of lignans and proteins was obtained at pH 9 at temperatures of 170 °C and 160 °C, respectively. Maximum recovery of carbohydrates was at pH 4 and 150 °C.

Hartonen et al. [56] compared the extraction of naringenin and other flavonoids from knotwood of aspen using SWE, ultrasonic extraction, Soxhlet extraction and reflux in methanol. They found SWE to be favourable. A high amount of oleuropein and olive biophenols were extracted from olive leaves using SWE. Eight phenolic compounds were extracted from potato peel using SWE [61]. A total of 32 phenolic compounds (19 anthocyanins, 6 phenolic acids, 3 flavonols, resveratrol,catechin, epicatechin and allagic acid) were extracted from wet and dried fruit berries and by-products using SWE [62]. Monrad et al. [63] extracted higher yields of polyphenolics such as anthocyanins and procyanidins using wet grape pomace with SWE at 140 °C. Petersson et al. [64] looked at extraction and degradation kinetic studies on anthocyanins from red onion using SW conditions. Pongnaravane et al. [65] compared the effectiveness of SWE of anthraquinones from *Morinda citrifolia* with that of other extraction methods, such as ethanol extraction in a stirred vessel, Soxhlet extraction and ultrasound-assisted extraction [46]. The results of their study showed that SWE extracts presented comparable antioxidant activities to those of Soxhlet extracts, and that SWE extracts were more effective than ethanol extracts and ultrasound-assisted extracts in terms of antioxidant activity.

Flavonoids are polyphenolic compounds. They reduce damage associated with conditions such as cardiovascular disease and cancer [66, 67]. Ko et al. [68] extracted non-polar flavonoids from 8 plants using SWE. They optimized different temperatures for various samples. Flavanones of hesperidine and narirutin from Citrus unshiu peel were extracted using SWE. Maximum yields of hesperidine and narirutin were obtained at an extraction temperature of 160 °C using an extraction time of only 10 min.

The effect of SWE parameters (termed by the authors as pressurized hot water extraction) such as temperature and extraction times were tested on the total phenolic content and DPPH radical scavenging capacity of kafir lime fruit peel extract [69]. 200 °C and 15 min of extraction time yielded the best results under the selected conditions. He et al. [70] tested total phenolic content and antioxidant capacities of pomegranate seed residues using varying SWE conditions. Their best conditions for extraction of phenolic compounds were found to be 220 °C, 30 min of extraction time and a 1:40 solid to water ratio.

4.3.3 Carotenoids

Carotenoids are naturally occurring fat-soluble pigments. There are about 600 well known compounds, which are divided into two main classes, xanthophylls and carotenes [8]. They are found mainly in fruits and vegetables. Conventional extraction methods for carotenoids use toxic/non-green organic solvents such as acetone, petroleum ether, diethyl ether, tetrahydrofuran, hexane, dichloromethane and methanol [71, 72]. However, carotenoids have been extracted from green algae, [73] carrots, green beans and broccoli [74] using SWE.

4.3.4 Pesticides and PAHs in Food

Consumers are very concerned about pesticides and polycyclic aromatic hydrocarbon (PAH) residues on food. Both groups of compounds are well known for their toxic and carcinogenic properties. Pan et al., determined 21 pesticides from black tea using SWE at 150 °C [75]. A rapid SWE method was developed for PAH determination using temperatures between 150 and 200 °C [76]. Martorell et al. [77] used SWE to find 16 PAHs from various foodstuffs. Edible vegetable oil was also found to contain PAHs [78]. SWE can be thought of as a fast, reliable green extraction technique for pesticides and PAHs from plant matrices.

4.3.5 Subcritical Water Chromatography (SWC)

Subcritical water chromatography (SWC) has received some attention in recent years. Subcritical water (SW) can be used to replace the traditional solvent mixtures in the mobile phase of HPLC in reverse phase chromatography [12]. Many HPLC systems already have a column oven. Increasing the temperature of SW generates an effective gradient for elution of the solvent as the polarity of the eluent decreases. Temperature plays an important role in this separation. Increasing temperature can decrease viscosity, which might cause band spreading due to the higher diffusion rate at elevated temperatures [79]. Subcritical water has been applied as an HPLC analytical solvent to extract and quantify caffeine, chlorophenols and anilines [80]. In another study, Rodriguez-Meizoso et al. [21] demonstrated that the combined use of SWE and high-performance liquid chromatography-diode array detection (HPLC-DAD) was a suitable protocol to obtain and characterize nutraceuticals from natural sources, i.e., oregano. They also reported that changing the SW temperature could be used as a means of fine tuning the extraction selectivity for the extraction of antioxidant compounds from oregano. Causon et al. [81] separated n-alcohols with subcritical water chromatography at high temperatures (exceeding 200 °C) using monolithic capillary columns. Subcritical water was compared as an eluent with methanol/water and acetonitrile/water mobile phases for reverse phase liquid chromatography [82]. SWE has been linked directly to SWC using a cold polystyrene-divinylbenzene trap to separate pharmaceuticals and antioxidants [83].

Some of the important disadvantages of HPLC systems are their expense and the fact that they are not environmentally friendly, using high purity toxic organic solvents. The pump in a SWC system does not need to have to have a degasser unit like many traditional HPLC systems. Therefore using SW in HPLC can decrease the cost of the system. In a laboratory it is fairly easy to build your own SWC system. Another advantage of the SWC is its compatibility with many commercial detectors such as UV, FID, fluorescence detectors, refractive index detectors, electrochemical detectors, light scattering detectors and even MS detectors [84].

4.3.6 *Microwave Subcritical Water Extraction*

Microwave extraction or microwave-assisted extraction is a relatively new extraction technique. Microwaves are applied during the extraction process to heat the solvents and sample matrix which increases the kinetics of extraction. Microwave extraction has a number of advantages over traditional methods of extraction of compounds from plant samples, e.g. shorter extraction times, use of less solvents and being more cost effective. There are various types of microwave extraction systems, such as open or closed ones. Many companies offer open or closed vessel microwave extraction systems with differing power levels. Microwave-assisted water extraction [85], microwave-assisted distillation [86] and microwave-assisted hydro-distillation [87] have been used in closed systems. Subcritical water extraction needs the water to be kept within the appropriate pressure ranges at different temperatures (as given in Table 4.3). Closed microwave systems do not need much pressure to reach subcritical water extraction conditions. For example, only pressures of 2.0, 3.6 and 6.1 bar are needed for extraction temperatures of 120, 140 and 160 °C, respectively (Table 4.3). Temperature is one of the most important factors contributing to the recovery yield when using microwave-assisted extraction and SWE techniques. Coelho et al. [88] and Passos and Coimbra [89] have recently used microwave superheated water extraction of carbohydrates from brewers' spent grain and spent coffee grounds. They also found that temperature is very important in extraction efficiency from their samples. Teo et al. [90] extracted stevioside and rebaudioside from *Stevia rebaudiana* using microwave-assisted extraction and [*sic*] pressurized hot water extraction. They have found comparable results using both techniques. People sometimes produce subcritical water conditions in closed microwave-assisted extraction systems without realising it by varying temperatures, pressures and amounts of sample loaded. This may well be helping with their extractions, however, their 'accidental' use of SWE does not receive the credit it should for the higher extraction efficiencies produced. Microwave-assisted extraction and SWE are very effective extraction techniques for plant matrices. It is probable that combining the two techniques into one may produce an even better extraction technique. We need to await results of future work to come to this conclusion.

4.4 Conclusions

SWE is a fast, reliable, clean, cheap, environmentally friendly and green sustainable technique. SW can extract polar, mid-polar and even non-polar compounds from plant samples. Phenolics, essential oils, flavonoids, pectins and proteins have been extracted using SWE. Plants are considered a very complex matrix. Temperature plays an important role in extraction efficiency of target compounds. Increasing temperature from 175 °C may cause degradation problems with many samples. Pressure is important to keep water in the liquid state. Green and sustainable processes are becoming more popular. The cost of the SWE process may appear more expensive in the laboratory scale, however, on an industrial scale it is competitive with commercial methods. SW is being used for newer applications such as in the HPLC mobile phase, extraction together with microwaves and also for pectin and protein extraction.

References

1. Smith RM (2002) Extractions with superheated water. J Chromatogr A 975(1):31–46. doi:10.1016/S0021-9673(02)01225-6
2. Clifford AA (2002) Extraction of natural products with superheated water. In: Clack JH, Macquarrie D (eds) Handbook of green chemistry and technology. Blackwell, Oxford
3. Ozel MZ, Clifford AA (2004) Superheated water extraction of fragrance compounds from Rosa canina. Flavour Frag J 19(4):354–359. doi:10.1002/Ffj.1317
4. Ozel MZ, Gogus F, Lewis AC (2006) Comparison of direct thermal desorption with water distillation and superheated water extraction for the analysis of volatile components of Rosa damascena Mill. using GCxGC-TOF/MS. Anal Chim Acta 566(2):172–177. doi:10.1016/j.aca.2006.03.014
5. Reuters T (1990–2013) Web of science. http://apps.webofknowledge.com/
6. Miller DJ, Hawthorne SB, Gizir AM, Clifford AA (1998) Solubility of polycyclic aromatic hydrocarbons in subcritical water from 298 K to 498 K. J Chem Eng Data 43(6):1043–1047. doi:10.1021/Je980094g
7. Kutlular O, Ozel MZ (2009) Analysis of essential oils of origanum onites by superheated water extraction using GCxGC-TOF/MS. J Essent Oil Bear Pl 12(4):462–470
8. Mustafa A, Turner C (2011) Pressurized liquid extraction as a green approach in food and herbal plants extraction: a review. Anal Chim Acta 703(1):8–18. doi:10.1016/j.aca.2011.07.018
9. Rodriguez-Meizoso I, Castro-Puyana M, Borjesson P, Mendiola JA, Turner C, Ibanez E (2012) Life cycle assessment of green pilot-scale extraction processes to obtain potent antioxidants from rosemary leaves. J Supercrit Fluid 72:205–212. doi:10.1016/j.supflu.2012.09.005
10. Miller DJ, Hawthorne SB (1998) Method for determining the solubilities of hydrophobic organics in subcritical water. Anal Chem 70(8):1618–1621. doi:10.1021/Ac971161x
11. Miller DJ, Hawthorne SB (2000) Solubility of liquid organic flavor and fragrance compounds in subcritical (hot/liquid) water from 298 K to 473 K. J Chem Eng Data 45(2):315–318. doi:10.1021/Je990278a
12. Smith RM (2006) Superheated water: the ultimate green solvent for separation science. Anal Bioanal Chem 385(3):419–421. doi:10.1007/s00216-006-0437-y
13. Carr AG, Mammucari R, Foster NR (2011) A review of subcritical water as a solvent and its utilisation for the processing of hydrophobic organic compounds. Chem Eng J 172(1):1–17. doi:10.1016/j.cej.2011.06.007
14. Teutenberg T, Lerch O, Gotze HJ, Zinn P (2001) Separation of selected anticancer drugs using superheated water as the mobile phase. Anal Chem 73(16):3896–3899. doi:10.1021/Ac0101860
15. Ozel MZ, Gogus F, Lewis AC (2003) Subcritical water extraction of essential oils from Thymbra spicata. Food Chem 82(3):381–386. doi:10.1016/S0308-8146(02)00558-7
16. Rovio S, Hartonen K, Holm Y, Hiltunen R, Riekkola ML (1999) Extraction of clove using pressurized hot water. Flavour Frag J 14(6):399–404. doi:10.1002/(Sici)1099-1026(199911/12)14:6<399::Aid-Ffj851>3.0.Co;2-A
17. Basile A, Jimenez-Carmona MM, Clifford AA (1998) Extraction of rosemary by superheated water. J Agric Food Chem 46(12):5205–5209. doi:10.1021/Jf980437e

18. Priego-Lopez E, de Castro MDL (2004) Superheated water extraction of linear alquilbenzene sulfonates from sediments with on-line preconcentration/derivatization/detection. Anal Chim Acta 511(2):249–254. doi:10.1016/j.aca.2004.02.005
19. Hashimoto S, Watanabe K, Nose K, Morita M (2004) Remediation of soil contaminated with dioxins by subcritical water extraction. Chemosphere 54(1):89–96. doi:10.1016/S0045-6535(03)00673-8
20. Ibanez E, Kubatova A, Senorans FJ, Cavero S, Reglero G, Hawthorne SB (2003) Subcritical water extraction of antioxidant compounds from rosemary plants. J Agric Food Chem 51(2):375–382. doi:10.1021/Jf025878j
21. Rodriguez-Meizoso I, Marin FR, Herrero M, Senorans FJ, Reglero G, Cifuentes A, Ibanez E (2006) Subcritical water extraction of nutraceuticals with antioxidant activity from oregano. Chemical and functional characterization. J Pharm Biomed 41(5):1560–1565. doi:10.1016/j.jpba.2006.01.018
22. Teo CC, Tan SN, Yong JWH, Hew CS, Ong ES (2010) Pressurized hot water extraction (PHWE). J Chromatogr A 1217(16):2484–2494. doi:10.1016/j.chroma.2009.12.050
23. IAPWS Release on the static dielectric constant of ordinary water substance for temperatures from 238 K to 873 K and pressures up to 1000 MPa. http://www.iapws.org.
24. Richter BE, Jones BA, Ezzell JL, Porter NL, Avdalovic N, Pohl C (1996) Accelerated solvent extraction: a technique for sample preparation. Anal Chem 68(6):1033–1039. doi:10.1021/Ac9508199
25. Franks F (1983) Water, 1st edn. The Royal Society of Chemistry, London
26. Fernandez-Prini RJ, Corti HR, Japas ML (1991) High-temperature aqueous solutions: thermodynamic properties. CRC Press, Boca Raton
27. Caffarena ER, Grigera JR (2004) On the hydrogen bond structure of water at different densities. Phys A 342(1–2):34–39. doi:10.1016/j.physa.2004.04.057
28. Hawthorne SB, Miller DJ (1994) Direct comparison of soxhlet and low-temperature and high-temperature supercritical CO_2 extraction efficiencies of organics from environmental solids. Anal Chem 66(22):4005–4012. doi:10.1021/Ac00094a024
29. Mockel HJ, Welter G, Melzer H (1987) Correlation between reversed-phase retention and solute molecular-surface type and area. 1. Theoretical outlines and retention of various hydrocarbon classes. J Chromatogr 388(2):255–266. doi:10.1016/S0021-9673(01)94487-5
30. Jimenez-Carmona MM, Ubera JL, de Castro MDL (1999) Comparison of continuous subcritical water extraction and hydrodistillation of marjoram essential oil. J Chromatogr A 855(2):625–632. doi:10.1016/S0021-9673(99)00703-7
31. Gamiz-Gracia L, de Castro MDL (2000) Continuous subcritical water extraction of medicinal plant essential oil: comparison with conventional techniques. Talanta 51(6):1179–1185. doi:10.1016/S0039-9140(00)00294-0
32. Gogus F, Ozel MZ, Lewis AC (2006) Extraction of essential oils of leaves and flowers of Achillea monocephala by superheated water. Flavour Frag J 21(1):122–128. doi:10.1002/Ffj.1541
33. Kubatova A, Lagadec AJM, Miller DJ, Hawthorne SB (2001) Selective extraction of oxygenates from savory and peppermint using subcritical water. Flavour Frag J 16(1):64–73. doi:10.1002/1099-1026(200101/02)16:1<64::Aid-Ffj949>3.3.Co;2-4
34. Ong ES, Len SM (2003) Evaluation of surfactant-assisted pressurized hot water extraction for marker compounds in Radix Codonopsis pilosula using liquid chromatography and liquid chromatography/electrospray ionization mass spectrometry. J Sep Sci 26(17):1533–1540. doi:10.1002/jssc.200301578
35. Ong ES, Len SM (2003) Pressurized hot water extraction of berberine, baicalein and glycyrrhizin in medicinal plants. Anal Chim Acta 482(1):81–89. doi:10.1016/S0003-2670(03)00196-X
36. Ong ES, Len SM (2004) Evaluation of pressurized liquid extraction and pressurized hot water extraction for tanshinone I and IIA in Salvia miltiorrhiza using LC and LC-ESI-MS. J Chromatogr Sci 42(4):211–216

37. Pineiro Z, Palma M, Barroso CG (2004) Determination of catechins by means of extraction with pressurized liquids. J Chromatogr A 1026(1–2):19–23. doi:10.1016/j.chroma.2003.10.096
38. Eller FJ, Taylor SL (2004) Pressurized fluids for extraction of cedarwood oil from Juniperus virginianna. J Agric Food Chem 52(8):2335–2338. doi:10.1021/Jf030783i
39. Deng CH, Yao N, Wang AQ, Zhang XM (2005) Determination of essential oil in a traditional Chinese medicine, Fructus amomi by pressurized hot water extraction followed by liquid-phase microextraction and gas chromatography-mass spectrometry. Anal Chim Acta 536(1–2):237–244. doi:10.1016/j.aca.2004.12.044
40. Deng CH, Yang XH, Zhang XM (2005) Rapid determination of panaxynol in a traditional Chinese medicine of Saposhnikovia divaricata by pressurized hot water extraction followed by liquid-phase microextraction and gas chromatography-mass spectrometry. Talanta 68(1):6–11. doi:10.1016/j.talanta.2005.04.040
41. Ozel MZ, Kaymaz H (2004) Superheated water extraction, steam distillation and Soxhlet extraction of essential oils of Origanum onites. Anal Bioanal Chem 379(7–8):1127–1133. doi:10.1007/s00216-004-2671-5
42. Zou SP, Wu YL, Yang MD, Li C, Tong JM (2010) Bio-oil production from sub- and supercritical water liquefaction of microalgae Dunaliella tertiolecta and related properties. Energ Environ Sci 3(8):1073–1078. doi:10.1039/C002550j
43. Crescenzi C, Di Corcia A, Nazzari M, Samperi R (2000) Hot phosphate-buffered water extraction coupled on line with liquid chromatography/mass spectrometry for analyzing contaminants in soil. Anal Chem 72(14):3050–3055. doi:10.1021/Ac000090q
44. Andersson T, Pihtsalmi T, Hartonen K, Hyotylainen T, Riekkola ML (2003) Effect of extraction vessel geometry and flow homogeneity on recoveries of polycyclic aromatic hydrocarbons in pressurised hot water extraction. Anal Bioanal Chem 376(7):1081–1088. doi:10.1007/s00216-003-2078-8
45. Carabias-Martinez R, Rodriguez-Gonzalo E, Revilla-Ruiz P, Hernandez-Mendez J (2005) Pressurized liquid extraction in the analysis of food and biological samples. J Chromatogr A 1089(1–2):1–17. doi:10.1016/j.chroma.2005.06.072
46. Herrero M, Cifuentes A, Ibanez E (2006) Sub- and supercritical fluid extraction of functional ingredients from different natural sources: Plants, food-by-products, algae and microalgae – a review. Food Chem 98(1):136–148. doi:10.1016/j.foodchem.2005.05.058
47. Shalmashi A, Golmohammad F, Eikani MH (2008) Subcritical water extraction of caffeine from black tea leaf of Iran. J Food Process Eng 31(3):330–338. doi:10.1111/j.1745-4530.2007.00156.x
48. Extraction CPCS (2003) http://www.criticalprocesses.com/Superheated%20water%20-%20More%20details.htm
49. Gogus F, Ozel MZ, Lewis AC (2005) Superheated water extraction of essential oils of Origanum micranthum. J Chromatogr Sci 43(2):87–91
50. Ayala RS, de Castro MDL (2001) Continuous subcritical water extraction as a useful tool for isolation of edible essential oils. Food Chem 75(1):109–113
51. Ozel MZ, Gogus F, Hamilton JF, Lewis AC (2005) Analysis of volatile components from Ziziphora taurica subsp taurica by steam distillation, superheated-water extraction, and direct thermal desorption with GCxGC-TOFMS. Anal Bioanal Chem 382(1):115–119. doi:10.1007/s00216-005-3156-x
52. Ozel MZ, Gogus F, Lewis AC (2006) Determination of Teucrium chamaedrys volatiles by using direct thermal desorption-comprehensive two-dimensional gas chromatography-time-of-flight mass spectrometry. J Chromatogr A 1114(1):164–169. doi:10.1016/j.chroma.2006.02.036
53. Kim JW, Mazza G (2006) Optimization of extraction of phenolic compounds from flax shives by pressurized low-polarity water. J Agric Food Chem 54(20):7575–7584. doi:10.1021/Jf0608221

54. Garcia-Marino M, Rivas-Gonzalo JC, Ibanez E, Garcia-Moreno C (2006) Recovery of catechins and proanthocyanidins from winery by-products using subcritical water extraction. Anal Chim Acta 563(1–2):44–50. doi:10.1016/j.aca.2005.10.054
55. Ardag H, Ozel MZ, Sen A (2011) Polycyclic aromatic hydrocarbons in water from the Menderes River, Turkey. Bull Environ Contam Toxicol 86(2):221–225. doi:10.1007/s00128-011-0199-x
56. Hartonen K, Parshintsev J, Sandberg K, Bergelin E, Nisula L, Riekkola ML (2007) Isolation of flavonoids from aspen knotwood by pressurized hot water extraction and comparison with other extraction techniques. Talanta 74(1):32–38. doi:10.1016/j.talanta.2007.05.040
57. Aliakbarian B, Fathi A, Perego P, Dehghani F (2012) Extraction of antioxidants from winery wastes using subcritical water. J Supercrit Fluid 65:18–24. doi:10.1016/j.supflu.2012.02.022
58. Choi MPK, Chan KKC, Leung HW, Huie CW (2003) Pressurized liquid extraction of active ingredients (ginsenosides) from medicinal plants using non-ionic surfactant solutions. J Chromatogr A 983(1–2):153–162. doi:10.1016/S0021-9673(02)01649-7
59. Shotipruk A, Kiatsongserm J, Pavasant P, Goto M, Sasaki M (2004) Pressurized hot water extraction of anthraquinones from the roots of Morinda citrifolia. Biotechnol Progr 20(6):1872–1875. doi:10.1021/Bp049779x
60. Lamm LJ, Yang Y (2003) Off-line coupling of subcritical water extraction with subcritical water chromatography via a sorbent trap and thermal desorption. Anal Chem 75(10):2237–2242. doi:10.1021/Ac020724o
61. Singh PP, Saldana MDA (2011) Subcritical water extraction of phenolic compounds from potato peel. Food Res Int 44(8):2452–2458. doi:10.1016/j.foodres.2011.02.006
62. King JW, Grabiel R, Wightman JD (2003) Subcritical water extraction of anthocyanins from fruit berry substrates. In: 6th international symposium on supercritical fluids. Versailles, France, 28–30 April 2003
63. Monrad JK, Srinivas K, Howard LR, King JW (2012) Design and optimization of a semi-continuous hot-cold extraction of polyphenols from Grape Pomace. J Agric Food Chem 60(22):5571–5582. doi:10.1021/Jf300569w
64. Petersson EV, Liu JY, Sjoberg PJR, Danielsson R, Turner C (2010) Pressurized hot water extraction of anthocyanins from red onion: a study on extraction and degradation rates. Anal Chim Acta 663(1):27–32. doi:10.1016/j.aca.2010.01.023
65. Pongnaravane B, Goto M, Sasaki M, Anekpankul T, Pavasant P, Pavasant P, Shotipruk A (2006) Extraction of anthraquinones from roots of Morinda citrifolia by pressurized hot water: antioxidant activity of extracts. J Supercrit Fluid 37(3):390–396. doi:10.1016/j.supflu.2005.12.013
66. Erlund I (2004) Review of the flavonoids quercetin, hesperetin naringenin. Dietary sources, bioactivities, and epidemiology. Nutr Res 24(10):851–874. doi:10.1016/j.nutres.2004.07.005
67. Jeon SM, Bok SH, Jang MK, Lee MK, Nam KT, Park YB, Rhee SJ, Choi MS (2001) Antioxidative activity of naringin and lovastatin in high cholesterol-fed rabbits. Life Sci 69(24):2855–2866. doi:10.1016/S0024-3205(01)01363-7
68. Ko MJ, Cheigh CI, Chung MS (2014) Relationship analysis between flavonoids structure and subcritical water extraction (SWE). Food Chem 143:147–155. doi:10.1016/j.foodchem.2013.07.104
69. Khuwijitjaru P, Chalooddong K, Adachi S (2008) Phenolic content and radical scavenging capacity of kaffir lime fruit peel extracts obtained by pressurized hot water extraction. Food Sci Technol Res 14(1):1–4. doi:10.3136/Fstr.14.1
70. He L, Zhang XF, Xu HG, Xu C, Yuan F, Knez Z, Novak Z, Gao YX (2012) Subcritical water extraction of phenolic compounds from pomegranate (Punica granatum L.) seed residues and investigation into their antioxidant activities with HPLC-ABTS (center dot^{+}) assay. Food Bioprod Process 90(C2):215–223. doi:10.1016/j.fbp.2011.03.003
71. Akhtar MH, Bryan M (2008) Extraction and quantification of major carotenoids in processed foods and supplements by liquid chromatography. Food Chem 111(1):255–261. doi:10.1016/j.foodchem.2008.03.071
72. Fikselova M, Silhar S, Marecek J, Francakova H (2008) Extraction of carrot (Daucus carota L.) carotenes under different conditions. Czech J Food Sci 26(4):268–274

73. Denery JR, Dragull K, Tang CS, Li QX (2004) Pressurized fluid extraction of carotenoids from Haematococcus pluvialis and Dunaliella salina and kavalactones from Piper methysticum. Anal Chim Acta 501(2):175–181. doi:10.1016/j.aca.2003.09.026
74. McInerney JK, Seccafien CA, Stewart CM, Bird AR (2007) Effects of high pressure processing on antioxidant activity, and total carotenoid content and availability, in vegetables. Innov Food Sci Emerg 8(4):543–548. doi:10.1016/j.ifset.2007.04.005
75. Pan Y, Yi X, Deng X, Zhao S, Chen S, Yang H, Han L, Zhu J (2012) Determination of multipesticides in black tea by subcritical water extraction and gas chromatography-tandem mass spectrometry. Chin J Chromatogr 30(11):1159–1165
76. Latawiec AE, Reid BJ (2010) Sequential extraction of polycyclic aromatic hydrocarbons using subcritical water. Chemosphere 78(8):1042–1048. doi:10.1016/j.chemosphere.2009.11.029
77. Martorell I, Perello G, Marti-Cid R, Castell V, Llobet JM, Domingo JL (2010) Polycyclic aromatic hydrocarbons (PAH) in foods and estimated PAH intake by the population of Catalonia, Spain: temporal trend. Environ Int 36(5):424–432. doi:10.1016/j.envint.2010.03.003
78. Welling P, Kaandorp B (1986) Determination of polycyclic aromatic-hydrocarbons (Pah) in edible vegetable-oils by liquid-chromatography and programmed fluorescence detection – comparison of caffeine complexation and Xad-2 chromatography sample cleanup. Z Lebensm Unters Forsch 183(2):111–115. doi:10.1007/Bf01041927
79. Chienthavorn O, Smith RM (1999) Buffered superheated water as an fluent for reversed-phase high performance liquid chromatography. Chromatographia 50(7–8):485–489. doi:10.1007/Bf02490746
80. Li B, Yang Y, Gan YX, Eaton CD, He P, Jones AD (2000) On-line coupling of subcritical water extraction with high-performance liquid chromatography via solid-phase trapping. J Chromatogr A 873(2):175–184. doi:10.1016/S0021-9673(99)01322-9
81. Causon TJ, Shellie RA, Hilder EF (2009) High temperature liquid chromatography with monolithic capillary columns and pure water eluent. Analyst 134(3):440–442. doi:10.1039/B815886j
82. Allmon SD, Dorsey JG (2010) Properties of subcritical water as an eluent for reversed-phase liquid chromatography-Disruption of the hydrogen-bond network at elevated temperature and its consequences. J Chromatogr A 1217(37):5769–5775. doi:10.1016/j.chroma.2010.07.030
83. Tajuddin R, Smith RM (2002) On-line coupled superheated water extraction (SWE) and superheated water chromatography (SWC). Analyst 127(7):883–885. doi:10.1039/B203298h
84. Smith RM, Chienthavorn O, Wilson ID, Wright B, Taylor SD (1999) Superheated heavy water as the eluent for HPLC-NMR and HPLC-NMR-MS of model drugs. Anal Chem 71(20):4493–4497. doi:10.1021/Ac9905470
85. Nkhili E, Tomao V, El Hajji H, El Boustani ES, Chemat F, Dangles O (2009) Microwave-assisted water extraction of green tea polyphenols. Phytochem Anal 20(5):408–415. doi:10.1002/Pca.1141
86. Mohamadi M, Shamspur T, Mostafavi A (2013) Comparison of microwave-assisted distillation and conventional hydrodistillation in the essential oil extraction of flowers Rosa damascena Mill. J Essent Oil Res 25(1):55–61. doi:10.1080/10412905.2012.751555
87. Liu YQ, Wang HW, Wei SL, Yan ZJ (2012) Chemical composition and antimicrobial activity of the essential oils extracted by microwave-assisted hydrodistillation from the flowers of two plumeria species. Anal Lett 45(16):2389–2397. doi:10.1080/00032719.2012.689905
88. Coelho E, Rocha MAM, Saraiva JA, Coimbra MA (2014) Microwave superheated water and dilute alkali extraction of brewers' spent grain arabinoxylans and arabinoxylo-oligosaccharides. Carbohydr Polym 99:415–422. doi:10.1016/j.carbpol.2013.09.003
89. Passos CP, Coimbra MA (2013) Microwave superheated water extraction of polysaccharides from spent coffee grounds. Carbohydr Polym 94(1):626–633. doi:10.1016/j.carbpol.2013.01.088
90. Teo CC, Tan SN, Yong JWH, Hew CS, Ong ES (2009) Validation of green-solvent extraction combined with chromatographic chemical fingerprint to evaluate quality of Stevia rebaudiana Bertoni. J Sep Sci 32(4):613–622. doi:10.1002/jssc.200800552

Chapter 5
Liquefied Dimethyl Ether: An Energy-Saving, Green Extraction Solvent

Peng Li and Hisao Makino

Abstract Extraction is an essential procedure in the fields of food, pharmacy, and renewable bio-fuels, and it affords the recovery of desired components and the removal of undesired components from the natural feedstock. Conventional extraction techniques involving organic solvents and supercritical fluids have been extensively studied and used. Generally, these techniques are either economically or environmentally unfavourable because of the use of toxic solvents and considerable heating and pressurizing. Recently, a new extraction technique involving the use of liquefied dimethyl ether (DME) as a green solvent has attracted tremendous attention. This technique is economically efficient and environmentally friendly by virtue of the unique physical and chemical properties of DME. Additionally, the DME method can extract/remove the desired/undesired components as well as dewater (dry) the wet materials simultaneously. These advantages render the DME method practicable in several industrial fields. This chapter attempts to outline the potential of liquefied DME as an extraction solvent by elucidating the operating principles, procedures, and some recent studies and results.

Keywords Extraction • Dimethyl ether • Dewatering

5.1 Introduction

Extraction techniques are widely employed for the isolation of various compositions from natural resources in industries pertaining to food and pharmacy, and is being considered employing in the field of renewable bio-fuels production from natural biomasses In some cases, the extraction technique is an essential step in

P. Li (✉) • H. Makino
Energy Engineering Research Laboratory, Central Research Institute of Electric Power Industry (CRIEPI), Yokosuka 240-0196, Japan
e-mail: yolipeng@criepi.denken.or.jp

eliminating the undesired components from the feedstock. Basically, extraction can be categorised based on the use of traditional solvents and critical fluids. The traditional solvent extraction is the most common because chemical solvents have high selectivity and solubility for the target compositions. In the traditional solvent extraction, the solvent used is in the liquid form at room temperature and ambient pressure. A wide range of organic solvents such as chloroform, methanol, hexane, and petroleum ether have been applied [1, 10]; the selection of solvent is usually dependent upon the type of target feedstock and target chemicals contained in the feedstock. The major factors that influence the efficiency and selectivity of solvent extraction are the physicochemical properties of the selected solvent such as polarity, and the operating conditions such as temperature. In addition, the economic and environmental concerns must be considered. The principle of the traditional chemical solvent extraction is as follows: an organic solvent is in contact with the extractable materials while the target compositions are continuously extracted into the organic phase because of the permeability effect of the solvent on the materials. The selection of the organic solvents is of tremendous relevance because it must possess biocompatibility, maximum solubility for target compositions, and extraction ability. The disadvantage of traditional solvent extraction is that most of the organic solvents cause health hazards and/or environmental pollution. Besides, pretreatment processes such as drying and cell disruption are normally required for the extraction of available components from natural resources, increasing the overall cost burden.

Supercritical fluid technology (SFT) is more efficient compared to traditional solvent extraction, and shows high selectivity. The basic principle of SFT involves the establishment of a certain phase (supercritical) which is beyond the critical point of the fluid, wherein the meniscus separating the liquid and vapour phases disappears, leaving behind a single homogeneous phase [27]. SFT is a suitable substitute for organic solvents for a range of industrial and laboratory processes. Carbon dioxide (CO_2) and water are the most commonly used supercritical fluids, which could be potentially used for the extraction of natural products. For example, supercritical CO_2 has several advantages over traditional solvents, especially for extracting less polar chemicals [11, 19]. The disadvantages of SFT are associated with the high costs of operation and safety related issues; for instance, in the case of supercritical CO_2, the operating temperature and pressure are above 304.25 K and 7.39 MPa, respectively. In order to reduce the costs, recent studies involving near- or sub- critical solvents such as dimethyl ether have been reported [2].

Central Research Institute of Electric Power Industry (CRIEPI) recently developed a new extraction technology that differs from both the traditional organic solvent extraction and SFT. This technique uses liquefied dimethyl ether (DME) as the solvent to extract the target compositions, and remove the water from the wet materials simultaneously. The advantages of this method are that it is highly cost-efficient and environmentally friendly. In fact, the use of DME as an extraction solvent has been approved by the European Union [6]. Thus far, this technique has been studied at both laboratory and bench scales for use in the (1) dewatering and extraction of organic components from vegetal biomasses [22];

(2) extraction of bioactive components from green tea leaves [18]; (3) extraction of lipids and hydrocarbons from high-moisture microalgae [14, 16, 17]; (4) removal of polychlorinated biphenyls (PCBs) from polluted soil and other materials [26]. A pilot-scale study on the use of DME for extraction purposes is also being carried out in CRIEPI.

This chapter presents a complete picture of the current knowledge and recent studies on the use of liquefied DME for extraction. It provides the necessary theoretical background and relevant details of the method including the technique, mechanism, and some recent results pertaining to the extraction of natural products.

5.2 Basic Principles

5.2.1 Properties of DME

The chemical structure of DME is shown in Fig. 5.1. DME is the simplest ether with the formula, CH_3OCH_3, which is in a gaseous state at room temperature and pressure. The standard boiling point of DME is -24.8 °C and its saturated vapour pressure has been previously mentioned and explained in detail [28]; for example, the pressure is 0.51–0.59 MPa over the normal temperature range of (20–25 °C). Like other organic solvents, DME has high affinity for organic compositions, but the difference is that it is also partially miscible in water. The phase equilibrium relationship for the liquefied DME/water system is known [9]. At room temperature, water is soluble in DME over the range of 7–8 wt%. For example, the weight of DME required for the extraction of water is 1/(0.07–0.08) times the weight of water. DME is almost non-toxic; the European Food Safety Authority has determined that there are no health concerns with regard to the use of DME as an extraction solvent in food processing [6].

5.2.2 Theoretical Principles of DME Extraction

A schematic depicting the steps involved in DME extraction is shown as Fig. 5.2. In the first step, the target components and/or water in the natural materials are extracted using liquefied DME; and as a result, a mixture of DME, water and organic components is formed. Secondly, the concentration of water and organic components in liquefied DME increases and reaches saturation, while the materials are dried. Thirdly, the DME in the mixture is vapourised, and then the organic components and water are separated. In the final step, the DME gas is again liquefied for the next circulation.

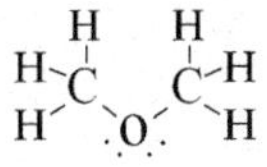

Fig. 5.1 Structure of DME molecule

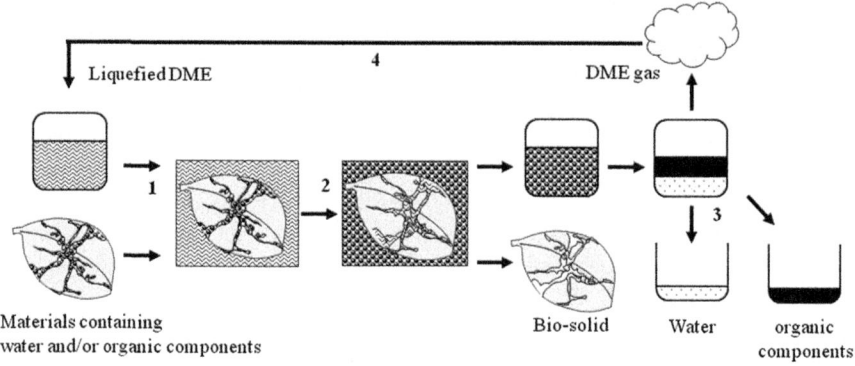

Fig. 5.2 Flow chart of process of DME extraction for wet natural feedstock

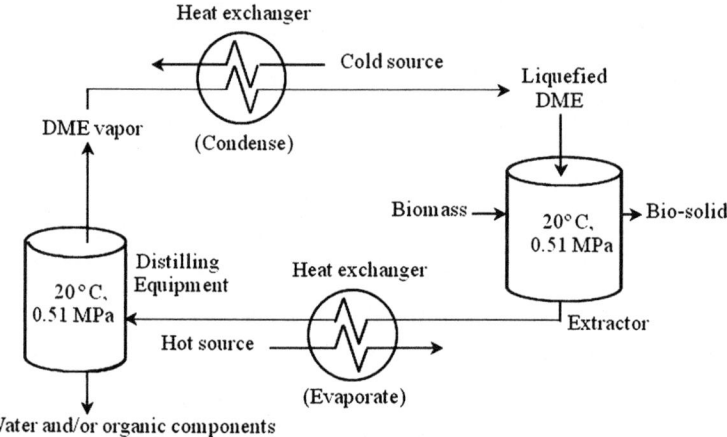

Fig. 5.3 Schematic illustration of the DME extraction system

The principle of energy saving in the aforementioned process is shown in Fig. 5.3, which represents a DME extraction system [22]. In this system, at ambient temperature, DME is mixed with organic components and water in an extractor, using which water and organic components are separated. By employing techniques such as filtration and sedimentation, the water-organic-DME mixture can be separated from the materials and ejected from the extractor. If the concentration of the organic components in the mixture is insufficient for further use, the mixture is recycled as the extractant. Next, the DME in the mixture is evaporated in a heat exchanger by a low-level waste heat. As a consequence, supersaturated water appears beneath the liquefied DME phase because the organic component dissolves more readily into the liquefied DME. The organic-DME mixture is then separated from the water layer, and most of the DME at this stage exists as a gas. The DME vapour is condensed in the heat exchanger using cold heat sources. The reutilization of DME, including its evaporation and liquefaction, permits the efficient use of low-level heat source. For large-scale applications, a heating source of unharnessed waste heat at about 40 °C is desirable for DME evaporation. DME gas is then liquefied again at a slightly lower temperature for recirculation. At this stage, a cooling source of about 10 °C such as geo-heat, which is present within the first 50 m of the earth's surface, is desirable [4].

5.3 Experimental

5.3.1 Laboratory-Scale DME Extraction Apparatus

Kanda et al., designed the laboratory-scale apparatus to evaluate the extraction efficiency of the proposed method [13]. The apparatus mainly consisted of three parts generally; however, in order to test the different materials based on their respective properties, the apparatus was slightly modified. As shown in Fig. 5.4, a vessel for storing the liquefied DME (TVS-1-100, Taiatsu Techno Corp., Saitama, Japan), a vessel as extraction column (HPG-10-5 Taiatsu Techno Corp.) and a storage vessel to hold the mixture of DME, water and/or extracted organic components (HPG-96-3, Taiatsu Techno Corp.) were connected in series. The test materials were loaded into the extraction column. Nitrogen (0.6 MPa) was used to push the liquefied DME through the extraction system.

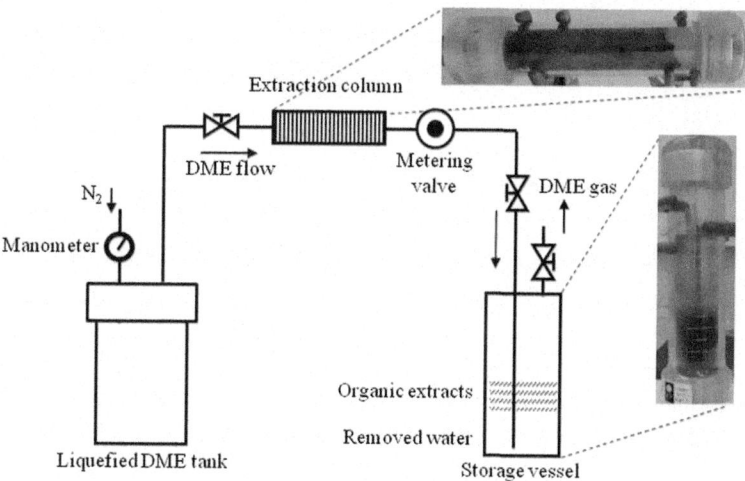

Fig. 5.4 Schematic diagram of the lab-scale DME extraction apparatus (the sample in the extraction column is green tea leaves)

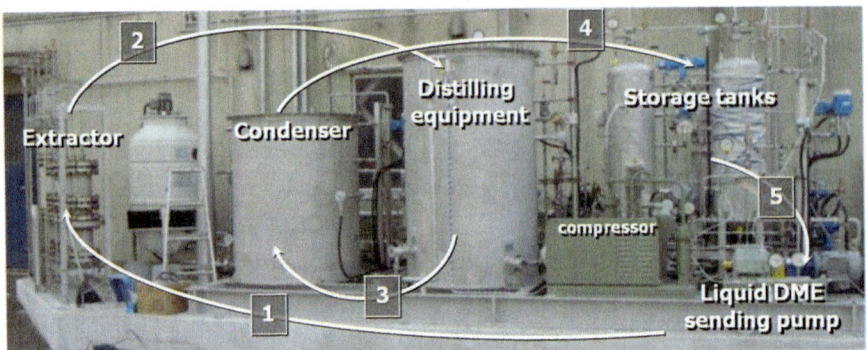

Fig. 5.5 The prototype of the DME extraction process

5.3.2 Bench-Scale DME Extraction Equipment

The world's first prototype of the DME extraction process was reported by Kanda and Makino as shown in Fig. 5.5 [12]. Briefly, the equipment consisted of a liquefied DME pump, extraction column (volume 0.01 m^3, inner diameter 0.15 m, length 0.55 m), evaporator, flash distillation tower (volume: 0.1 m^3), and condenser, connected in series to form a closed loop. In this extraction system, the operation pressure was 0.51 MPa, and temperature in the extractor and distillation tower was around 20 °C. Liquefied DME was mixed with wet test materials in the extractor, and water and organic compounds were extracted. The mixture of water, organic compounds and DME was separated from the test materials and ejected from the extractor. Next, DME in the mixture was evaporated in the heat exchanger at 30 °C, and the water and organic components were separated from DME in the distillation tower. The DME vapour was then condensed in the heat exchanger at 15 °C.

5.4 Results and Discussion

5.4.1 Extraction and Dewatering of Vegetal Biomass

Li et al., reported the extraction of three representative common vegetal biomasses including spent coffee grounds, green tea waste, and orange peels for validating the performance of DME extraction method. These materials are the main sources of industrial food waste generated on a huge scale worldwide, warranting their effective utilisation. For example, the world annually produces around 6 million tons of spent coffee grounds from the beverage factories [24]. According to a credible report, the spent coffee grounds contain around 10–15 % of oily substances, depending on the coffee species, which can be easily converted into biodiesel [20].

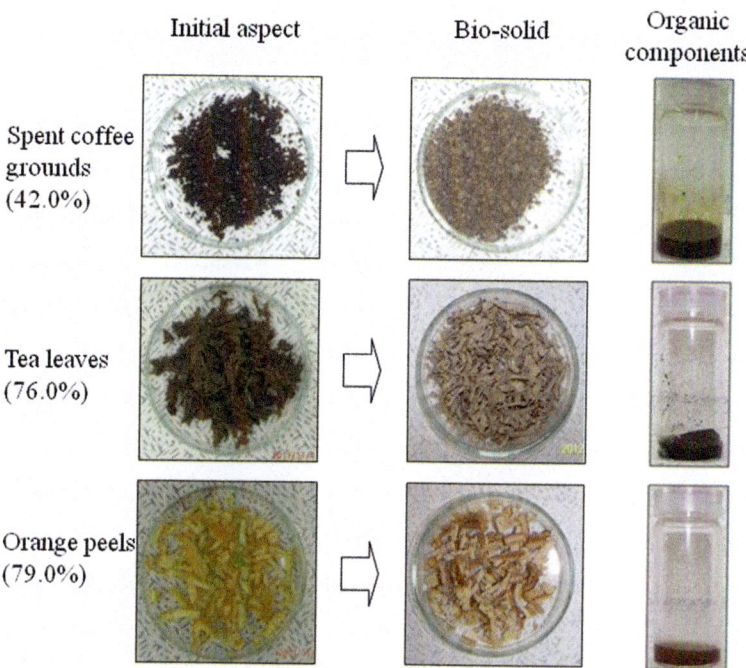

Fig. 5.6 Pictures of original samples, dewatered bio-solids and extracted organic components. The value in parentheses is the initial water content of biomass

Green tea is consumed worldwide as one of the most popular, traditional beverages, particularly in Asian countries such as China, India, and Japan. The annual global production of tea was about 4.51 million tons in 2010 [7].

As shown in Fig. 5.6, the surfaces of these vegetal biomasses became relatively brighter after DME extraction owing to a substantial decrease in the water and pigment contents of these samples. Figure 5.7 shows the changes in the water content of the test samples and amounts of the organic extracts in the storage vessel. As the amount of liquefied DME passing through the extraction system increased, the water content in samples decreased while the amount of organic extracts in the storage vessel significantly increased. When the amounts of consumed DME were 218.1, 196.9, and 200.5 g, the corresponding water contents in the samples were 5.0 %, 10.0 %, and 11.9 %, respectively. On the other hand, to obtain the maximum amounts of organic extracts from the spent coffee grounds and green tea waste, approximately 276.4 and 277.1 g of DME were consumed. The difference in the DME consumption for dewatering and extraction of organic components for spent coffee grounds and green tea waste may be due to the difference in the biologic properties of these biomasses. The results also implied that the dewatering velocity of the liquefied DME exceeded its extraction velocity for organic components, at least for spent coffee grounds and green tea waste.

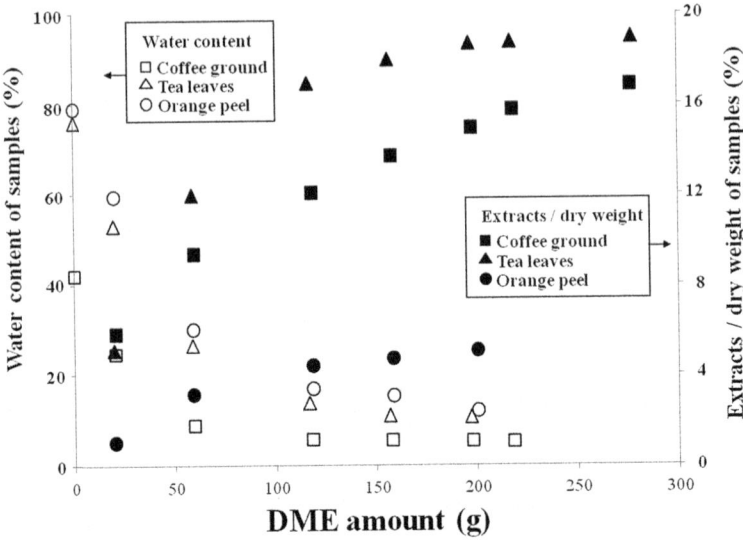

Fig. 5.7 Changes of water content in the samples and yield of organic extracts

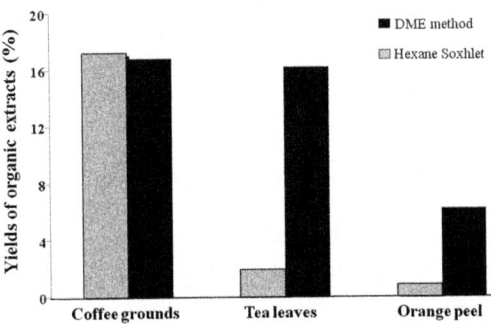

Fig. 5.8 Organic extracts yield using the DME (*black*) and hexane Soxhlet (*gray*) methods

The extraction yields of the organic components obtained using the DME extraction method were compared to those obtained from the widely-used conventional Soxhlet method involving hexane. As shown in Fig. 5.8, with the Soxhlet method, the extraction yields of the spent coffee grounds, green tea waste, and orange peels were 17.2 ± 0.2 %, 1.9 ± 0.1 % and 0.9 ± 0.05 %, respectively; however, the extraction yields using the DME method were 16.8 ± 1.0 %, 16.2 ± 1.5 %, and 6.2 ± 0.5 %, respectively. The difference in the extraction yields between the two methods was because of the difference in the chemical compositions of the tested biomasses and the chemical properties between DME and hexane.

Table 5.1 Sample details of tested algae varieties

	Genus	Cell form	Water content (%)	Location
N-595	*Oscillatoria agardhii*	Filar	85	Northern Ireland
N1263	*Oscillatoria agardhii*	Filar	85	Germany
ONC	*Microcystis aeruginosa*	Granular	93.4	Okinawa island (Japan)
GSK	*Microcystis aeruginosa*	Granular	91.1	Okinawa island (Japan)
GK12	*Monoraphidium chlorophyta*	Granular	78.2	–
Kanogawa	*Cymbela*	Acicular	93	Lake Kanogawa Ozu (Japan)
Hirosawa[a]	*Microcystis*	Granular	91	Hirosawa mere Kyoto (Japan)

[a]Microalgae sample was mixed-species

5.4.2 Extraction of Bio-oils from Microalgae

Fossil fuel depletion and global warming have impelled researchers to work on bio-fuel production from biomasses such as crops, animal fat, and microalgae [25]. Among these, microalgae have attracted significant attention as the newest generation of biofuel resource [8]. Compared to terrestrial plants, microalgae have high oil content and growth rate; mass algal cultivation can be performed on unexploited lands using systems supplying nutrients, thus avoiding competition with limited arable lands [21].

In the conventional process, the recovery of bio-oils from microalgae generally requires multiple solid-liquid separation steps. These processes involve drying, cell wall disruption, and solvent extraction [23]. The extraction of bio-oils is usually performed with toxic organic solvents such as hexane, chloroform, and methanol, which means these processes are highly energy-intensive and damaging to the environment [23]. For example, on the laboratory scale, bio-oil extraction with hexane is normally carried out using the Soxhlet method at 70 °C for 18 h [5]. Such long duration of extraction and heating is a key drawback. The most rapid and effective conventional extraction method for bio-oils is the Bligh-Dyer's method [1], which involves drying, cell disruption, and solvent (chloroform-methanol) extraction. This standard method has been indispensable, not only for bio-oil extraction from microalgae but also for the quantification of crude oil derived from biological materials [10].

Kanda et al., investigated the extraction of bio-oils using liquefied DME on several natural blue-green microalgae, and the results were compared to those obtained using the Bligh-Dyer's method [16]. The sample details of the tested algae are listed in Table 5.1. The extraction volumes achieved using liquefied DME and the Bligh-Dyer's method are shown in Fig. 5.9. White columns represent the bio-oil extraction yield using liquefied DME on the dry weight of the microalgae while the black columns represent the results of the Bligh-Dyer's method. Both NIES-595 and NIES-1263 belong to *Oscillatoria agardhii*, but their extraction yields were

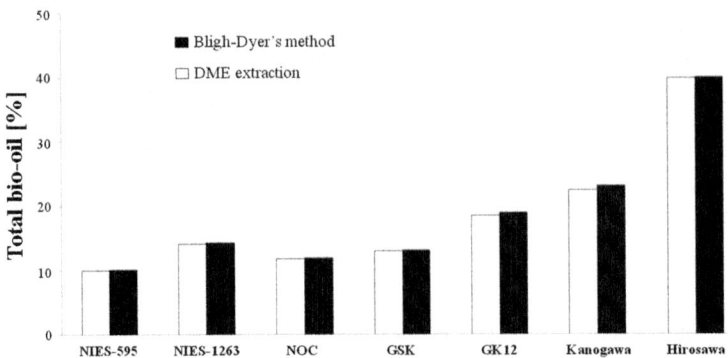

Fig. 5.9 Bio-oil extraction ratio using the DME and BD methods for several species of natural microalgae

9.9 ± 1 % and 14.0 ± 1 %, respectively. Conversely, the extraction rates of ONC 11.0 ± 2 % and GSK 12.0 ± 1.5 % were similar. The extraction yield of GK12 was 18.5 ± 2 % while that of the mixed-species of microalgae collected at Lake Kanogawa was 22.5 ± 1 %. The extraction yield of Hirosawa Mere showed the highest extraction rate at 40.1 ± 2 %. The extraction yield of all the species was more than 97.0 % as determined by the Bligh-Dyer's method. This implied that the extraction yields obtained using the DME extraction method were comparable to those obtained using the Bligh-Dyer's method.

In another recent study Kanda et al., proposed the use of this method to directly extract hydrocarbons from the wet *botryococcus braunii*, which is a fine energy source containing a considerable amount of hydrocarbons [17]. Our results indicated that the extraction yields of hydrocarbons using DME were approximately identical to those obtained using the Soxhlet extraction method involving hexane.

5.4.3 Other Recent Studies Involving DME

The use of DME not only affords the extraction of available components from natural feedstock but also the removal of water and undesired components. Oshita et al., successfully investigated the removal of PCBs and water from the river sediments [26]. The maximum extraction efficiencies of liquefied DME for PCBs and water were 99 % and 97 %, respectively. Only about 2 % of PCBs remained in the sediment after DME treatment. Kanda et al., proposed this method to remove the odorous components and water from the slurry of biosolids [15]. In their study, the moisture content of the test sludge cake was reduced from 78.9 to 8.0 %. The amounts of the odorous components in the dewatered sludge including hydrogen sulphide, methyl mercaptan, methyl sulphide, and acetaldehyde were reduced significantly.

Table 5.2 Balancing contents of caffeine and catechins

Amount (μg)	Post-HWE green tea leaves	DME extraction			
		Tea leaves residue	Organic compounds	Water	Loss
Total (%)	100.0	28.4	1.6	70.0	–
Caffeine (μg/g)	1,210	Nd	47	498	665
Catechins (μg/g)					
C	105	41	3	35	26
EC	798	201	16	342	239
GC	587	175	7	239	166
EGC	3,568	1,256	32	1,395	885
CG	79	42	4	28	5
ECG	1,574	740	42	510	282
GCG	362	187	7	110	58
EGCG	7,437	4,164	95	2,163	1,015

Nd not detected, *DME* dimethyl ether, *C* catechin, *EC* epicatechin, *GC* gallocatechin, *EGC* epigallocatechin, *CG* catechin gallate, *ECG* epicatechin gallate, *GCG* gallocatechin gallate, *EGCG* epigallocatechin gallate

5.4.4 Properties of Extracted Components and Dewatered Bio-solids

The concentration and chemical compositions of the extracts derived from different extraction solvents were different sometimes despite the slight changes in the extraction conditions. Herein, two recent results obtained using the DME method on the extraction of natural products were presented. In the first example, the extractions of caffeine and eight catechins including catechin (C), epicatechin (EC), gallocatechin (GC), epigal- locatechin (EGC), catechin gallate (CG), epicatechin gallate (ECG), gallocatechin gallate (GCG), and epigallocatechin gallate (EGCG) from green tea leaves were studied [18]. Table 5.2 shows the concentrations of caffeine and catechins in the residue, organic extracts, and removed water. The amounts (mass) of residue, organic extracts, and removed water from 100.0 % of green tea waste were found to be 28.4 %, 1.6 %, and 70.0 %, respectively. Kanda et al., evaluated the distribution of caffeine and catechins in the samples after DME extraction. Probably, the losses in the contents of caffeine and catechins, as shown in Table 5.2, were due to the differences in the analysis methods used for wet green tea leaves (and its residue), and organic compounds and water. As shown in Table 5.2, no caffeine was detected in the residue, indicating the good extractive ability of the DME method for removing caffeine from such biological materials. Of the removed caffeine, 41.2 % was present in water, which implied that the removed water might not be suitable for low-caffeine applications without additional treatment. All the catechins were detected in the residue, extracts, and water. However, we found that the catechins remained in the residue and water rather than in the organic extracts, probably because they migrated to the water layer from the upper DME-organic layer with the evaporation of DME (refer to Fig. 5.4). Approximately 29.1–42.9 %

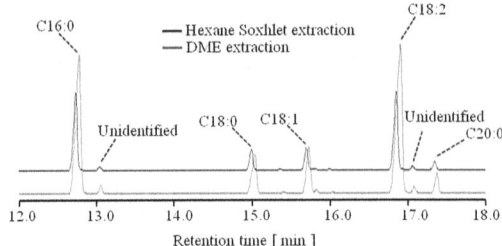

Fig. 5.10 Gas chromatogram of lipids derived from coffee grounds via DME and hexane Soxhlet extraction

Table 5.3 Proximate analysis and main elemental compositions of extracted lipids and dewatered solids from vegetal biomass

Analysis (wt.% dry basis)	Coffee grounds		Microalgae (*Kanogaw*)	
	Lipids	Solids	Lipids	Solids
Proximate analysis				
Ash yield	Nd	1.9	1.1	7.5
Volatile matter	–	81.1	–	80.1
Fixed carbon	–	17.0	–	12.4
Ultimate analysis				
C	77.0	51.5	70.9	46.9
H	11.4	6.97	10.0	6.65
N	1.96	2.43	2.62	10.7
O	9.60	37.1	15.0	27.8
S	–	0.15	0.16	0.41
HHV (MJ kg^{-1})	38.9	21.1	33.8	18.3

of the catechins were extracted into the water fraction while 25.2–56.0 % remained in the residue after DME extraction. Here, it was noteworthy that 56.0 % of the most important catechin, EGCG, still remained in the residue after DME extraction.

In the case of the extraction of lipids from spent coffee grounds, the chemical compositions of the lipids extracted using liquefied DME were determined and compared to those extracted using the Soxhlet method involving hexane. As shown in Fig. 5.10, the gas chromatograms of the lipids obtained via DME extraction resembled those obtained using the Soxhlet extraction with hexane; the carbon number of the detected lipids was in the range of C16–C18, which consisted of both saturated and unsaturated fatty acids. This outcome was almost identical to a previous report on the lipids derived from spent coffee grounds [20].

As mentioned earlier, this DME based technology can produce not only organic extracts but also dried bio-solids as byproducts. For example, the properties of the extracts and bio-solids derived from the spent coffee grounds and microalgae are shown in Table 5.3. The concentrations of carbon and hydrogen in the lipids of both spent coffee grounds and algae are higher, while nitrogen and oxygen are lower compared to those in the solids. The higher heating values (HHVs) of the lipids derived from either spent coffee grounds or microalgae are equivalent to those of

5 Liquefied Dimethyl Ether: An Energy-Saving, Green Extraction Solvent

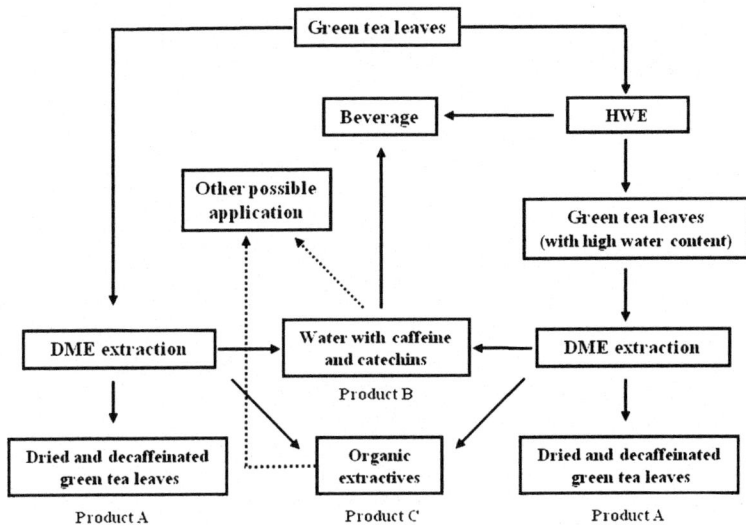

Fig. 5.11 Proposed utilisation of the DME method in the tea industry

the first-generation biodiesel, and are essentially the same as the traditional fossil oils [3]. In addition to the lipids, the bio-solids derived from both the spent coffee grounds and microalgae via DME extraction also retained sufficient calorific density to render themselves as potential carbon neutral fuels.

5.4.5 Future Possible Applications of DME Extraction

Owing to the unique properties of DME, this technique could be applied in the extraction of natural products in several industrial fields. Here, two possible applications of this technique were proposed for the tea industry and renewable bio-fuel production from vegetal biomass. The utilisation of the DME method in the tea industry has been depicted in Fig. 5.11. The right side in the figure has been conceptualised according to the outcomes of a recent study [18]. Here, the high-moisture green tea leaves after hot water extraction (HWE) are treated with liquefied DME. As mentioned earlier, it is different from the conventional method for that the DME method can simultaneously dewater (i.e. drying) and directly extracts the organic constituents from the natural feedstock at room temperature. This means that the heating of the extractant and the downstream hot-drying are both unnecessary. Furthermore, DME is a safe solvent and does not remain stable at room temperature. As a result, DME can be used for food processing. Therefore, the product B can be used for beverage production either with or without pre-treatment. The dried and decaffeinated product A can be used for the production of powdered

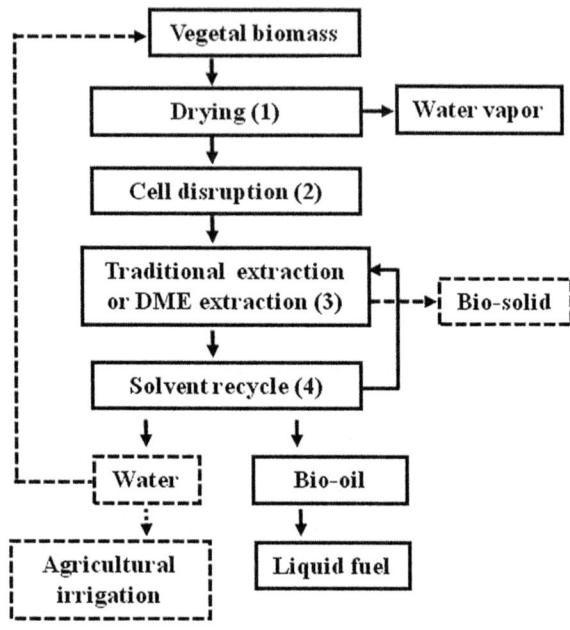

Fig. 5.12 Proposed utilisation of the DME method in the algae bio-fuel production compared to traditional method

green tea. The left side in the figure is a prospective concept derived from this study. Here, the green tea leaves could be extracted directly using the DME method without HWE. Thereupon, the amounts of caffeine and catechins in product B and catechins in product A should be much higher than those remaining post-HWE. Finally, the chemical compositions of product C from green tea leaves either with or without HWE should be further studied carefully because such condensed organic extracts from tea leaves usually contain other bio-active components, which may be of commercial value.

For the production of renewable bio-fuels, the DME technique is advantageous over the traditional extraction technologies [14, 16, 17, 22]. Herein, both traditional and DME approaches were integrated in a graphic illustration as shown in Fig. 5.12. There are four main steps: drying, cell disruption, solvent extraction, and solvent recycling, numbered as steps 1, 2, 3, and 4, respectively. In the traditional method, all four steps are normally required. The product obtained from the traditional method is only bio-oil, which can be chemically converted into liquid bio-fuels such as biodiesel. The water contained in the biomass is vaporized in step 1 in the traditional method, and the disposal of the biomass residues needs to be considered. In the DME approach, only steps 3 and 4 are required because of the dewatering ability and penetrability of DME. Besides the main product of bio-oil, the byproducts, namely dewatered bio-solids and removed water, could also be meaningfully utilised. For example, if the removed water from the vegetal feedstock did not contain any DME at room temperature, then such water could be used for agricultural irrigation.

5.5 Conclusions and Future Applications

In comparison to the traditional solvent extraction technology and SFT, the DME extraction has many advantages, particularly for natural product extraction from high moisture containing natural feedstock. As an organic solvent, liquefied DME is water-soluble; therefore, water can be removed from the feedstock simultaneously with the organic components. The boiling point of DME is close to the room temperature; therefore, the circulation of DME (gasification and liquefaction) is efficiency in energy consumption. The safety of DME allows for the application of this technique to the pharmacy and food industries. Above all, the uniqueness of this technique lies in the fact that it couples both dewatering and extraction, and therefore, both extracted chemicals and residue are obtained as products. However, as a new technique, some certain phenomenon such as the excellent penetrability of DME, needs to be clarified further. Additional efforts should also be directed toward testing other sources of natural feedstock, and expanding the possible application of this technique to other fields. A pilot-scale study should also be carried out to make this technique industrially practicable as soon as possible.

References

1. Bligh EG, Dyer WJ (1959) A rapid method of lipid extraction and purification. Can J Biochem Physiol 37:911–917
2. Catchpole O, Ryan J, Zhu Y, Fenton K, Grey J, Vyssotski M, Mackenzie A, Nekrasov E, Mitchell K (2010) Extraction of lipids from fermentation biomass using near-critical dimethylether. J Supercrit Fluids 53:34–41
3. Czernik S, Bridgwater AV (2004) Overview of applications of biomass fast pyrolysis oil. Energ Fuels 18:590–598
4. Delrue F, Setier P-A, Sahut C, Cournac L, Roubaud A, Peltier G, Froment AK (2012) An economic, sustainability, and energetic model of biodiesel production from micro algae. Bioresour Technol 111:191–200
5. Demirbas A, Science S, Turkey T (2009) Production of biodiesel from algae oils. Energ Sources A 31:163–168
6. EFSA (2009) Scientific opinion of the panel on food contact materials, enzymes, flavourings and processing aids (CEF) on dimethyl ether as an extraction solvent. EFSA J 84:1–13. http://dx.doi.org/10.2903/j.efsa.2009.984
7. Food and Agriculture Organization of the United Nations-Production (FAOSTAT) (2010) http://faostat.fao.org/DesktopDefault.aspx?PageID=567&lang=en#ancor
8. Gallagher BJ (2011) The economics of producing biodiesel from algae. Renew Energy 36:158–162
9. Holldorff H, Knapp H (1988) Binary vapour-liquid-liquid equilibrium of dimethyl ether-water and mutual solubilities of methyl chloride and water: experimental results and data reduction. Fluid Phase Equilib 44:195–209
10. Jae-Yon L, Chan Y, So-Young J, Chi-Yong A, Hee-Mock O (2010) Comparison of several methods for effective lipid extraction from micro algae. Bioresour Technol 101:75–77
11. Jaime L, Mendiola JA, Ibáñez E, Martin-Álvarez PJ, Cifuentes A, Reglero G, Señoráns FJ (2007) β-Carotene isomer composition of sub- and supercritical carbon dioxide extracts. Antioxidant activity measurement. J Agric Food Chem 55:10585–10590

12. Kanda H, Makino H (2009) Clean up process for oil-polluted materials by using liquefied DME. J Environ Eng 4:356–361
13. Kanda H, Makino H (2010) Energy-efficient coal dewatering using liquefied dimethyl ether. Fuel 89:2104–2109
14. Kanda H, Li P (2011) Simple extraction method of green crude from natural blue-green microalgae by dimethyl ether. Fuel 90:1264–1266
15. Kanda H, Morita M, Makino H, Takegami K, Yoshikoshi A, Oshita K, Takaoka M, Morisawa S, Takeda N (2011) Deodorization and dewatering of biosolids by using dimethyl ether. Water Environ Res 83:23–25
16. Kanda H, Li P, Ikehara T, Yasumoto-Hirose M (2012) Lipids extracted from several species of natural blue-green microalgae by dimethyl ether: extraction yield and properties. Fuel 95: 88–92
17. Kanda H, Li P, Yoshimura T, Okada S (2013a) Wet extraction of hydrocarbons from *Botryococcus braunii* by dimethyl ether as compared with dry extraction by hexane. Fuel 105:535–539
18. Kanda H, Li P, Makino H (2013b) Production of decaffeinated green tea leaves using liquefied dimethyl ether. Food Bioprod Process. http://dx.doi.org/10.1016/j.fbp.2013.02.001
19. Kitada K, Machmudah S, Sasaki M, Goto M, Nakashima Y, Kumamoto S, Hasegawa T (2009) Supercritical CO_2 extraction of pigment components with pharmaceutical importance from Chlorella vulgaris. J Chem Technol Biotechnol 84:657–661
20. Kondamude N, Mohapatra SK, Misra M (2008) Spent coffee grounds as a versatile source of green energy. J Agric Food Chem 56:11757–11760
21. Liam B, Philip O (2010) Biofuels from microalgae – a review of technologies for production processing, and extraction of biofuels and co-products. Renew Sust Energ Rev 14:557–577
22. Li P, Kanda H, Makino H (2014) Simultaneous production of bio-solid fuel and bio-crude from vegetal biomass using liquefied dimethyl ether. Fuel 116:370–376
23. Molina Grima E, Belarbi E-H, Acién Fernández FG, Robles Medina A, Chisti Y (2003) Recovery of microalgal biomass and metabolites: process options and economics. Biotechnol Adv 20:491–515
24. Mussatto SI, Carneiro LM, Silva JPA, Roberto IC, Teixeira JA (2011) A study on chemical constituents and sugars extraction from spent coffee grounds. Carbohydr Polym 83:368–374
25. Meng X, Yang J, Xu X, Zhang L, Nie Q, Xian M (2009) Biodiesel production from oleaginous microorganisms. Renew Energy 34:1–5
26. Oshita K, Takaoda M, Kitade S, Takeda N, Kanda H, Makino H, Matsumoto T, Morisawa S (2010) Extraction of PCBs and water from river sediment using liquefied dimethyl ether as an extractant. Chemosphere 78:1148–1154
27. Sawangkeaw R, Bunyakiat K, Ngamprasertsith S (2010) A review of laboratory-scale research on lipid conversion to biodiesel with supercritical methanol. J Supercrit Fluids 55:1–13
28. Wu J, Zhou Y, Lemmon EW (2011) An equation of state for the thermodynamic properties of dimethyl ether. J Chem Eng Data 40:023104

Chapter 6
Ethyl Lactate Main Properties, Production Processes, and Applications

Carla S.M. Pereira and Alírio E. Rodrigues

Abstract Petroleum is still the basis for the production of chemicals. Nevertheless, alternatives such as biomass and waste have been developed due to both environmental impacts of petroleum production and use, and uncertainty about the longevity and stability of petroleum supplies.

Ethyl lactate is derived from nature-based feedstocks (it is synthesized trough the esterification reaction between ethanol and lactic acid, both reactants generated from biomass raw materials), and can be used in place of several environment-damaging halogenated and toxic solvents, including ozone depleting chlorofluorocarbons, carcinogenic methylene chloride, and toxic ethylene glycol ethers and chloroform. This chapter presents an overview regarding ethyl lactate main properties, its synthesis and production processes, with particular emphasis to reactive/separation processes based on innovative technologies, as reactive distillation, membrane reactors and simulated moving bed reactors, and its applications (mainly for extraction of bioactive compounds from natural sources).

6.1 Introduction

The environmental regulations as well as the increase in crude oil prices raised stringent and compelling demands for the design and implementation of greener products and processes.

Green solvents were developed as a more environmentally friendly alternative to petrochemical solvents, being the most popular water (aqueous biphasic),

---

C.S.M. Pereira • A.E. Rodrigues (✉)
LSRE – Laboratory of Separation and Reaction Engineering – Associate Laboratory
LSRE/LCM, Faculdade de Engenharia, Universidade do Porto, Rua Dr. Roberto Frias,
4200-465 Porto, Portugal
e-mail: arodrig@fe.up.pt

Table 6.1 Ethyl lactate major benefits [4] (Reproduced by permission of The Royal Society of Chemistry)

100 % biodegradable	Renewable – biomass derived
FDA approved	EPA approved SNAP solvent
Non carcinogenic	Non corrosive
Great penetration characteristics	Stable in solvent formulations until exposed to water
Rinses easily with water	High solvency power for resins, polymers and dyes
High boiling point	Easy and inexpensive to recycle
Low VOC	Not a ozone depleting chemical
Low vapor pressure	Not a hazardous air pollutant

FDA Food and Drug Administration, *EPA* Environmental Protection Agency, *SNAP* Significant New Alternatives Policy, *VOC* Volatile Organic Compound

supercritical carbon dioxide and ionic liquids [1]. An increasing interest is also being given to bio-based solvents (produced from biomass or waste) as the lactate ester family solvents, which included ethyl lactate [2].

Anastas and Warner developed "the twelve principles of green chemistry", which are a list of suggestions on how to design greener processes and/or greener products [3]. Ethyl lactate is in accordance with at least eight of these principles [4]:

1. Ethyl lactate is produced from renewable raw materials (by the reaction of ethanol with lactic acid; both reactants can be obtained by fermentation of biomass): *7th principle "Use of Renewable Feedstocks"*.
2. Ethyl lactate is 100 % biodegradable, easy to recycle, non-corrosive, non-carcinogenic, non-toxic (U.S. Food and Drug Administration approved its use in food products) and non-ozone depleting [5]: *3rd principle "Less Hazardous Chemical Syntheses", 4th principle "Designing Safer Chemicals" and 10th principle "Design for Degradation"*.
3. Ethyl lactate can be produced using heterogeneous catalysts and without using an excess of any of the reactants; the elimination of homogenous catalysts (usually mineral acids) avoids the presence of corrosive catalysts and, as consequence, eliminates a further step of their neutralization: *1st principle "Prevention" and 9th principle "Catalysis"*.
4. Ethyl lactate can be produced by using hybrid technologies where reaction and separation of the products take place in a single unity eliminating the use of solvents, reducing the capital cost (less separation units are needed) and requiring less energy consumption: *5th principle "Safer Solvents and Auxiliaries" and 6th principle "Design for Energy Efficiency"*.

Due to the recognition of ethyl lactate as an environmentally benign chemical together with other ethyl lactate benefits, summarized in Table 6.1, several applications of this green solvent are addressed in the literature, as pharmaceutical preparations, fragrances, for inks and coatings industries, food additives, and more recently, in organic synthesis and as extractive solvent of bioactive components from natural sources.

6 Ethyl Lactate Main Properties, Production Processes, and Applications

Considering the increasing importance of this compound, this chapter provides: (1) ethyl lactate main properties, where temperature dependent properties, as viscosity, vapor pressure, heat of vaporization and heat capacity are addressed; (2) ethyl lactate synthesis by the esterification reaction between ethanol and lactic acid and its production processes by using multifunctional reactors, where reaction and separation steps are integrated into a single unit; and (3) ethyl lactate applications with particular emphasis to the use of this compound as extraction solvent of bioactive components.

6.2 Ethyl Lactate Properties

Ethyl lactate (CAS No.: 97-64-3, IUPAC name Ethyl (S)-2-hydroxypropanoate), with molecular formula $C_5H_{10}O_3$, is a clear to slightly yellow liquid, that when dilute presents a mild, buttery, creamy, whit hints of fruit and coconut odor. It can be found naturally in small quantities in a variety of foods, as chicken, wine and some fruits or it can be derived from renewable resources as corn or sugar crops (see Sect. 6.3.1 – "Raw Materials").

Ethyl lactate is 100 % biodegradable, non-corrosive [6], non-carcinogenic and non-ozone depleting, indeed it is so benign that it is approved by Food and Drug Administration (FDA) to be used as food flavor additive.

Ethyl lactate has a high solvency power; in Table 6.2, the solvating properties of ethyl lactate and the petrochemical N-methyl pyrrolidone are presented. It also has high boiling point, low vapor pressure, and low surface tension. Some physical and thermodynamic properties of ethyl lactate are presented in Tables 6.3 and 6.4.

The values of density, viscosity, vapor pressure, heat capacity, heat of vaporization and thermal conductivity at different temperatures (Table 6.4) were calculated by using Eqs. 6.1, 6.2, 6.3, 6.4, 6.5, 6.6, respectively, with the constants presented in Table 6.5 (based on appropriate experimental data available in literature).

Table 6.2 Solvating properties of ethyl lactate and N-methyl pyrrolidone ([4] – Reproduced by permission of The Royal Society of Chemistry)

	Ethyl lactate	N-methyl pyrrolidone
Kauri butanol value	>1,000	350
Solubility parameters		
Hildebrand parameter	21.3	23.1
Disperse Hansen parameter	7.8	8.8
Polar Hansen parameter	3.7	6.0
Hydrogen-bonding Hansen parameter	6.1	3.5
Miscibility	Miscible in water and hydrocarbons	Miscible in water and hydrocarbons

Table 6.3 Basic properties of ethyl lactate (Data from [7])

Properties	Ethyl lactate
Molecular weight – M (g/mol)	118.133
Melting temperature – T_f (K)	248.25
Normal boiling temperature – T_b (K)	426.15–427.15
Critical temperature – T_c (K)	588.00
Critical pressure – P_c (bar)	38.60
Critical volume – V_c (cm^3/mol)	354.0
Acentric factor – ω	0.793

Table 6.4 Ethyl lactate properties at different temperatures

	278.15 K	298.15 K	318.15 K
ρ (g.cm^3)	1.06	1.03	0.99
η (cP)	4.55	2.21	1.26
P_{vp} (mmHg)	1.07	3.75	1.12×10^1
C_p (Jmol^{-1} K^{-1})	2.46×10^{-2}	2.55×10^{-2}	2.64×10^{-2}
ΔH^V (Jmol^{-1})	6.17×10^4	6.01×10^4	5.84×10^4
λ (Wm^{-1} K^{-1})	1.73×10^{-1}	1.68×10^{-1}	1.63×10^{-1}

The density was determined by the Rackett equation, with ρ_L (g.cm^{-3}) and T (K), given by:

$$\rho_L = AB^{-\left(1-\frac{T}{T_c}\right)^n} \tag{6.1}$$

The liquid viscosity dependency on temperature was described by the following correlation, with η_L(cP) and T (K):

$$\log_{10}\eta_L = A + \frac{B}{T} + CT + DT^2 \tag{6.2}$$

The correlation of vapor pressure as a function of temperature was determined by the Antoine-type equation with extended term:

$$\log_{10} P_{vp} = A + \frac{B}{T} + C\log_{10}T + DT + ET^2 \tag{6.3}$$

with P_{vp}(mmHg) and T (K).

The liquid heat capacity correlation was calculated by the expression:

$$C_p = A + BT + CT^2 + DT^3 + ET^4 \tag{6.4}$$

with C_p(Jmol^{-1}K^{-1}) and T (K).

The heat of vaporization as a function of temperature was determined by:

$$\Delta H^V = A\left[1 - T_r\right]\left(B + CT_r + DT_r^2\right) \tag{6.5}$$

with $T_r = T/T_c$ and ΔH^V(JK^{-1}mol^{-1}).

6 Ethyl Lactate Main Properties, Production Processes, and Applications 111

Table 6.5 Constants used in the calculation of ethyl lactate temperature dependent properties

Constants	Density (Eq. 6.1) [7]	Viscosity (Eq. 6.2) [7]	Vapour pressure (Eq. 6.3) [7]	Heat capacity (Eq. 6.4) [7]	Heat vaporization (Eq. 6.5) [8]	Thermal conductivity (Eq. 6.6) [8]
A	0.33372	−20.0105	32.0863	−46.239	8.0260×10^7	2.8358×10^{-1}
B	0.21190	3.2123×10^3	-2.9164×10^3	2.1823	4.0930×10^{-1}	-3.5110×10^{-4}
C	–	4.1891×10^2	−9.5666	-5.9832×10^{-3}	–	–
D	–	-3.2733×10^{-5}	6.5114×10^{-3}	6.8683×10^{-6}	–	–
E	–	–	4.5645×10^{-13}	–	–	–
n	0.45530	–	–	–	–	–
T_{min} (K)	247.15	247	247	248	247.15	247.15
T_{max} (K)	T_c	T_c	T_c	529	588.00	427.65

The correlation of the thermal conductivity was calculated by:

$$\lambda = A + BT + CT^2 + DT^3 + ET^4 \tag{6.6}$$

with $\lambda(\text{Wm}^{-1}\text{K}^{-1})$ and T (K).

All the presented properties make of ethyl lactate a suitable compound for several applications, as can be observed in Sect. 6.4.

6.3 Production Processes

6.3.1 Raw Materials

Common biorefinery building blocks as ethanol and lactic acid can be used to produce ethyl lactate.

Ethanol is an important raw material in the chemical industry and it is the most widely used biofuel for transportation. It can be produced from several biomass crops, as sugar crops (*e.g.*, sugar cane and sugar beet), starch crops (*e.g.*, corn and cassava) or cellulosic feedstocks (*e.g.*, wood, grasses and agricultural residues).

The worldwide production of ethanol is growing every year. According to Merchant Research and Consulting Report [9], from 2007 to 2012 the global ethanol production increased by 56 %.

USA is the leader in ethanol market, with 59 % share of the global production, followed by Brazil with 24 % share.

Lactic acid, an important chemical platform for the economy of renewable compounds, can be produced by the fermentation of different carbohydrates, such as glucose (from starch), maltose (produced by specific enzymatic starch conversion), sucrose (from syrups, juices, and molasses), or lactose (from whey) [10, 11]. Other feedstocks, particularly from wastes, are being investigated [12, 13].

According to a recent Report by Global Industry Analysts, Inc. (2012), the lactic acid global market is forecast to reach 328.9 thousand metric tons by the year 2015 [14]. This market growth is driven by a rise in demand from existing end-use markets (mainly for the production of biodegradable polylactic acid, a well-known sustainable bioplastic material [15, 16]) and development of new product applications.

6.3.2 Synthesis

The ethyl lactate synthesis involves a liquid phase reversible reaction between ethanol and lactic acid, catalyzed by an acid catalyst, and having water as a by-product:

$$\text{Ethanol (Eth)} + \text{Lactic Acid (La)} \xrightleftharpoons{H^+} \text{Ethyl Lactate (EL)} + \text{Water (W)}$$

6 Ethyl Lactate Main Properties, Production Processes, and Applications

Lactic acid is an α-hydroxy acid; it contains a hydroxyl group adjacent to the carboxylic acid functional group. A review on the chemistry of this compound can be found in the literature [17]. The lactic acid bifunctional nature promotes intermolecular esterification in aqueous solutions above 20 wt.% to form linear dimer, and higher oligomer acids [18, 19]. An 88 wt.% lactic acid solution comprises 43.5 mol % of monomer, 9.2 mol% of dimer, 1.8 mol% of trimer and about 45 mol% of water, while an 20 wt.% aqueous lactic acid solution is constituted only by monomer and water, with a monomer molar percentage of about 5.6 mol% [20].

The degree of self-esterification increases with increasing acid concentration, which compromises the use of lactic acid as reactant for the synthesis of ethyl lactate; the use of high lactic acid concentration implies the presence of oligomers that, during the esterification, will be converted into the corresponding esters. These esters will simultaneously undergo hydrolysis and transesterification leading to a mixture of acid and ester monomers and oligomers, according to:

$2La_1 \iff La_2 + W$ (lactic acid dimer formation)
$La_1 + La_2 \iff La_3 + W$ (lactic acid trimer formation)
...
$La_1 + La_{n-1} \iff La_n + W$ (lactic acid oligomer formation) with $n \geq 2$
$La_1 + Eth \iff EL_1 + W$ (ethyl lactate formation)
$La_2 + Eth \iff EL_2 + W$ (ethyl lactate dimer formation)
$La_3 + Eth \iff EL_3 + W$ (ethyl lactate trimer formation)
...
$La_n + Eth \iff EL_n + W$ (ethyl lactate oligomer formation)

where:

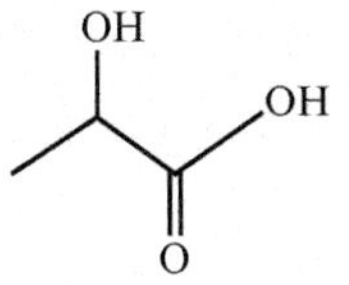

Lactic acid (La_1) 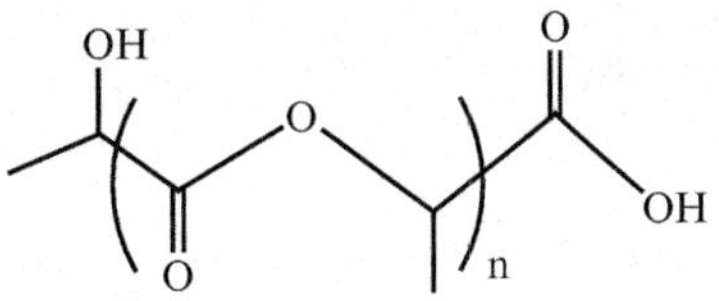   Lactic acid oligomers (La_{n+1})

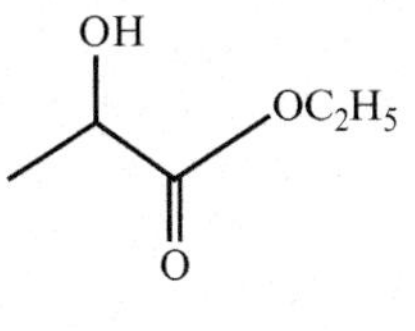

Ethyl lactate (EL_1) 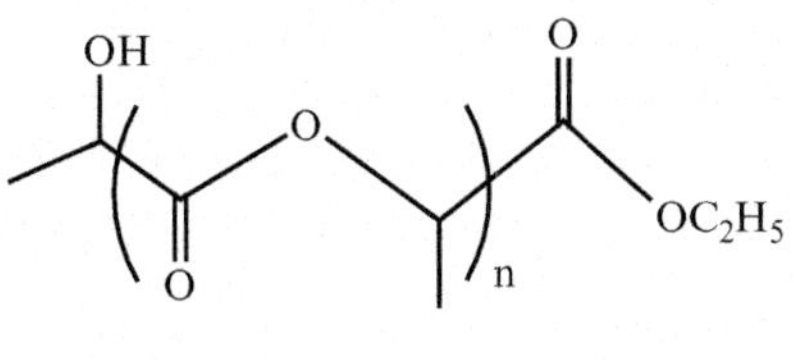   Ethyl lactate oligomers (EL_{n+1})

In the ethyl lactate reaction kinetics studies, some authors use 20 wt.% lactic acid solution as reactant in order to avoid the oligomers formation; however, even when high lactic acid concentrations are used, the formation of oligomers is usually neglected. As far as our knowledge goes, just two works take into account the oligomers formation; nevertheless, their amount, at equilibrium, is less than 5 % [20, 21].

The water concentration is a strong factor on the extent of the oligomers formation, but the amount of ethanol is also important. For example, for a lactic acid solution of 88 wt.%, the oligomers composition at equilibrium is 2.4 molar%, when using a molar ratio between ethanol and lactic acid of 1, and is 0.4 molar% when using an ethanol to lactic acid molar ratio of 3 [22].

A summary of the kinetic studies performed for the lactic acid esterification with ethanol is presented in Table 6.6 [4]. As can be observed, most of the studies consider heterogeneous catalysts, which is easily explained by their significant advantages over the homogeneous ones, i.e.: easy to separate from the reaction mixture; long life time; higher purity of products (side reactions can be eliminated or are less significant); and elimination of the corrosive environment caused by the discharge of acid containing waste. Some works also study the self-catalyzed reaction (without the use of catalyst), but the use of catalyst is favorable, especially in this esterification reaction as the self-catalyzed reaction rate is extremely slow.

Few authors take into account, in the kinetic model, the non-ideality of the reaction mixture using activities instead of concentration; the kinetic model is mainly expressed in terms of species concentration. In spite of the number of kinetic studies available for this system, only one presents the thermodynamic equilibrium constant defined as function of the species liquid activities, described by the following equation: $\ln(k) = 2.9625 - 515.13/T(K)$ [22]. There is another study, regarding the vapor-liquid reactive equilibrium for the ethyl lactate synthesis, where the thermodynamic equilibrium constant was determined; however, the values predicted by the proposed equilibrium constant expression are not in very good agreement with the experimental ones [23].

6.3.3 Ethyl Lactate Production by Multifunctional Reactors

The conventional way to produce ethyl lactate is in a batch reactor, where the esterification reaction between ethanol (usually in excess) and lactic acid is carried out until equilibrium; then the equilibrium mixture is fed to various separation units (mainly energy intensive distillation steps) in order to recover ethyl lactate with the desired purity, to remove water, and to recycle the unconverted ethanol and lactic acid back to the reactor. The disadvantage of this process is in its economics, since it represents high energy costs and investment in several reaction and separation units.

The most feasible engineering solution for the production of this type of compounds that involve equilibrium limited reactions is using multifunctional reactors, where reaction and separation steps are combined in a single unit. This process

Table 6.6 Summary of the kinetic studies of the esterification reaction between ethanol and lactic acid ([4] – Reproduced by permission of The Royal Society of Chemistry)

Refs.	Catalyst	Temperature range (°C)	Lactic acid solution (wt. %)	Oligomers presence	Kinetic model	Expression of the components	Activation energy (kJ/mol)
Troupe and DiMilla [24]	Sulfuric acid	25–100	85; 44	Neglected	Empirical equation	Concentrations	62.47
Tanaka et al. [21]	Amberlyst 15	90–92	91	Considered	Simple nth-order rev. rate expressions	Concentrations	47.00[a]
Benedict et al. [25]	Without catalyst	95	88	Neglected	Homogeneous	Concentrations	—
	Amberlyst XN-1010	75–95	88	Neglected	Based on single-site mechanisms	Concentrations	30.54
Engin et al. [26]	Heteropoly acid supported on Lewatit® S100	70	92	Neglected	Simple nth-order rev. rate expressions	Concentrations	—
Zhang et al. [27]	002	60–88	20	Neglected	LH	Activities[b]	51.58
	NKC				LH	Activities[b]	52.26
Asthana et al. [20]	Amberlyst 15	62–90	20; 50; 88	Considered	Simple nth-order rev. rate expressions	Concentrations	48.00[a]
Delgado et al. [28]	Amberlyst 15	55–86	20	Neglected	LH	Activities[c]	52.29
	Without catalyst	55–85	20	Neglected	Homogeneous	Activities[c]	62.50
Pereira et al. [22]	Amberlyst 15	50–90	88	Neglected	LH	Activities[c]	49.98
Bamoharram et al. [29]	Preyssler acid	70–85	20	Neglected	Simple nth-order rev. rate expressions	Concentrations	47.11

LH Langmuir-Hinshelwood, *rev.* reversible
[a] activation energy of the lactic acid monomer esterification
[b] activities coefficients calculated by the UNIFAC model
[c] activities coefficients calculated by the UNIQUAC model

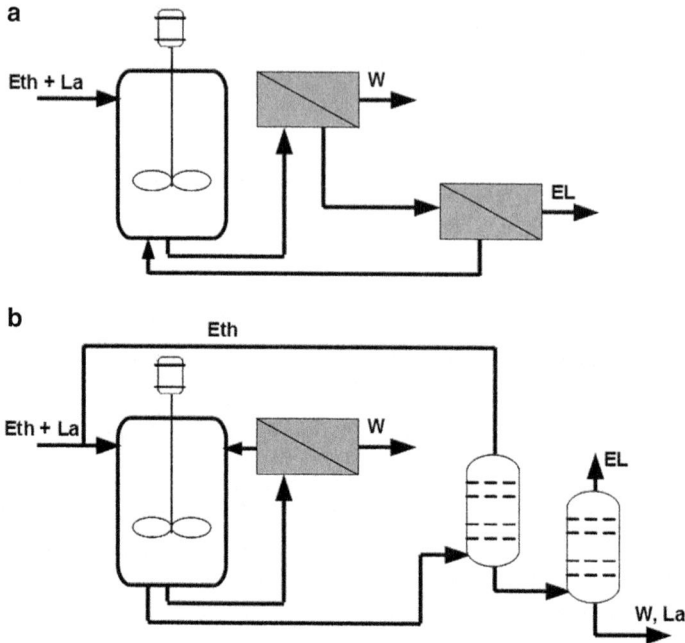

Fig. 6.1 Schematic representation of the ethyl lactate production process patented by Argonne National Laboratory [30]: reactor coupled with a pervaporation membrane unit and followed by: (**a**) pervaporation unit; (**b**) two distillation columns

intensification methodology by process integration brings significant advantages when compared with conventional process, as better energy efficiency (the same resources are used to perform reaction and separation steps), conversion beyond the equilibrium value (the products are removed from the reaction medium as they are formed), productivity improvement, lower solvent consumption and of course this integration results in compacter production plants.

Some multifunctional reactors were already studied aiming the sustainable production of ethyl lactate, which are presented next.

6.3.3.1 Membrane Reactors

A process based in a reactor with an external pervaporation membrane unit for reaction medium dehydration and followed by separation of the reaction mixture in a plurality of pervaporation steps or, alternatively, followed by two consecutive distillation columns is represented in Fig. 6.1. This process was patented by Argonne National Laboratory [30] and is currently applied by VERTEC BIOSOLVENTS™ Company in the ethyl lactate production.

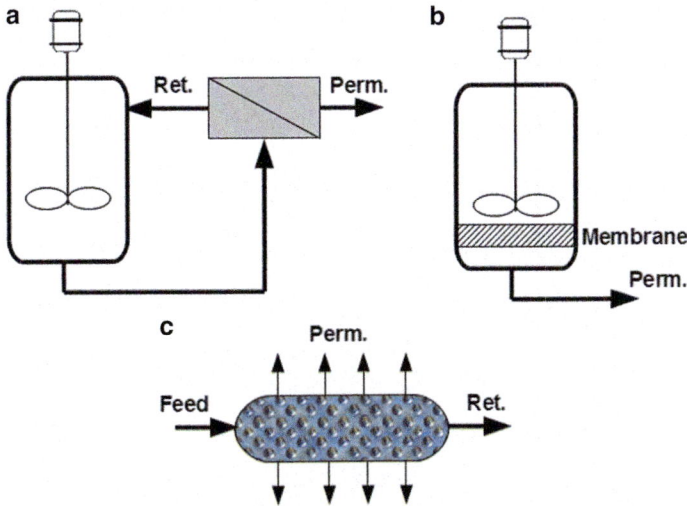

Fig. 6.2 Layout of a pervaporation membrane reactor: (**a**) batch reactor coupled with a pervaporation unit; (**b**) membrane and reactor in the same unit; (**c**) continuous integrated membrane reactor

Other authors also focused their studies on the ethyl lactate synthesis using pervaporation and/or vapor-permeation membrane reactors. Three configurations were assessed: batch reactor, where the lactic acid esterification reaction takes place, followed by a membrane for water removal, and recycle of the retentate to the reactor [30, 25, 31–33] (Fig. 6.2a); membrane inside a batch reactor [34, 21, 35] (Fig. 6.2b) and; tubular membrane packed in the lumen side with a heterogeneous catalyst (Amberlyst 15-wet) [36] (Fig. 6.2c).

The type of hydrophilic membranes tested were polymeric [25, 31, 30], ceramic [32] and organic–inorganic hybrid membranes [33, 34] for pervaporation and zeolites for vapor permeation [35, 21].

The main results obtained for the ethyl lactate production by using membrane reactors are summarized in Table 6.7.

6.3.3.2 Reactive Distillation

The reactive distillation (RD) technology, where reaction is integrated with separation by distillation (Fig. 6.3), was successfully implemented by Asthana and collaborators [37]; for a bottom temperature of 128 °C, and a feed comprising a mixture of ethanol and 88 wt.% lactic acid solution (ethanol/lactic acid molar ratio of 3.6:1), it was obtained a 95 % lactic acid conversion and 95 % ethyl lactate purity (ethanol free basis).

Table 6.7 Membrane reactors for the ethyl lactate synthesis ([4] – Reproduced by permission of The Royal Society of Chemistry)

Refs.	Membrane	Temperature	Membrane water flux (kg/m^{2}h)[a]	Catalyst	Eth/La Molar ratio	La Conversion	EL Purity
Rathin and Shih-Perng [30]	GFT PerVap 1005	95 °C	1.20	Amberlyst XN-1010	2.0	99 %	76 %
Jafar et al. [34]	Zeolite A	70 °C	0.18	p-toluene sulphonic acid	2.0	95 %	–
Tanaka et al. [21]	Zeolite T	120 °C	0.33	Amberlyst 15	2.4	99 %	–
Benedict et al. [25, 31]	GFT PerVap 1005	95 °C	–	Amberlyst XN-1010	1.2	71 %[b]	–
Budd et al. [35]	Zeolite/polyelectrolyte multilayer	100 °C	0.60[c]	p-toluene sulfonic acid	2.0	90 %	–
Ma et al. [33]	Chitosan–TEOS	80 °C	0.19[d] 0.27	Amberlyst 15	3.0	80 %[e]	–
Pereira et al. [36]	Microporous silica	70 °C	2.55	Amberlyst 15	1	98 %	96 %

[a] for water/ethanol liquid mixture (≈10/90 wt %)
[b] after 8 h
[c] sheet membrane (70 °C)
[d] tube membrane (70 °C)
[e] after 9 h

Fig. 6.3 Typical reactive distillation column applied to ethyl lactate synthesis

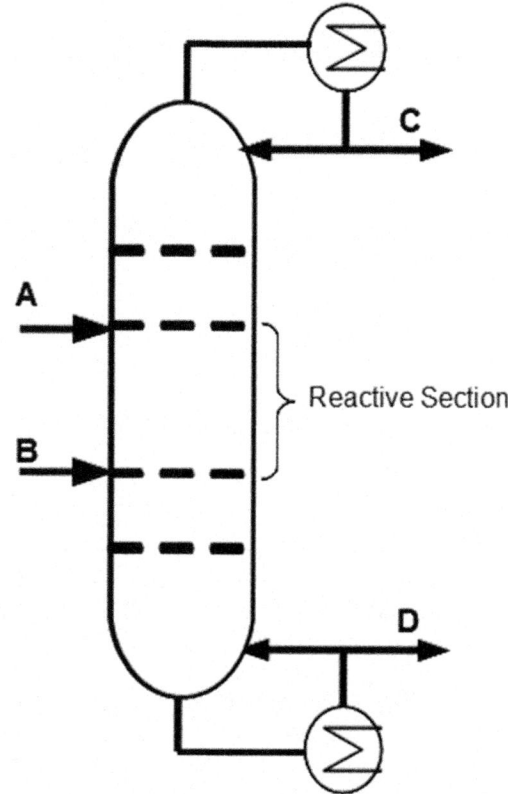

This process was also studied by Gao and co-workers [38]; however, when using a bottom temperature of about 115 °C and a feed ratio of ethanol to lactic acid of 4:1, an ethyl lactate yield of just 53 % was achieved.

In the most recent study [39], simulation work was performed considering the fermentation step for the production of lactic acid from sucrose integrated with the ethyl lactate synthesis using RD technology.

6.3.3.3 Chromatographic Reactors Based Technologies

The Simulated Moving Bed Reactor (SMBR), which combines chemical reaction with continuous counter-current chromatography, was also evaluated for the production of ethyl lactate [40]. It is reported an ethyl lactate productivity of 18.06 $Kg_{EL}/(L_{ads}\cdot day)$, a desorbent consumption of 4.75 L_{Eth}/Kg_{EL} and a ethyl lactate purity of 95 % (ethanol free basis), when using this technology at 50 °C, pure ethanol as desorbent and Amberlyst-15 wet resin as catalyst and selective

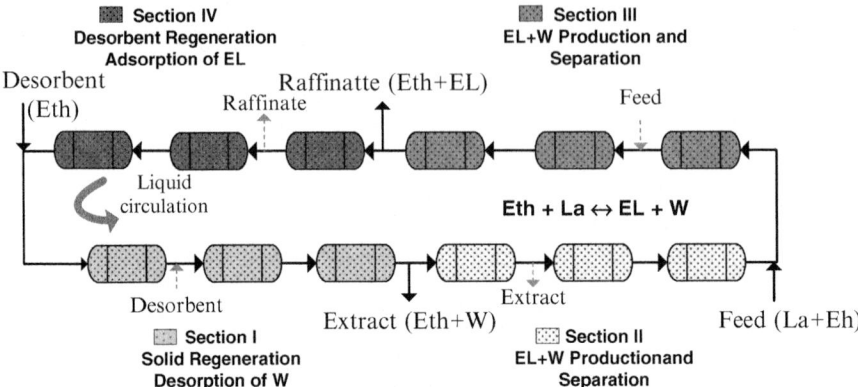

Fig. 6.4 Schematic representation of the ethyl lactate synthesis by SMBR

adsorbent to water (see Fig. 6.4). The maximum ethyl lactate productivity attained was 31.7 $Kg_{EL}/(L_{ads}.day)$, but it was accompanied by a high consumption of ethanol (7.6 L_{Eth}/Kg_{EL}), which implies diluted outlet streams; in the SMBR two outlet streams diluted in the desorbent used are obtained: the extract and the raffinate, requiring, therefore, additional separation steps.

In order to achieve high ethyl lactate productivity without the penalty on the consumption of ethanol (desorbent), the SMBR was integrated with hydrophilic membranes, which resulted in a new hybrid technology: the Simulated Moving Bed Membrane Reactor (PermSMBR) [41, 42].

Once the physical PermSMBR unit does not exist yet, this technology was just theoretically assessed, but using mathematical models that strongly rely on experimental data (the SMBR [40] and the pervaporation performance of the compounds involved in ethyl lactate synthesis [36] were experimentally assessed) [41, 42]. The PermSMBR revealed a high performance with high productivity and low solvent consumption which proves this technology potential for the sustainable synthesis of ethyl lactate even when compared with other intensified processes. For example, for an ethyl lactate productivity of around 16 $Kg_{EL}/(L_{resin}.day)$, at 50 °C, the SMBR ethanol consumption is 165 % higher than that of the PermSMBR. For a productivity of about 41 $Kg_{EL}/(L_{resin}.day)$, the RD process developed by Asthana et al. [32] (bottom temperature of 128 °C) requires a larger amount of ethanol by 152 % than the PermSMBR at 70 °C [42]. In all cases, an ethyl lactate purity of 95 % (ethanol free basis) was obtained. Nevertheless, it should be mentioned that a fair comparison among these technologies must be performed in terms of economical evaluation, where capital and operational costs are considered.

A schematic representation of a PermSMBR based plant for the production of ethyl lactate is shown in Fig. 6.5.

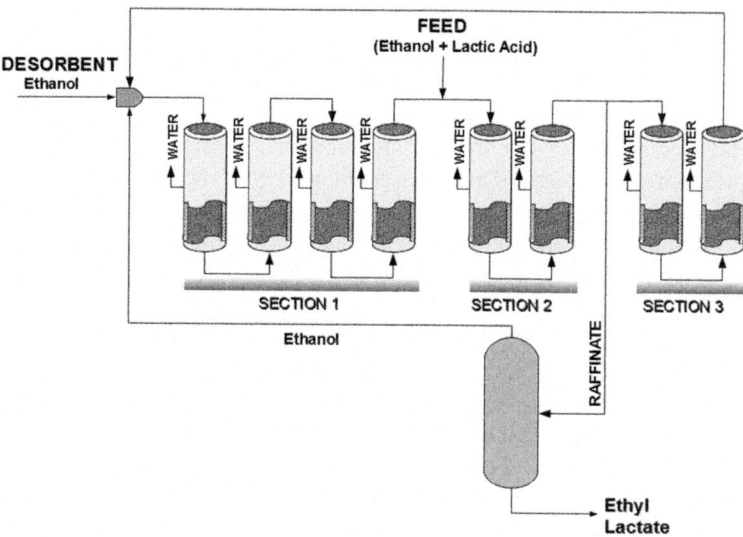

Fig. 6.5 PermSMBR process scheme for ethyl lactate production

6.4 Applications

Ethyl lactate has many applications such as food additive (flavoring agent), fragrances, pharmaceutical (as dissolving/dispersing excipient [43, 44]), and agricultural (for instance in copper [45] or cadmium [46] removal from contaminated soils).

Nevertheless, its main application is as green solvent; it can be used for chemical reactions [1, 47] (for instance, in the greener synthesis of aryl aldimines [48], synparvolide B [49], varitriol [50], and spiro-oxindole derivatives [51]), in magnetic tape coatings replacing the hazardous air pollutants MEK, MIBK and toluene [52], as paint stripper and graffiti remover, as cleaning agent for the polyurethane industry and for metal surfaces, efficiently removing greases, oils, adhesives and solid fuels. Moreover, it has the ability to replace a range of environment damaging halogenated and toxic solvents as is the case of N-methyl Pyrrolidone [53], acetone and xylene, in their numerous applications.

Among the solvent applications, one that is recently attracting an increasing attention is the use of ethyl lactate for the extraction of bioactive components from natural sources, probably motivated by the fact that an important trend in the bioactive compounds extraction field is the search for new environmentally green and food grade solvents.

The extraction of phytosterols from wet corn fiber by using ethyl lactate, which provides an oil product with free phytosterols and free fatty acids, was proposed by Abbas et al. [54]. Sclareol, a highly water-insoluble plant natural product, was selectively extracted from Clary sage using ethyl lactate and recovered from the liquid solution by a CO_2 gas anti-solvent methodology [55]. Ishida and Chapman [56] reported ethyl lactate as an excellent solvent to extract carotenoids from different sources: lycopene isomers from tomatoes, and lutein and β-carotene from corn and carrots. A keto-carotenoid, namely astaxanthin, was extracted using ethyl lactate instead of the traditionally non-environmentally friendly solvents used, as chloroform, acetone and petroleum ether [57]; when compared with the conventional extraction procedures, the reported method gave better results (higher extraction efficiency within short extraction time). The potential application of ethyl lactate to recover squalene from pretreated olive oil deodorizer and tocopherol from olive oil was assessed by Hernández et al. [58] and Vicente et al. [59], respectively. Manic et al. [60] studied the solubility of high-value compounds as caffeine, vanillic acid, ferulic acid, caffeic acid and thymol in ethyl lactate in order to infer ethyl lactate suitability as extractive solvent for these species. The solubility results obtained in the previous study motivated the authors to evaluate the actual extraction of caffeine from vegetal sources, such as green coffee beans and green tea leaves, by using ethyl lactate [61]. The use of this solvent instead of ethyl acetate (traditionally used) improved the caffeine recovery. The extraction of γ-limolenic acid from Spirulina microalgae using ethyl lactate was also employed [62].

6.5 Conclusions

Ethyl lactate is considered a green solvent that due to its unique advantages such as biodegradable, non-toxic, high solvency power, excellent miscibility with organics, among others, has already several applications as demonstrated by the examples addressed in this chapter. Besides, new applications are underway, as the cases of the use of ethyl lactate in organic synthesis and in the extraction of bioactive components from natural sources. The feasibility of ethyl lactate for the extraction of high-value products such as carotenoids, caffeine and γ-limolenic acid, among others, was already demonstrated.

The synthesis of ethyl lactate involves a thermodynamic limited reaction between ethanol and lactic acid, having water as by-product. Therefore, in order to improve the production of ethyl lactate, making it more commercially attractive and competitive when compared with petro based chemicals, the use of reactive/separation technologies, where at least one of the products is continuously removed from the reaction mixture to lead to the depletion of the limiting reactant, increasing ethyl lactate yield and purity, is beneficial. In this chapter, a literature survey related to intensified ethyl lactate production processes, as membrane reactors, reactive distillation and chromatographic reactors was presented. In all cases, it was possible to achieve conversion far beyond the equilibrium value (≥ 95 %) and high ethyl lactate purity; however, each process is operated at different conditions and requires different capital investment and so to make a decision on what is the most suitable ethyl lactate production process, an economical evaluation taking into account each process capital and operational costs should be performed.

References

1. Sheldon RA (2005) Green solvents for sustainable organic synthesis: state of the art. Green Chem 7(5):267–278
2. Aparicio S, Halajian S, Alcalde R, García B, Leal JM (2008) Liquid structure of ethyl lactate, pure and water mixed, as seen by dielectric spectroscopy, solvatochromic and thermophysical studies. Chem Phys Lett 454(1–3):49–55
3. Anastas PT, Warner JC (1998) Green chemistry: theory and practice. Oxford Science, New York
4. Pereira CSM, Silva VMTM, Rodrigues AE (2011) Ethyl lactate as a solvent: properties, applications and production processes – a review. Green Chem 13(10):2658–2671
5. Clark JH, Tavener SJ (2006) Alternative solvents: shades of green. Org Process Res Dev 11(1):149–155. doi:10.1021/op060160g
6. Clary JJ, Feron VJ, van Velthuijsen JA (1998) Safety assessment of lactate esters. Regul Toxicol Pharm 27(2):88–97. http://dx.doi.org/10.1006/rtph.1997.1175
7. Yaws CL (1999) Chemical properties handbook. McGraw-Hill Education, New York
8. Rowley RL, Wilding WV, Oscarson JL, Giles NF (2009) DIPPR-801 data compilation of pure chemical properties, design institute for physical properties. Brigham Young University, Provo, Utah. http://dippr.byu.edu
9. Ethanol (EtOH): 2013 World Market Outlook and Forecast up to 2017 (2013) Merchant research and consulting. http://mcgroup.co.uk/researches/ethanol. Accessed 19 Nov 2013
10. Benninga H (1990) A history of lactic acid making: a chapter in the history of biotechnology. Kluwer Academic, Dordrecht
11. Corma Canos A, Iborra S, Velty A (2007) Chemical routes for the transformation of biomass into chemicals. Chem Rev 107(6):2411–2502
12. Helen Shiphrah V, Sahu S, Ranjan Thakur A, Ray Chaudhuri S (2013) Screening of bacteria for lactic acid production from whey water. Am J Biochem Biotechnol 9(2):118–123
13. Gao M-T, Hirata M, Toorisaka E, Hano T (2007) Lactic acid production with the supplementation of spent cells and fish wastes for the purpose of reducing impurities in fermentation broth. Biochem Eng J 36(3):276–280. http://dx.doi.org/10.1016/j.bej.2007.02.030
14. Lactic acid – A Global Strategic Business Report (2012) Global Industry Analysts, Inc. http://www.strategyr.com/Lactic_Acid_Market_Report.asp. Accessed 19 Nov 2013
15. Datta R, Tsai SP, Bonsignore P, Moon SH, Frank JR (1995) Technological and economic potential of poly(lactic acid) and lactic acid derivatives. FEMS Microbiol Rev 16(2–3):221–231
16. Wee YJ, Kim JN, Ryu HW (2006) Biotechnological production of lactic acid and its recent applications. Food Technol Biotechnol 44(2):163–172
17. Holten CH (1971) Lactic acid. Verlag Chemie, Copenhagen
18. Montgomery R (1952) Acidic constituents of lactic acid-water systems. J Am Chem Soc 74(6):1466–1468
19. Vu DT, Kolah AK, Asthana NS, Peereboom L, Lira CT, Miller DJ (2005) Oligomer distribution in concentrated lactic acid solutions. Fluid Phase Equilib 236(1–2):125–135
20. Asthana NS, Kolah AK, Vu DT, Lira CT, Miller DJ (2006) A kinetic model for the esterification of lactic acid and its oligomers. Ind Eng Chem Res 45(15):5251–5257
21. Tanaka K, Yoshikawa R, Ying C, Kita H, Okamoto KI (2002) Application of zeolite T membrane to vapor-permeation-aided esterification of lactic acid with ethanol. Chem Eng Sci 57(9):1577–1584
22. Pereira CSM, Pinho SP, Silva VMTM, Rodrigues AE (2008) Thermodynamic equilibrium and reaction kinetics for the esterification of lactic acid with ethanol catalyzed by acid ion exchange resin. Ind Eng Chem Res 47:1453–1463
23. Delgado P, Sanz MT, Beltran S (2007) Isobaric vapor-liquid equilibria for the quaternary reactive system: Ethanol + water + ethyl lactate + lactic acid at 101.33 kPa. Fluid Phase Equilib 255(1):17–23

24. Troupe RA, DiMilla E (1957) Kinetics of the ethyl alcohol-lactic acid reaction. Ind Eng Chem 49(5):847–855. doi:10.1021/ie50569a028
25. Benedict DJ, Parulekar SJ, Tsai SP (2003) Esterification of lactic acid and ethanol with/without pervaporation. Ind Eng Chem Res 42(11):2282–2291
26. Engin A, Haluk H, Gurkan K (2003) Production of lactic acid esters catalyzed by heteropoly acid supported over ion-exchange resins. Green Chem 5(4):460–466
27. Zhang Y, Ma L, Yang J (2004) Kinetics of esterification of lactic acid with ethanol catalyzed by cation-exchange resins. React Funct Polym 61(1):101–114
28. Delgado P, Sanz MT, Beltran S (2007) Kinetic study for esterification of lactic acid with ethanol and hydrolysis of ethyl lactate using an ion-exchange resin catalyst. Chem Eng J 126(2–3):111–118
29. Bamoharram FF, Heravi MM, Ardalan P, Ardalan T (2010) A kinetic study of the esterification of lactic acid by ethanol in the presence of Preyssler acid an eco-friendly solid acid catalyst. React Kinet Mech Catal 100(1):71–78
30. Rathin D, Shih-Perng T (1998) Esterification of fermentation-derived acids via pervaporation. US Patent 5,723,639
31. Benedict DJ, Parulekar SJ, Tsai SP (2006) Pervaporation-assisted esterification of lactic and succinic acids with downstream ester recovery. J Membr Sci 281(1–2):435–445
32. Wasewar K, Patidar S, Agarwal VK (2009) Esterification of lactic acid with ethanol in a pervaporation reactor: modeling and performance study. Desalination 243(1–3):305–313
33. Ma J, Zhang M, Lu L, Yin X, Chen J, Jiang Z (2009) Intensifying esterification reaction between lactic acid and ethanol by pervaporation dehydration using chitosan-TEOS hybrid membranes. Chem Eng J 155(3):800–809
34. Jafar JJ, Budd PM, Hughes R (2002) Enhancement of esterification reaction yield using zeolite: a vapour permeation membrane. J Membr Sci 199(1):117–123
35. Budd PM, Ricardo NMPS, Jafar JJ, Stephenson B, Hughes R (2004) Zeolite/polyelectrolyte multilayer pervaporation membranes for enhanced reaction yield. Ind Eng Chem Res 43(8):1863–1867. doi:10.1021/ie034142o
36. Pereira CSM, Silva VMTM, Pinho SP, Rodrigues AE (2010) Batch and continuous studies for ethyl lactate synthesis in a pervaporation membrane reactor. J Membr Sci 361(1–2):43–55
37. Asthana N, Kolah A, Vu DT, Lira CT, Miller DJ (2005) A continuous reactive separation process for ethyl lactate formation. Org Process Res Dev 9(5):599–607
38. Gao J, Zhao XM, Zhou LY, Huang ZH (2007) Investigation of ethyl lactate reactive distillation process. Chem Eng Res Des 85(4A):525–529
39. Lunelli BH, De Morais ER, Maciel MRW, Filho RM (2011) Process intensification for ethyl lactate production using reactive distillation. Chem Eng Trans 24:823–828
40. Pereira CSM, Zabka M, Silva VMTM, Rodrigues AE (2009) A novel process for the ethyl lactate synthesis in a simulated moving bed reactor (SMBR). Chem Eng Sci 64(14):3301–3310
41. Silva VMTM, Pereira CSM, Rodrigues AE (2009) Simulated moving bed membrane reactor, new hybrid separation process and used thereof. PT Patent 104,496; WO Patent 2010/116,335
42. Silva VMTM, Pereira CSM, Rodrigues AE (2010) PermSMBR – a new hybrid technology: application on green solvent and biofuel production. AIChE J 57:1840–1851
43. Muse J, Colvin HA (2005) Use of ethyl lactate as an excipient for pharmaceutical compositions. US Patent 2005/0287,179 A1
44. McConville JT, Carvalho TC, Kucera SA, Garza E (2009) Ethyl lactate as a pharmaceutical-grade excipient and development of a sensitive peroxide assay. Pharm Technol 33(5):74–82
45. Guo H, Wang W, Sun Y, Li H, Ai F, Xie L, Wang X (2010) Ethyl lactate enhances ethylenediaminedisuccinic acid solution removal of copper from contaminated soils. J Hazard Mater 174(1–3):59–63
46. Li J, Sun Y, Yin Y, Ji R, Wu J, Wang X, Guo H (2010) Ethyl lactate-EDTA composite system enhances the remediation of the cadmium-contaminated soil by Autochthonous Willow (Salix × aureo-pendula CL 'J1011') in the lower reaches of the Yangtze River. J Hazard Mater 181(1–3):673–678
47. Sheldon RA (2007) The E Factor: fifteen years on. Green Chem 9(12):1273–1283

48. Bennett JS, Charles KL, Miner MR, Heuberger CF, Spina EJ, Bartels MF, Foreman T (2009) Ethyl lactate as a tunable solvent for the synthesis of aryl aldimines. Green Chem 11(2):166–168
49. Sabitha G, Rao AS, Yadav JS (2011) Stereoselective synthesis of (−)-synparvolide B. Tetrahedron Asymmetry 22(8):866–871
50. Srinivas B, Sridhar R, Rao KR (2010) Stereoselective total synthesis of (+)-varitriol. Tetrahedron 66(44):8527–8535
51. Dandia A, Jain AK, Laxkar AK (2013) Ethyl lactate as a promising bio based green solvent for the synthesis of spiro-oxindole derivatives via 1,3-dipolar cycloaddition reaction. Tetrahedron Lett 54(30):3929–3932
52. Nikles SM, Piao M, Lane AM, Nikles DE (2001) Ethyl lactate: a green solvent for magnetic tape coating. Green Chem 3(3):109–113
53. Reisch M (2008) Solvent users look to replace NMP. Chem Eng News 86(29):32
54. Abbas C, Rammelsberg AM, Beery K (2003) Extraction of phytosterols from corn fiber using green solvents. US Patent 20,030,235,633
55. Tombokan XC, Aguda RM, Danehower DA, Kilpatrick PK, Carbonell RG (2008) Three-component phase behavior of the sclareol-ethyl lactate-carbon dioxide system for GAS applications. J Supercrit Fluids 45(2):146–155
56. Ishida BK, Chapman MH (2009) Carotenoid extraction from plants using a novel, environmentally friendly solvent. J Agric Food Chem 57(3):1051–1059
57. Wu W, Lu M, Yu L (2011) A new environmentally friendly method for astaxanthin extraction from Xanthophyllomyces dendrorhous. Eur Food Res Technol 232(3):463–467
58. Hernández EJ, Luna P, Stateva RP, Najdanovic-Visak V, Reglero G, Fornari T (2011) Liquid-liquid phase transition of mixtures comprising squalene, olive oil, and ethyl lactate: application to recover squalene from oil deodorizer distillates. J Chem Eng Data 56(5):2148–2152
59. Vicente G, Paiva A, Fornari T, Najdanovic-Visak V (2011) Liquid-liquid equilibria for separation of tocopherol from olive oil using ethyl lactate. Chem Eng J 172(2–3):879–884
60. Manic MS, Villanueva D, Fornari T, Queimada AJ, MacEdo EA, Najdanovic-Visak V (2012) Solubility of high-value compounds in ethyl lactate: measurements and modeling. J Chem Thermodyn 48:93–100
61. Bermejo DV, Luna P, Manic MS, Najdanovic-Visak V, Reglero G, Fornari T (2012) Extraction of caffeine from natural matter using a bio-renewable agrochemical solvent. Food and Bioproducts Processing 91(4):303–309
62. Golmakani MT, Mendiola JA, Rezaei K, Ibáñez E (2012) Expanded ethanol with CO_2 and pressurized ethyl lactate to obtain fractions enriched in γ-Linolenic Acid from Arthrospira platensis (Spirulina). J Supercrit Fluids 62:109–115

Chapter 7
Ionic Liquids as Alternative Solvents for Extraction of Natural Products

Milen G. Bogdanov

Abstract Ionic liquids (ILs) have been proved as promising substituents of the flammable, volatile, and toxic organic solvents in numerous processes. This chapter considers the role of ILs in the extraction of natural products from their native sources and represents a comprehensive overview on the recent achievements in the IL-assisted solid-liquid extractions of secondary metabolites from plant matrices. By analyzing the similarities and differences between the ILs and molecular solvents, important factors that influence the extraction efficiency are discussed, and some general conclusions regarding the advantages and disadvantages of the use of ILs are emphasized. The effect of the IL structure on the extraction efficiency and the possible extraction mechanism and the approaches for both IL recycling and solute recovery after extraction are also discussed.

7.1 Introduction

Plants, animals, and microorganisms represent a sustainable source of natural products useful to human beings [1]. Particularly, the plant kingdom offers a variety of species, which have been used for millennia as remedies for numerous diseases in different world areas [2]. Therefore, diverse natural species are still the main source of ideas toward the development of new drugs, functional foods, and food additives. Bioactive natural compounds are secondary metabolites, generated through various biological pathways in secondary metabolism processes [3], and typically, their manufacturing from the natural sources proceeds according to well-established procedures [4], which usually begin with exhaustive extraction with molecular solvents (VOCs), e.g., saturated hydrocarbons, alcohols, chloroalkanes, etc., and

M.G. Bogdanov (✉)
Faculty of Chemistry and Pharmacy, University of Sofia "St. Kl. Ohridski",
1, James Bourchier Blvd, 1164 Sofia, Bulgaria
e-mail: ohmb@chem.uni-sofia.bg

F. Chemat and M. Abert Vian (eds.), *Alternative Solvents for Natural Products Extraction*, Green Chemistry and Sustainable Technology,
DOI 10.1007/978-3-662-43628-8_7, © Springer-Verlag Berlin Heidelberg 2014

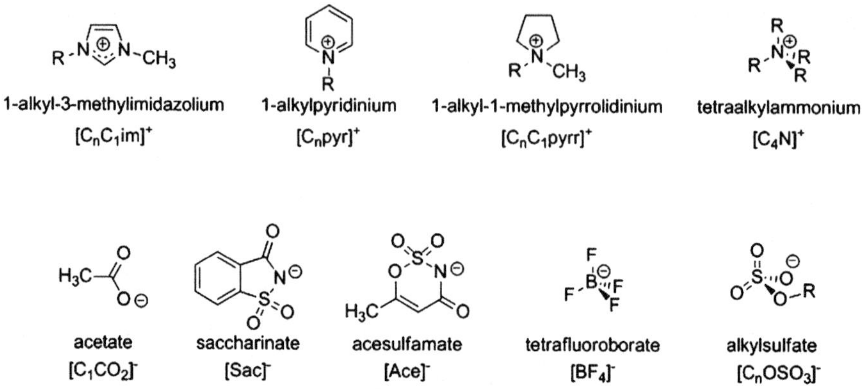

Fig. 7.1 Structure, name, and abbreviation of commonly used cations and anions in ionic liquids

followed by additional chemical treatment of the extracts in order for the compounds of interest to be isolated in a pure form. These procedures are laborious, time and energy consuming, and require complicated equipment. Moreover, the organic solvents employed in the production of natural products are flammable, volatile, and toxic, which is in a contradiction with the universally accepted nowadays 12 principles of the green chemistry [5]. Thus, the need for extractants of improved characteristics from safety, ecological, and toxicological standpoint can be put forward.

Room-temperature ionic liquids (RTILs) are promising candidates that could meet the above requirements [6]. Consisting entirely of ions (usually charge-stabilized organic cation and inorganic or organic anion, *cf.* Fig. 7.1), they are liquids at ambient temperature and display a wide range of unique properties, such as negligible vapor pressure, nonflammability, high thermal stability, low chemical reactivity, etc. [7]. These unique properties, together with the fine-tunable density, viscosity, polarity, and miscibility with other common solvents [8–10], favor their application in diverse fields such as synthesis [11], catalysis [12], electrochemistry [13], and analytical chemistry [14], to name just a few. Furthermore, harmful VOCs have been successfully replaced by RTILs in different separation processes including liquid-liquid and solid-liquid extractions [15].

This chapter considers the role of the ILs in the extraction of secondary metabolites from natural sources. It begins with a short description of the ILs as solvents, emphasizing the similarities and differences with the molecular solvents, and proceeds with a comprehensive overview on the recent achievements in the IL-assisted solid-liquid extractions of natural products by means of different extractive techniques. Factors that influence the extraction efficiency are discussed, and some general conclusions regarding the advantages and disadvantages of the extraction methods employed are drawn. The effect of the IL structure on the extraction efficiency and the possible extraction mechanism and the approaches for both IL recycling and solute recovery after extraction are also discussed. This chapter is designed, on the one hand, in a way to provide detailed information to the experienced researchers and, on the other hand, to give some clues to the researchers who are just entering in this still unexplored area to help them to avoid potential pitfalls and to identify the best method and the most suitable IL for a particular extraction process.

7.2 Ionic Liquids as Solvents for Extraction

The knowledge of solubility data for the compounds of interest is essential for a successful performance of the extraction and separation processes. Due to the low solubility of natural compounds in water, other solvents such as alcohols, ethers, chloroalkanes, and normal alkanes are commonly used. However, considering the substance intended to be extracted, a particular solvent could be selected if literature data for the compound class under investigation is known. Where such an information is unavailable, one should follow the principle "like dissolves like," i.e., that the solvent used should have a similar polarity to the compound(s) of interest. The widely accepted and understood concept of polarity is based on the definition that polarity is a sum of all possible (specific and nonspecific) intermolecular interactions between a solvent and any potential solute, excluding these interactions resulting in a chemical reaction [16]. This can be considered both as a physical and a chemical phenomenon that comprises Coulombic interactions, dipole-dipole interactions, hydrogen-bonding interactions, and donor-acceptor acid-base interactions. Regarding IL polarities, they depend on the nature of the IL components and are typically in the range from dipolar non-hydrogen-bond-donating solvents (DMF, DMSO, acetonitrile) up to polar hydrogen-bond-donating ones (primary alcohols, water) [17, 18]. This similarity, together with the others discussed below, suggests ILs as good candidates for VOC substituents in dissolving natural products of different polarity.

Besides solubility, which is a key feature in obtaining a crude extract with any solvent, there are additional criteria for the proper solvent selection. Among them, melting and boiling points, density, viscosity, and surface tension of the solvent are of a significant importance. The melting temperature of the solvent should preferably be lower than the ambient. RTILs meet those criteria, thus allowing easy handling. Furthermore, because of their high thermal and chemical stability, the use of ILs allows extractions to be conducted at higher temperatures than the one offered by the common VOCs. The densities reported for ILs to date vary between 1 and 1.6 g cm^{-3} [19] and appear to be the least sensitive physicochemical property to variations in temperature and impurity content. High viscosity of a solvent is not desirable in the extraction processes because it hinders the mass transfer of the solute of interest. Compared to the common molecular solvents, ILs demonstrate higher viscosity, ranging from 10 to 500 mPa s at room temperature [19], and this can be considered as a drawback for their use as extractants. Nevertheless, this property of ILs is strongly dependent on the temperature and water content [20], thus allowing the above shortcoming to be overcome by temperature elevation or by the use of IL-molecular solvents mixtures. Low surface tension is important for solvent penetration into the plant matrix by promoting better wetting of the solids. Unlike most molecular solvents, which exhibit surface tensions at room temperature around or below 22 mN m^{-1}, the employed ILs in the solid-liquid extraction processes present surface tension values ranging from 20 mN m^{-1} up to 50 mN m^{-1} [21]. Consequently, some ILs could be considered as substituents of commonly used for extraction primary alcohols such as methanol, ethanol, isopropanol, etc., and could be employed as additives to water in order to reduce its surface tension.

It is noteworthy that pure ILs can be considered as self-assembly amphiphiles which form H-bonded-polymeric network, the latter being a general structural pattern for both solid and liquid phases [22, 23]. However, the introduction of other molecules into the pure ILs disrupts the H-bonded network and generates a secondary nanostructure with polar and nonpolar domains. In case of dilution by solvents, depending on the solvent polarity and H-bonding ability, supramolecular aggregates, triple ions, contact ion pairs, or solvent-separated ions can be formed [22]. For water, it has been shown to be located in the polar domain and that the degree of its interactions with the IL ions is strongly dependent on the anion nature [24]. It has been also found that the aqueous solutions of imidazolium-based ILs can form aggregates, especially as the alkyl group attached to the imidazolium core becomes longer, and a number of studies [25–30] demonstrated that the mesophase structure of IL-aqueous mixtures could be "tuned" simply by careful selection of the anion and adjustment of the water concentration, thereby tailoring the system for a selective interactions with the solute of a particular interest.

Besides physicochemical properties of ILs that affect the extraction outcome, some other requirements regarding the whole process of economical and environmental impact should also be taken into account. A high selectivity of a solvent enables fewer technological stages to be used, and in the case of a complex mixture, where multiple components could be extracted, a group selectivity is desirable. Availability and costs are also important. The solvent should be readily available, and it is not its price that is important, but the annual cost due to the inevitable operation losses. Although the ILs perform as excellent extractants, they are still expensive compared to the conventional molecular solvents. Therefore, efficient recycling is another important issue that addresses the economics of their use. To this end, the recovery and reusability of the ILs after extraction is a key issue that should be thought over in more details in the near future. From an environmental standpoint, the appropriate solvent should be as less as possible volatile, flammable, corrosive, and toxic. Corrosive and flammable solvents increase the process' demands not only because of the sophisticated equipment required but might also result in a more expensive pre- and posttreatment of the waste products. The removal of solvents from residual plant material can also cause serious problems, and posttreatment may be necessary to reduce the residue level. In food and pharmaceutical processing, only nontoxic solvents should be taken into consideration, since any hazard associated with the solvent inevitably requires extra safety measures.

Based on the above reasoning, a quite obvious conclusion could be drawn that the ILs, mainly due to the versatility of possible ion combinations and fine-tunable physicochemical properties, could be considered as potential substituents of the volatile, flammable, and toxic organic solvents commonly employed in solid-liquid extraction processes. Nevertheless, one should keep in mind that consideration of ILs in general as nonflammable, nontoxic, biodegradable, noncorrosive, and all properties related to their "greenness" are true in the same extent as they are not. For example, it sounds somehow confusing when $[C_nC_1im][BF_4]$ and $[C_nC_1im][PF_6]$ are chosen as best extractants among others, since it is well known nowadays that both $[BF_4]^-$ and $[PF_6]^-$ anions hydrolyze in the presence of water [31] to give the highly corrosive HF.

7.3 Solid-Liquid Extraction with Ionic Liquids

Plants are complex matrices containing a range of secondary metabolites which differ in their functional groups and polarities, thereby leading to the simultaneous dependence of the extraction of these metabolites on the plant material type, solutes nature, and extractant properties. Therefore, as much as possible, factors controlling the partitioning of the compounds of interest should be taken into account in order for best extraction outcome to be achieved and reproducible results to be enabled (*cf.* Fig. 7.2).

Regarding the raw material, the particle size and moisture content are important, and their influence on the extraction efficiency should be always considered, especially in the case when a novel method for quantification is going to be developed.

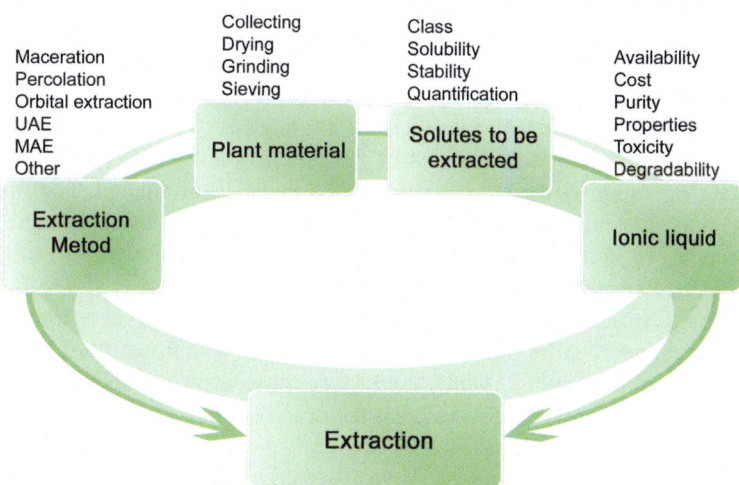

Fig. 7.2 Some important factors to be considered prior to extraction

Besides this, different plant parts such as leaves, flowers, branches, bark, seeds, fruits, or rhizomes and roots could be extracted, and the plant parts used should be clearly denoted, since all of them differ by morphology and chemical composition [32]. Furthermore, the chemical composition of the plant might vary with season, maturity, and growing area, so it is of a great importance that all ingredients of the batch sample, particularly when they belong to the same compound class, are known. This is necessary because the presence of additional unknowns might compromise the analysis and thereby result in a wrong interpretation of the data obtained. Another issue worth to be addressed is the nomenclature used for naming the plant material. Various nomenclature systems have been applied to plants to date. These include common names (e.g., black pepper), which are well accepted in our everyday lives; botanical names (e.g., *Piper nigrum*), which are widely used in the scientific community; and pharmaceutical names (e.g., *Piperis nigri fructus*), which are used to denote unequivocally the medicinal plant parts. Because of their easy recognizability, someone would prefer to use common names, but should eventually keep in mind that they vary from region to region and from language to language. Moreover, different plants may be known by the same common name and vice versa, thus causing a lack of clarity and confusion if they are not used in a proper way. Therefore, the botanical nomenclature, because of its wide acceptance and accuracy in naming biological species, should be preferred when the results are disseminated to an international audience. However, the common names of a particular plant could be also used for the sake of simplicity, but in this case the botanical name as well as the plant parts extracted should be clearly denoted.

Since the solid-liquid extraction may be affected by a large variety of factors, an appropriate optimization should be performed. The procedure for adjusting these factors comprises a series of apparently simple steps; however, the ultimate success of this type of research depends on the attention devoted to each aspect of the work. The variables mainly evaluated in the solid-liquid extractions with ILs depend on the extraction techniques employed (*cf.* Sect. 7.3.1) and could be summarized as follows: IL structure (both anion and cation type), IL concentration, sample moisture content, pH, preliminary soaking time, extraction temperature, irradiation power, irradiation time, pressure, solid-liquid ratio, particle size, and number of extraction cycles. It is noteworthy that in most of the articles published to date, otherwise important operation parameters are neglected and so not studied. For instance, limited number of articles controls the moisture of the plant material or examines the influence of pH on the extraction efficiency – two factors that affect the efficiency and robustness of the extraction procedure in a great manner.

Besides IL structure, four to six variables are mainly selected for optimization in the papers summarized in this chapter (*cf.* Table 7.1). The majority of the authors prefer the unvaried method, which comprises consecutive variables changing, in order for the influence of each particular factor to be assessed and thus the highest extraction yield to be achieved step-by-step. In some cases, in order to prove that the total amount of the solutes of interest is recovered from the batch sample, a comparison with other representative techniques such as Soxhlet extraction had been performed. Even though this approach could ensure an exhaustive extraction, it is noteworthy that it does not consider the simultaneous influence of multiple factors on the extraction efficiency, thus not corresponding to the overall optimized conditions. In order for these relationships to be assessed properly and the most significant process parameters to be found, more sophisticated statistical approaches such as orthogonal design and response surface methodology are also employed (Table 7.1).

7.3.1 Extraction Procedures

A range of techniques, which differ in their cost and complexity, can be used for extraction of natural products from plant materials. In the ideal case, the extraction method selected should be exhaustive, i.e., to yield as much of the desired metabolites or as many compounds as possible. It should be simple, fast, safe, economical, environmentally friendly, and reproducible. Regarding ILs, the classical extraction under stirring at ambient conditions or elevated temperature (HRE) has been shown effective and economical, but more sophisticated extraction technologies, such as microwave-assisted extraction (IL-MAE), ultrasound-assisted extraction (IL-UAE), joint application of UAE and MAE, IL-assisted ultrahigh pressure extraction (IL-UPE), and IL-based negative-pressure cavitation-assisted extraction (IL-NPCE) are also employed. All these methods possess specific requirements, and in some cases, a comparative analysis between them has been performed. The following section describes the recent achievements concerning IL-assisted extractions by means of different extraction procedures. Factors that influence the extraction efficiency are discussed, and some general conclusions regarding the advantages and disadvantages of the extraction methods employed are drawn.

7.3.1.1 Classical Extraction

The extraction of value-added chemicals from plants can in some instances be accelerated by conducting the process at high temperature, since elevated temperature simultaneously increases the compounds' solubility inside the solid matrix and facilitates their diffusion into the extractant. Indeed, classical extraction procedures such as maceration, percolation, or batch extraction are often carried out at a temperature higher than the ambient. Among the classical methods, batch extraction seems preferred by many authors and has been applied as an extraction technique for recovery of alkaloids, phenolic compounds, and lipids.

Considering the fact that some ILs are able to dissolve cellulose, Jin et al. [56] explored the ability of pure $[C_4C_1im]Cl$ to improve the release of phenolic aldehyde paeonol from the roots of *Cynanchum paniculatum*. The authors studied the influence of several parameters on the extraction efficiency and found that at the optimized conditions (70 °C, 8 h, s/l ratio 1:7.3), the IL-assisted extraction

Table 7.1 Recent application of ionic liquids in solid-liquid extraction of secondary metabolites

Plant source[a]	Extracted compound(s)	ILs used[b]	Extraction method[c] and optimal conditions	References
Alkaloids				
Nelumbo nucifera (Indian lotus) Seed	Liensinine Isoliensinine Neferine	[C_4C_1im]X, {X = Cl, Br, [PF_6]}; [$C_nC_1$im][BF_4], (n = 2, 4, 6, 8)	IL-MAE 1.5 M [C_4C_1im][BF_4] or 1 M [C_6C_1im][BF_4], 280 W, irradiation 1.5 min, s/l ratio 1:15 and 1:10 [g/mL], respectively	[33]
Nelumbo nucifera (Indian lotus) Leaf	*N*-Nornuciferine *O*-Nornuciferine Nuciferine	[C_4C_1im]X, {X = Cl, [BF_4], [PF_6]}; [$C_nC_1$im]Br, (n = 2, 4, 6, 8)	IL-MAE 1 M [C_6C_1im]Br, 280 W, irradiation 2 min, s/l ratio 1:30 [g/mL]	[34]
Camptotheca acuminata Samara	Camptothecin 10-Hydroxycamptothecin	[C_4C_1im]X, {X = Cl, [BF_4], [NO_3], [ClO_4], [OTs], [HSO_4]}; [$C_nC_1$im]Br, (n = 2, 4, 6, 8)	IL-MAE 0.8 M [C_8C_1im]Br, pre-soaking 2 h, 105 °C, 385 W, irradiation 2 × 8 min (2 cycles), s/l ratio 1:12 [g/mL]	[35]
Stephania tetrandra Root	Fangchinoline Tetrandrine	[C_4C_1im][BF_4]	IL-UAE 1.5 M [C_4C_1im][BF_4], pH = 9.8, 150 W, irradiation 40 min, s/l ratio 1:20 [g/mL]	[36]
Piper nigrum (White pepper) Seed	Piperine	[C_4C_1im]X, {X = Br, [BF_4], [PF_6], [H_2PO_4]}; [C_6C_1im][BF_4]; [(HO_3S)^4C_4C_1im]Br	IL-UAE 2 M [C_4C_1im][BF_4], 500 W, irradiation 30 min, s/l ratio 1:15 [g/mL]	[37]
Catharanthus roseus (Madagascar rosy periwinkle) Leaf	Vindoline Catharanthine Vinblastine	[C_4C_1im]X, {X = Cl, I, [BF_4], [ClO_4], [OTs], [HSO_4], [NO_3]}; [(C_1 = C_2)C_1im]Br [$C_nC_1$im]Br, (n = 2, 4, 6, 8);	IL-UAE 0.5 M [(C_1 = C_2)C_1im]Br, pre-soaking 2 h, 250 W, irradiation 3 × 30 min (3 cycles), s/l ratio 1:10 [g/mL]	[38]

Plant	Compound	IL / Conditions	Method	Ref.
Camptotheca acuminata Samara	Camptothecin	[C_4C_1im]X, {Cl, [BF_4], [NO_3], [ClO_4], [HSO_4]}; [$C_nC_1$im]Br, (n = 2, 3, 4, 6, 8); [($C_1=C_2$)C_1im]Br; [ChC_1im]Br; [BzC_1im]Br	IL-UAE 0.75 M [C_8C_1im]Br, 250 W, irradiation 3 × 35 min (3 cycles), s/l ratio 1:12 [g/mL]	[39]
10-Hydroxycamptothecin				
Glaucium flavum (Yellow horned poppy) Aerial parts	Glaucine	[C_4C_1im]X, {X = Cl, Br, [Sac]; [Ace]}; [$C_nC_1$im][Ace], (n = 4, 6, 8, 10)	HRE 1 M [C_4C_1im][Ace], 25 °C, stirring 1 h or 80 °C, stirring 20 min, s/l ratio 1:40 [g/mL]	[40, 41]
Paullinia cupana (Guaraná) Seed	Caffeine	[C_4C_1im]X, {Cl, [OTs]}; [C_2C_1im]Cl; [(HO)$^2C_2C_1$im]Cl; [C_2C_1im][C_1CO_2]; [C_4C_1pyrr]Cl	HRE 2.34 M [C_4C_1im]Cl, 70 °C, stirring 30 min, s/l ratio 1:10 [g/mL]	[42]
Piper nigrum (Black pepper) Seed	Piperine	[$C_{12}C_1$im]X, {X = Br, [OTf], [N(CN)$_2$], [C_1CO_2]}; [$C_nC_1$im]Cl, {n = 10, 12, 14}; [C_{12}betaine]Cl	HRE 0.05 M [C_{12}betaine]Cl, 25 °C, stirring 3 h, s/l ratio 1:19 [g/mL]	[43]
Phenolic compounds and acids				
Fallopia japonica (Japanese knotweed) Rhizome	*trans*-resveratrol	[C_4C_1im]X, {Cl, Br, [BF_4]}	IL-MAE 2.5 M [C_4C_1im]Br, 60 °C, irradiation 10 min, size 0.30–0.45 mm, s/l ratio 1:20 [g/mL]	[44]
Smilax china (China root) Tubers	*trans*-resveratrol	[C_4C_1im]X, {Cl, [BF_4], [N(CN)$_2$], [H_2PO_4]}; [$C_nC_1$im]Br, (n = 2, 4, 6); [C_4C_1im]$_2$[SO_4]; [C_4pyr]Cl; [(C_1)$_4$N]Cl	IL-MAE 2.5 M [C_4C_1im]Br, 60 °C, irradiation 10 min, size 0.45–0.90 mm, s/l ratio 1:20 [g/mL]	[45]
Quercetin				

(continued)

Table 7.1 (continued)

Plant source[a]	Extracted compound(s)	ILs used[b]	Extraction method[c] and optimal conditions	References
Psidium guajava (Apple guava) Leaf	Gallic acid Ellagic acid Quercetin	$[C_4C_1im]X$, {Cl, $[BF_4]$, $[N(CN)_2]$, $[H_2PO_4]$}; $[C_nC_1im]Br$, ($n = 2, 4, 6$); $[C_4C_1im]_2[SO_4]$; $[C_4pyr]Cl$; $[(C_1)_4N]Cl$	IL-MAE 2.5 M $[C_4C_1im]Br$, 70 °C, irradiation 10 min, size 0.30–0.45 mm, s/l ratio 1:20 [g/mL]	[45]
Myrica rubra Leaf	Myricetin Quercetin	$[C_4C_1im]X$, {Cl, Br, $[BF_4]$, $[N(CN)_2]$, $[H_2PO_4]$, $[HSO_4]$}; $[C_2C_1im]$ Br or $[BF_4]$; $[C_4C_1im]_2[SO_4]$; $[HO_2C)C_1C_1im]Cl$; $[C_6C_1im]Br$; $[(C_1)_4N]Cl$; $[C_4pyr]Cl$	IL-MAE 2.0 M $[C_4C_1im][HSO_4]$, 70 °C, irradiation 10 min, s/l ratio 1:30 [g/mL]	[46]
Toona sinensis	Quercetin	$[C_4C_1im]X$, {Cl, Br, $[BF_4]$, $[OTs]$}; $[C_3C_1im]$ Br or $[BF_4]$; $[(C_1 = C_2)C_1im]$ Cl or $[BF_4]$	IL-MAE 0.1 M $[C_4C_1im]Br$, 60 °C, irradiation 8 min, s/l ratio 1:20 [g/mL]	[47]
Toona sinensis	Quercetin Kaempferol	$[C_4C_1im]X$, {X = Cl, $[OTs]$, $[C_1SO_3]$, $[HSO_4]$}; $[(C_1 = C_2)C_1im]Cl$ or $[BF_4]$; $[C_nC_1im]Br$ or $[BF_4]$, ($n = 2, 4, 6, 8, 10$)	IL-MAE 2.0 M $[C_4C_1im]Br$, 60 °C, irradiation 20 min, s/l ratio 1:30 [g/mL]	[48]
Rosa chinensis	Quercetin Kaempferol	$[C_4C_1im]X$, {X = Cl, $[OTs]$, $[C_1SO_3]$, $[HSO_4]$}; $[(C_1 = C_2)C_1im]Cl$ or $[BF_4]$; $[C_nC_1im]Br$ or $[BF_4]$, ($n = 2, 4, 6, 8, 10$)	IL-MAE 2.5 M $[C_8C_1im]Br$, 70 °C, irradiation 20 min, s/l ratio 1:40 [g/mL]	[48]
Cynanchum paniculatum Root	Paeonol	$[C_4C_1im]Cl$	IL-MAE $[C_4C_1im]Cl$ (pure), 136 W, irradiation 8 min, size ≤ 0.25 mm, s/l ration 1:7.3 [g/mL]	[49]

Source	Compounds	Ionic Liquid	Method/Conditions	Ref.
Dryopteris Fragrans Aerial part	Phloroglucinols	[C$_4$C$_1$im]X, {Cl, OH, [BF$_4$], [H$_2$PO$_4$]}; [C$_n$C$_1$im]Br, (n = 2, 4, 6, 8)	IL-MAE 0.75 M [C$_8$C$_1$im]Br, 50 °C, 600 W, irradiation 7 min, s/l ratio 1:12 [g/mL]	[50]
Cajanus cajan (Pigeon pea) Leaf	Apigenin, Formononetin, Stilbenes, Coumarins	[C$_4$C$_1$im]X, {X = Cl, [BF$_4$], [HSO$_4$], [H$_2$PO$_4$]}; [C$_n$C$_1$im]Br, (n = 2, 4, 6, 8)	IL-MAE 0.75 M [C$_4$C$_1$im]Br, 60 °C, 300 W, irradiation 7 min, s/l ratio 1:12 [g/mL]	[51]
Pertusaria pseudocorallina Lichen	Norstictic acid, Depsidones	[C$_1$C$_1$im][C$_1$OSO$_3$]; [C$_2$C$_1$im][C$_2$OSO$_3$]; [(HO)^{2}C$_2$C$_1$im][NTf$_2$]; [C$_3$C$_1$im][NTf$_2$]; [C$_4$C$_1$im][NTf$_2$]	IL-MAE [C$_1$C$_1$im][C$_1$OSO$_3$] (pure), 100 °C, irradiation 5 min, s/l ratio 1:20 [g/mL]	[52]
Pertusaria amara Lichen	Picrolichenic acid and derivatives	[C$_1$C$_1$im][C$_1$OSO$_3$]	IL-MAE [C$_1$C$_1$im][C$_1$OSO$_3$] (pure), 100 °C, irradiation 5 min, s/l ratio 1:20 [g/mL]	[52]
Ochrolechia parella Lichen	Variolaric acid, α-Alectoronic acid, Lecanoric acid	[C$_1$C$_1$im][C$_1$OSO$_3$]	IL-MAE [C$_1$ C$_1$im][C$_1$OSO$_3$] (pure), 100 °C, irradiation 5 min, s/l ratio 1:20 [g/mL]	[52]
Magnolia officinalis Bark	Magnolol, Honokiol	[C$_4$C$_1$im][PF$_6$]	IL-UAE 2.0 M [C$_4$C$_1$im][PF$_6$] in ethanol, pH = 7.15, 200 W, irradiation 30 min, s/l ratio 1:20 [g/mL]	[53]
Rheum spp. (Rhubarb)	Emodin, Chrysophanol, Rhein, Aloe-emodin physcion	[C$_4$C$_1$im]X, {Cl, Br, [BF$_4$]}	IL-UMAE 2.0 M [C$_4$C$_1$im]Br, 500 W, irradiation 2 min, s/l ratio 1:15 [g/mL]	[54]

(continued)

Table 7.1 (continued)

Plant source[a]	Extracted compound(s)	ILs used[b]	Extraction method[c] and optimal conditions	References
Lonicera japonica (Honeysuckle) Flower bulb	Chlorogenic acid	[C$_4$C$_1$im][BF$_4$]	IL-UAE 0.75 M [C$_4$C$_1$im][BF$_4$], pH = 1.2, 60 °C, 200 W, irradiation 40 min, size ≤ 0.4 mm, s/l ratio 1:20 [g/mL]	[55]
Cynanchum paniculatum Root	Paeonol	[C$_4$C$_1$im]Cl	HRE [C$_4$C$_1$im]Cl (pure), pre-soaking 8 h, 70 °C, size ≤ 0.25 mm, s/l ratio 1:7.3 [g/mL]	[56]
Apium graveolens (Celery)	Luteolin Apigenin	[C$_4$C$_1$im]X, {Cl, Br, [BF$_4$], [C$_1$OSO$_3$]}	IL-UAE 1 M [C$_4$C$_1$im][C$_1$OSO$_3$], pH = 1.0, 200 W, irradiation 90 min, s/l ratio 1:10 [g/mL]	[57]
Chamaecyparis obtusa (Japanese cypress) Leaf	Dihydrokaempferol Quercitrin Myricetin Amentoflavone	[C$_{10}$C$_1$im]X, {X = Cl, [BF$_4$], [PF$_6$], [NTf$_2$]}; [C$_n$C$_1$im]Br, (n = 2, 4, 6, 8, 10, 12)	IL-UAE 2.5 M [C$_{10}$C$_1$im]Br in methanol, 200 °C, 8 h, s/l ratio 1:13 [g/mL]	[58]
Larix gmelinii Bark	Procyanidins	[C$_4$C$_1$im]X, {X = Cl, [OH], [BF$_4$], [NO$_3$], [HSO$_4$], [C$_1$CO$_2$]}; [C$_n$C$_1$im]Br; (n = 2, 4, 6, 8, 10)	IL-UAE 1.25 M [C$_4$C$_1$im]Br, pre-soaking 3 h, 150 W, irradiation 30 min, s/l ratio 1:10 [g/mL]	[59]
Cajanus cajan (Pigeon pea) Root	Genistin Genistein Apigenin (flavonoids)	[C$_4$C$_1$im]X, {X = Cl, [BF$_4$], [HSO$_4$], [H$_2$PO$_4$]}; [C$_n$C$_1$im]Br, (n = 2, 4, 6, 8)	IL-NPCE, lab scale and pilot scale 0.53 M [C$_8$C$_1$im]Br, 74 °C, 15 min, pressure = −0.07 MPa s/l ratio 1:20 [g/mL]	[60]

Species	Compound	Ionic Liquid	Method / Conditions	Ref.
Cynanchum bungei Root	Acetophenones	[C$_n$C$_1$im][BF$_4$], (n = 2, 3, 4, 6); [C$_3$C$_1$im]X {X = Br, I}	IL-UAE 0.6 M [C$_4$C$_1$im][BF$_4$], 175 W, 25 °C, irradiation 50 min, size 177–250 mm, s/l ratio 1:35 [g/mL]	[61]
Cajanus cajan (Pigeon pea) Leaf	Flavonoid glycosides	[C$_4$C$_1$im]X, {X = Cl, [OH], [BF$_4$], [HSO$_4$], [H$_2$PO$_4$]}; [C$_n$C$_1$im]Br, (n = 2, 4, 6, 8)	IL-MAE 1.0 M [C$_4$C$_1$im]Br, 60 °C, irradiation 13 min, s/l ratio 1:20 [g/mL]	[62]
Glycosides and saponins				
Picrorhiza scrophulariflora Rhizome	Iridoid glycosides Phenylethanoid glycosides, cucurbitacin glycosides	[C$_4$C$_1$im][BF$_4$]	IL-UAE 1.5 M [C$_4$C$_1$im][BF$_4$], 500 W, irradiation 30 min, s/l ratio 1:500 [g/mL]	[63]
Fraxinus rhynchophylla (Qin Pi) Bark	Aesculin Aesculetin	[C$_4$C$_1$im]X, {Cl, I, [BF$_4$], [ClO$_4$], [HSO$_4$], [OTs]}; [C$_n$C$_1$im]Br, (n = 2, 4, 6, 8, 10, 12)	IL-UAE 0.86 M [C$_4$C$_1$im]Br, pre-soaking 4 h, 250 W, irradiation 44 min, s/l ratio, 1:11 [g/mL]	[64]
Glycyrrhiza spp. (Licorice) Root	Isoliquiritigenin (halcone) Liquiritin (flavanone gly.) Glycyrrhizic acid (saponin)	[C$_4$C$_1$im]X, {Cl, [OH], [BF$_4$], [NO$_3$], [ClO$_4$], [HSO$_4$], [C$_1$CO$_2$]}; [C$_n$C$_1$im]Br, (n = 2, 4, 6, 8, 10, 12)	IL-UAE 0.5 M [C$_4$C$_1$im]Br, pre-soaking 2 h, 214.91 W, irradiation 38.84 min, s/l ratio 1:12 [g/mL]	[65]
Acanthopanax senticosus (Siberian Ginseng) Root	Eleutheroside B and E (glycosides)	[C$_4$C$_1$im]X, {Cl, [OH], [BF$_4$], [NO$_3$], [ClO$_4$], [HSO$_4$], [C$_1$CO$_2$]}; [C$_n$C$_1$im]Br, (n = 2, 4, 6, 8, 10, 12)	IL-UAE 0.64 M [C$_4$C$_1$im]Br, pre-soaking 2 h, 250 W, irradiation 30 min, s/l ratio 1:25 [g/mL]	[66]

(continued)

Table 7.1 (continued)

Plant source[a]	Extracted compound(s)	ILs used[b]	Extraction method[c] and optimal conditions	References
Rhodiola rosea (Golden root) Root	Rhodiosin Rhodionin	[C$_4$C$_1$im]X, {X = Cl, [OH], [BF$_4$], [HSO$_4$]}; [C$_n$C$_1$im]Br, (n = 2, 4, 6, 8)	IL-UAE-SPT 2.0 M [C$_2$C$_1$im]Br, 360 W, irradiation 25 min, solvent flow 0.8 mL/min	[67]
Pueraria lobata (Kudzu) Root	Puerarin (isoflavone 8-C-glucoside)	[C$_4$C$_1$im]X, {X = Br, [BF$_4$]}; [(HO)^{2}C$_2$C$_1$im]X, {X = Br, [BF$_4$]}; [(HO$_2$C)C$_1$C$_1$im][BF$_4$]	IL-UAE 1.0 M [C$_4$C$_1$im]Br, 480 W, irradiation 27.43 min, s/l ratio 1:23[g/mL]	[68]
Camellia sinensis (Tea) Aerial part	Saponins Polyphenols	[C$_n$C$_1$im]Cl, (n = 2, 4, 6, 8); [(C$_1$ = C$_2$)C$_1$im]Cl; [(HO)^{2}C$_2$(C$_1$)$_3$N]X, {X = Cl, [C$_1$CO$_2$], [C$_6$CO$_2$]}; [C$_2$C$_1$im]X, {[N(CN)$_2$], [C$_2$OSO$_3$], [OTf], [(OH)^{1}C$_2$CO$_2$], [C$_1$CO$_2$]};	HRE 30 wt% [(HO)^{2}C$_2$(C$_1$)$_3$N]Cl, 60 °C, stirring 2 h, s/l ratio 1:10 [g/mL]	[69]
Ilex paraguariensis (Mate) Aerial part	Saponins Polyphenols	[C$_n$C$_1$im]Cl, (n = 2, 4, 6, 8); [(C$_1$ = C$_2$)C$_1$im]Cl; [(HO)^{2}C$_2$(C$_1$)$_3$N]X, {X = Cl, [C$_1$CO$_2$], [C$_6$CO$_2$]}; [C$_2$C$_1$im]X, {[N(CN)$_2$], [C$_2$OSO$_3$], [OTf], [(OH)^{1}C$_2$CO$_2$], [C$_1$CO$_2$]};	HRE 30 wt% [(HO)^{2}C$_2$(C$_1$)$_3$N]Cl, 60 °C, stirring 2 h, s/l ratio 1:10 [g/mL]	[69]
Panax ginseng (Ginseng) Root	Ginsenosides (saponins)	[C$_3$C$_1$im]X, {X = I, [BF$_4$]}; [C$_n$C$_1$im]Br, (n = 2, 3, 4, 6)	IL-UAE 0.3 M [C$_3$C$_1$im]Br, 250 W, irradiation 20 min, s/l ratio 1:10 [g/mL]	[70]

7 Ionic Liquids as Alternative Solvents for Extraction of Natural Products

Source	Product	Ionic Liquid	Method / Conditions	Ref.
Dioscorea nipponica Rhizome	Diogenin	[C$_4$C$_1$im]X, {[HSO$_4$], [H$_2$PO$_4$]} [C$_6$C$_1$im]X, {[HSO$_4$], [C$_1$SO$_3$], [OTf], [OTs]}, [(HO$_3$S)^{3}C$_3$C$_1$im]X, {[HSO$_4$], [H$_2$PO$_4$]}	IL-UAE 2.0 M [(HO$_3$S)^{3}C$_3$C$_1$im][HSO$_4$], sonic. 30 min, then 100 °C, 5 h, size 0.15–0.38 mm, s/l ratio 1:20 [g/mL]	[71]
Terpenoids, lipids, and essential oil				
Salvia miltiorrhiza (Danshen)	Tanshinones	[C$_4$C$_1$im]X, {Cl, Br, [BF$_4$], [NO$_3$], [SbF$_6$], [NTf$_2$]}; [C$_n$C$_1$im][PF$_6$], (n = 4, 6, 8, 12, 16)	IL-UPE 0.5 M [C$_8$C$_1$im][PF$_6$] in ethanol, pressure 300 Mpa, 2 min (1 cycle), < 0.251 mm, s/l ratio 1:20 [g/mL]	[72]
Illicium verum (Star anise) Fruit	Essential oil	[C$_4$C$_1$im][PF$_6$]	IL-MAE [C$_4$C$_1$im][PF$_6$] (pure), 440 W, 100 °C, size < 0.4 mm, s/l ratio 20:1.5 [g/mL]	[73]
Cuminum cyminum (Cumin) Fruit	Essential oil	[C$_4$C$_1$im][PF$_6$]	IL-MAE [C$_4$C$_1$im][PF$_6$] (pure), 440 W, 100 °C, irradiation 20 min, size < 0.4 mm, s/l ratio 20:1.5 [g/mL]	[73]
Rosmarinus officinalis (Rosemary) Leaf	Essential oil Carnosic acid Rosmarinic acid	[C$_4$C$_1$im]X, {X = Cl, [BF$_4$], [NO$_3$]}; [C$_n$C$_1$im]Br, (n = 2, 4, 6, 8, 10)	IL-MAE 1.0 M [C$_8$C$_1$im]Br, 700 W, irradiation 15 min, s/l ratio 1:12 [g/mL]	[74]

(continued)

Table 7.1 (continued)

Plant source[a]	Extracted compound(s)	ILs used[b]	Extraction method[c] and optimal conditions	References
Schisandra chinensis Fruit	Essential oil	[C_4C_1im]X, {Cl, [OH], [BF_4], [NO_3], [ClO_4], [HSO_4], [C_1CO_2]}; [$C_nC_1$im]Br, (n = 2, 4, 6, 8, 10, 12)	IL-MAE 0.25 M [$C_{12}C_1$im]Br, pre-soaking 4 h, 385 W, irradiation 40 min, s/l ratio 1:12 [g/mL]	[75]
Cinnamomum Spp. (Cinnamon) Bark	Essential oil, proanthocyanidins	[C_4C_1im]X, {X = Cl, [BF_4], [NO_3], [ClO_4], [C_1CO_2], [HSO_4]}; [$C_nC_1$im]Br, (n = 2, 4, 6, 8, 10)	IL-MAE 0.5 M [C_4C_1im]Br, 230 W, irradiation 15 min, s/l ratio 1:10 [g/mL]	[76]
Forsythia suspensa Seed	Essential oil	[C_4C_1im]X, {Cl, Br}; [C_2C_1im][C_1CO_2], [(C_1 = C_2)C_1im]Cl	IL-MAE 76 wt% [C_2C_1im][C_1CO_2], 300 W, 86 °C, irradiation 4.3 min, size 0.442–0.853 mm, s/l ratio 1:10 [g/mL]	[77]
Jatropha spp. Seed	Lipids Carbohydrates	[C_2C_1im]X, {[C_1OSO_3], [C_1CO_2]}	HRE 70 wt% [C_2C_1im][C_1CO_2] in methanol 120 °C, stirring 5 h, s/l ratio 1:4 [g/g]	[78]
Carthamus tinctorius (Safflower) Seed	Lipids Carbohydrates	[C_2C_1im]X, {[C_1OSO_3], [C_1CO_2]}	HRE 70 wt% [C_2C_1im][C_1CO_2] in methanol 120 °C, stirring 5 h, s/l ratio 1:4 [g/g]	[78]
Jatropha spp. Seed	Lipids Phorbol esters	[C_2C_1im]X, {[C_1OSO_3], [C_1CO_2]}	HRE 30 wt% [C_2C_1im][C_1CO_2] in methanol, 64 °C, stirring 22 h, s/l ratio 1:17 [g/g]	[79]

Chlorella vulgaris Alga	Lipids	[C_2C_1im]X, {Cl, [C_1CO_2], [BF_4], [HSO_4], [C_2OSO_3] [($C_2O)_2PO_2$], [SCN], [NTf_2], [C_1SO_3], [$AlCl_4$]}; [C_4C_1im]Cl; [($C_1 = C_2$)C_1im]Cl	HRE 1:1 w/w [C_2C_1im][C_1CO_2]-[C_2C_1im][NTf_2], 120 °C, stirring 2 h, s/l ratio 1:19 [g/g]	[80]
Chlorella vulgaris Alga	Lipids	[C_2C_1im]X, {[C_1CO_2], [HSO_4], [($C_2O)_2PO_2$], [SCN], [NTf_2]};	HRE 1:5 w/w [C_2C_1im][C_1CO_2]-$FeCl_3 \cdot 6H_2O$, 90 °C, stirring 1 h, s/l ratio 1:19 [g/g]	[81]
Chlorella vulgaris Alga	Lipids	[C_2C_1im]X, {Cl, Br, [C_1CO_2], [C_1OSO_3]}; [C_4C_1im]X, {Cl, [OTf], [BF_4], [C_1OSO_3], [NTf_2], [C_1SO_3], [PF_6]}	HRE 1:1 mL/mL [C_4C_1im][OTf] in methanol, 65 °C, stirring 18 h, s/l ratio 1:10 [g/mL]	[82]
Dunaliella spp. Microalgae	Lipids	[C_2C_1im][C_1OSO_3]	HRE 45 wt% [C_2C_1im][C_1OSO_3] in methanol, 65 °C, stirring 18 h, s/l ratio 1:16 [g/g]	[83]
Chlorella spp. Microalgae	Lipids	[C_2C_1im][C_1OSO_3]	HRE 45 wt% [C_2C_1im][C_1OSO_3] in methanol, 65 °C, stirring 18 h, s/l ratio 1:16 [g/g]	[83]
Brassica spp. (Rape) Seed	Lipids	[C_2C_1im][C_1OSO_3]	HRE 45 wt% [C_2C_1im][C_1OSO_3] in methanol, 65 °C, stirring 18 h, s/l ratio 1:16 [g/g]	[83]
Jatropha spp. Seed	Lipids	[C_2C_1im][C_1OSO_3]	HRE 45 wt% [C_2C_1im][C_1OSO_3] in methanol, 65 °C, stirring 18 h, s/l ratio 1:16 [g/g]	[83]

(continued)

Table 7.1 (continued)

Plant source[a]	Extracted compound(s)	ILs used[b]	Extraction method[c] and optimal conditions	References
Calophyllum inophyllum (Alexandrian laurel) Seed	Lipids	[C$_2$C$_1$im][C$_1$OSO$_3$]	HRE 45 wt% [C$_2$C$_1$im][C$_1$OSO$_3$] in methanol, 65 °C, stirring 18 h, s/l ratio 1:16 [g/g]	[83]
Millettia pinnata (Indian beech) Seed	Lipids	[C$_2$C$_1$im][C$_1$OSO$_3$]	HRE 45 wt% [C$_2$C$_1$im][C$_1$OSO$_3$] in methanol, 65 °C, stirring 18 h, s/l ratio 1:16 [g/g]	[83]
Others				
Artemisia annua	Artemisinin	[(HO)^{2}C$_2$(C$_1$)$_2$NH][C$_7$CO$_2$]; [(C$_1$OC$_2$)$_2$H$_2$N][NTf$_2$]	HRE [(HO)^{2}C$_2$(C$_1$)$_2$NH][C$_7$CO$_2$] (pure), 25 °C, 0.1 Mpa, 30 min, s/l ratio 1: 6.3 [g/g]	[84]
Illicium verum (Star anise) Pod	Shikimic acid	[C$_2$C$_1$im]X, {X = Cl, [C$_1$CO$_2$], [OTf], [NTf$_2$], [BF$_4$], [PF$_6$]}	IL-MAE [C$_2$C$_1$im][C$_1$CO$_2$] (pure), 100 °C, irradiation 10 min, s/l ratio 1:9 [g/mL]	[85]
Ligusticum chuanxiong (Szechuan lovage)	Senkyunolide I Senkyunolide H	[(NC)^{2}C$_2$(C$_1$)$_2$HN][C$_2$CO$_2$]; [(HO)^{2}C$_2$OC$_2$(C$_1$)$_2$HN][C$_2$CO$_2$]	IL-MAE [(HO)^{2}C$_2$OC$_2$(C$_1$)$_2$HN][C$_2$CO$_2$] (pure), 100 °C, irradiation 10 min, s/l ratio 1:2.5 [g/mL]	[86]

Botanical name; (common name); plant part (if available)[a]				
Schisandra chinensis (Five flavor berry) Fruit	Z-Ligustilide Schizandrins Schisantherin A Deoxyschizandrin γ-Schizandrin	[C₄C₁im]X, {[OH], [BF₄], [NO₃], [ClO₄], [HSO₄], [C₁CO₂]}; [CₙC₁im]Br, (n = 2, 4, 6, 8, 10, 12)	IL-UAE 0.8 M [C₁₂C₁im]Br, pre-soaking 4 h, 186.69 W, irradiation 30.56 min, s/l ratio 1:12 [g/mL]	[87]

[a]*Botanical name*; (common name); plant part (if available)

[b]Cations: [CₙC₁im] (1-alkyl-3-methylimidazolium); [C₁=C₂)C₁im] (1-allyl-3-methylimidazolium); [ChC₁im] (1-cyclohexyl-3-methylimidazolium); [BzC₁im] (1-benzyl-3-methylimidazolium); [(HO)²C₂C₁im] (1-(2-hydroxyethyl)-3-methylimidazolium); [(HO₂C)C₁C₁im] (1-acetic-3-methylimidazolium); [(HO₃S)³C₃C₁im] (1-methyl-3-(3-sulfopropyl)-imidazolium); [(HO₃S)⁴C₄C₁im] (1-methyl-3-(4-sulfobuthyl)-imidazolium); [C₄C₁pyrr] (1-butyl-1-methylpyrrolidinium); [C₁₂betaine] (2-(dodecyloxy)-N,N,N-trimethyl-2-oxoethanaminium); [C₄pyr] (N-butylpyridinium) [(C₁)₄N] (N,N,N,N-tetramethylammonium); [(HO)²C₂(C₁)₃N] (cholinium); [(HO)²C₂(C₁)₂HN] (N,N-dimethylethanolammonium); [(C₁OC₂)₂H₂N] (bis(2-methoxyethyl)ammonium); [(NC)²C₂(C₁)₂HN] (2-cyano-N,N-dimethylethanaminium); [(HO)²C₂OC₂(C₁)₂HN] (2-(2-hydroxyethoxy)-N,N-dimethylethanaminium). Anions: Cl (chloride); Br (bromide); I (iodide); [BF₄] (tetrafluoroborate); [PF₆] (hexafluorophosphate); [NO₃] (nitrate); [ClO₄] (perchlorate); [OTs] (tosylate); [HSO₄] (hydrogen sulfate); [H₂PO₄] (dihydrogen phosphate); [Sac] (saccharinate); [Ace] (acesulfamate); [CₙCO₂] (alkyl carboxylate); [OTf] (trifluoromethanesulfonate); [NTf₂] (bis(trifluoromethanesulfonyl) imide or triflimide); [N(CN)₂] (dicyanamide); [SO₄] (sulfate); [CₙSO₃] (alkyl sulfonate); [CₙOSO₃] (alkyl sulfate); [OH] (hydroxide); [SbF₆] (hexafluorostibanuide); [SCN] (thiocyanate); [AlCl₄] (tetrachloroaluminate); [(CₙO)₂PO₂] (dialkyl phosphate); [(OH)¹C₂CO₂] (lactate)

[c]IL-MAE (microwave-assisted extraction); IL-UAE (ultrasound-assisted extraction); IL-UPE (ultrahigh pressure extraction); IL-NPCE (IL-based negative-pressure cavitation-assisted extraction); HRE (classical extraction under stirring at ambient conditions or elevated temperature); IL-UMAE (joint application of UAE and MAE)

gave higher yield than that obtained by Soxhlet extraction. Additionally, they found that the extraction efficiency toward paeonol was strongly dependent on the pretreatment time, temperature, and solid-liquid ratio. It was observed that temperatures higher than 70 °C and soaking times longer than 8 h result in significantly reduced extraction efficiency, this being attributed to the increase in viscosity due to increased cellulose dissolution. Thus, the use of pure ILs could be considered as a drawback if compounds other than cellulose are intended to be extracted. This shortcoming could be overcome if IL mixtures with water, organic solvents, other ILs, or molten salts are employed instead of pure ILs. Taking into account the latter, Bogdanov et al. [40] studied the extraction ability of a series of water solutions of 1-alkyl-3-methylimidazolium-based ILs toward the aporphine alkaloid glaucine from aerial parts of *Glaucium flavum* (yellow horned poppy). The authors demonstrated that at same conditions (80 °C, 1 h, s/l ratio 1:30), 0.5 M [C_4C_1im][Ace] possesses higher extraction efficiency than that obtained by HRE in pure water (*ca.* 50 % increased) or by IL-UAE with 30 min preliminary soaking (*ca.* 25 % increased). Further, it was shown that the extraction efficiency is highly dependent both on the solid-liquid ratio and on the concentration of the IL applied. However, at the optimized conditions (1 M [C_4C_1im][Ace], 80 °C, 1 h, s/l ratio 1:40), a quantitative extraction outcome (compared to Soxhlet extraction with methanol) was achieved. In a subsequent study, Bogdanov and Svinyarov [41] conducted a detailed kinetic analysis on the same system and showed that the total amount of the target alkaloid could be extracted in less than 60 min, regardless of the temperature (20 min at 80 °C and 40–60 min at 20 °C). The latter result suggests the immense advantage of the use of IL-aqueous mixtures over pure ILs, since it allows such extractive systems to be employed to recover value-added chemicals at mild conditions by means of maceration or percolation – technologies commonly used in the industry. In another study, Cláudio et al. [42] demonstrated that aqueous solution of [C_4C_1im]Cl is a suitable extractive solvent for batch extraction of caffeine from the seeds of *Paullinia cupana* (guaraná). At the optimized extraction conditions (2.34 M [C_4C_1im]Cl, 70 °C, 30 min, s/l ratio 1:10), the authors reported more than 50 % enhanced extraction efficiency of the IL-aqueous system, compared to the one obtained by Soxhlet extraction with dichloromethane, and tagged that the particle size plays an important role in achieving higher yields. Recently, Ressmann et al. [43] introduced a novel, cost-efficient, and high-yield extraction media, consisting of surface-active ILs, for the extraction of active ingredients from natural sources. It was found in this study that IL-aqueous micellar systems might be employed to extract piperine from the seeds of *Piper nigrum* (black pepper) at concentrations one order of magnitude lower than those normally employed in IL-assisted extractions and that the extraction efficiency is strongly dependent on the CMC of the respective ILs. Using this strategy, the target alkaloid piperine was extracted quantitatively for 3 h at room temperature by means of 50 mM [C_{12}betaine]Cl-aqueous solution. In another study, Ribeiro et al. [69] employed water solutions of cholinium-based ILs to extract polyphenols and saponins from aerial parts of *Camellia sinensis* (tea) and *Ilex paraguariensis* (mate) by simple orbital extraction. The authors observed that 30 % aqueous solution of [$(HO)^2C_2(C_1)_3$N]Cl (choline chloride) gives comparable extraction efficiency with that obtained with 30 % ethanol and that the extraction outcome is independent on the temperature, thus allowing the process to be conducted under more benign conditions. An interesting issue that deserves additional attention appears from this study. Namely, that the ammonium salt used possesses melting temperature higher than 100 °C, which means that it could not be considered as an ionic liquid by definition. Considering that any "ionic liquid" used in solution loses the unique properties characteristic of its pure state, the question arises whether the term "quaternary salts" is not more appropriate to be adopted in such cases.

The batch extraction with ILs seems to be the preferred extraction technique for the recovery of lipids from various sources. Young et al. [83] first reported the use of IL-molecular cosolvent systems to extract bio-oil from the biomass of *Dunaliella* spp. and *Chlorella* spp. microalgae and the seeds of *Jatropha* spp., *Brassica* spp. (rape), *Calophyllum inophyllum* (Alexandrian laurel), and *Millettia pinnata* (Indian beech). It was shown in this study that $[C_2C_1im][C_1OSO_3]$-methanol (1:1.2, w/w) mixture is able to extract lipids in satisfactory yields after 18 h at 65 °C. Moreover, it was found that the extracted lipids auto-partitioned to a separate immiscible phase, thus allowing easy harvesting. Considering this as an advantage, Severa et al. [79] employed the same IL-methanol cosolvent system in a subsequent study for simultaneous extraction and separation of phorbol esters and lipids from the seeds of *Jatropha* spp. Similarly, Kim et al. [82] used a system consisting of $[C_4C_1im][OTf]$ and methanol to extract lipids from both commercial and cultivated microalgae *Chlorella vulgaris* and showed that a broad range of fatty acids are successfully extracted. In another study, Choi et al. [80] used IL-IL mixtures to improve the lipid extraction yields from algae biomass. Particularly, it was found that the $[C_2C_1im][HSO_4]$-$[C_2C_1im][SCN]$ mixture (1:1, w/w) gives nearly sixfold higher extraction outcome compared to $[C_2C_1im][HSO_4]$ in pure form. A similar synergistic effect, leading to an increased extraction yields, was also documented by Choi et al. [81] in the extraction of lipids from *Chlorella vulgaris* by means of mixtures of ILs with molten salts, such as $Zn(NO_3)_2.6H_2O$, $Mg(ClO_4)_2.6H_2O$, and $FeCl_3.6H_2O$.

Summarizing, the batch extractions with ILs do not require special equipment and are normally performed at a wide temperature range 20–120 °C. This factor and the soaking time and the solid-liquid ratio depend both on the type of the plant matrix and the type of compounds to be extracted. The extraction times are rather broad, ranging from 20 min to 18 h, depending on the temperature applied and the compounds intended to be extracted. Both pure ILs and their mixtures with water, organic solvents, other ILs and molten salts could be successfully employed in this process.

7.3.1.2 Ultrasound-Assisted Extraction

Ultrasound-assisted extraction (UAE) shows certain advantages in comparison to other extraction methods, since it is easy to perform, does not require complicated equipment, and significantly reduces the extraction times and solvent consumption [88]. In addition, UAE provides more effective mixing and faster energy transfer and allows operation at ambient conditions, which can be advantageous if thermally unstable compounds are to be extracted. Ultrasonic waves facilitate solvent penetration and swelling of the plant material due to the formation and further collapse of gas bubbles into the bulk of a solvent. This phenomenon causes enlargement of the matrix pores and occasionally cell tissue disruption, thereby promoting easier convection of the compounds of interest.

As can be seen from Table 7.1, IL-UAE appears to be the preferred extraction technique in IL-assisted extraction of glycosides, but alkaloids [36–39] and phenolic compounds [53–55, 57–59] are also frequently recovered by this method. Zhang et al. [36] studied the extraction ability of $[C_4C_1im]Br$ toward two bisbenzylisoquinoline alkaloids, namely, fangchinoline and tetrandrine, from the roots of *Stephania tetrandra*. They found that at the optimized conditions (1.5 M $[C_4C_1im]Br$, pH = 9.8, 150 W, irradiation 40 min, s/l ratio 1:20), ultrasound can improve the extraction yield (30–50 % increased), compared to other reference methods such as UAE and HRE with ethanol. Piperine, the alkaloid responsible for the pungency of black pepper (*Piper nigrum*), was also efficiently extracted by IL-UAE by means of 2 M $[C_4C_1im][BF_4]$ for 30 min, but at a significantly higher irradiation power – 500 W [37]. In another study, Yang et al. [38] extracted the antimicrotubule drug vinblastine, together with its precursors vindoline and catharanthine, from the leaves of *Catharanthus roseus* by means of 0.5 M $[(C_1=C_2)C_1im]Br$. Despite the fact that $[(C_1=C_2)C_1im]Br$ has proved to be an effective solvent, preliminary soaking for 2 h was found necessary for sufficient extraction in this case. Moreover, it was reported that at least three extraction cycles had to be performed to achieve quantitative yields. The same tendency was observed by Ma et al. [39] in the extraction of the quinoline alkaloids camptothecin and 10-hydroxycamptothecin from the seeds of *Camptotheca acuminata* by means of 0.75 M $[C_8C_1im]Br$. It is noteworthy that strong temperature dependence of extraction efficiency on time and ultrasonic power was observed when susceptible to oxidation and thermal degradation phenolic compounds such as anthraquinones [54], acetophenones [61], procyanidins [59], and chlorogenic acid [55] were extracted from their natural sources by IL-UAE. By contrast, less sensitive to oxidation phenolic compounds such as lignans and flavonoids have been extracted without such problems from the bark of *Magnolia officinalis* [53] and *Apium graveolens* [57], respectively. The latter suggests that the influence of as much as possible factors should be taken into account in order to establish the optimal extraction conditions for a particular process.

IL-UAE has also proved to be an efficient method for the extraction of a variety of glycosides. Cao et al. [63] demonstrated that IL-UAE with 1.5 M $[C_4C_1im][BF_4]$ simultaneously reduces the extraction time from 6 h to 30 min and enhances the extraction yields by factor of 35–40 compared to the conventional HRE in ethanol in the extraction of iridoid, phenylethanoid, and cucurbitacin glycosides from the rhizome of *Picrorhiza scrophulariiflora*. The enhanced effect of IL-UAE is also reported by Fan et al. [68] in the extraction of benzodiazepine site antagonist puerarin (isoflavone 8-C-glucoside) from the roots of *Pueraria lobata*. In this study, the influence of sonication time and power on the extraction efficiency was found significant, since ultrasound treatment longer than 30 min and power higher than 480 W results in a decrease of extraction yields, most likely due to degradation of the target compound. The same time dependence on the extraction efficiency was observed by Yang et al. [65] in the simultaneous extraction of isoliquiritigenin (halcone), liquiritin (flavanone glycoside), and glycyrrhizic acid (saponin) from the roots of *Glycyrrhiza* spp. and by Lin et al. [70] in the extraction of group of saponins named ginsenosides from the roots of *Panax ginseng*. Another important factor – preliminary soaking – was studied by Yang et al. [66] in the extraction of glycosides from the roots of *Acanthopanax senticosus*. The authors showed that IL-UAE of eleutheroside B and E with 0.64 M [C_4C_1im]Br is *ca.* 55–60 % more effective in comparison with pure water or sodium chloride solution at the same extraction conditions (sonication for 30 min at 250 W) and is comparable in extraction yields with methanol HRE but for a fourfold reduced extraction time, from 2 h to 30 min. The extraction ability of the IL-based extractant used seems to be overestimated in this case, since 2 h preliminary soaking was found necessary in order for quantitative yields to be achieved by IL-UAE, so 2.5 h seem to describe the whole extraction process in a more proper way. The importance of the preliminary soaking was also documented by Yang et al. [64] when the extraction of aesculin (coumarin glucoside) and aesculetin (phenolic coumarin) from the bark of *Fraxinus rhynchophylla* was studied. In this case, soaking time of 4 h prior to 44 min sonication at 250 W was found optimal. It was also found that 90 % extraction efficiency could be reached if at least two successive extractions were performed. Nevertheless, the efficiency of 0.86 M [C_4C_1im]Br aqueous solution was shown to be higher than that provided by pure water, acetone, ethanol, and methanol at the same conditions. Recently, Zhu et al. [67] developed an IL-based online ultrasonic extraction combined with solid-phase trapping (IL-OUAE-SPT) to extract selectively the flavonoid glycosides rhodiosin and rhodionin from the roots of *Rhodiola rosea*. Briefly, the target compounds were first extracted under sonication at 25 °C from the plant material by continuous pumping of the IL-based extractive system through the extraction cell and then selectively trapped with polyamide resin. Following this procedure, higher extraction yields *ca.* 15–30 % were achieved for significantly reduced extraction times in comparison with the UAE with methanol or maceration with water. Moreover, this approach was shown to reduce the level of other polyphenols in the extracts, thus protecting the analytical equipment used for quantification.

Summarizing, the optimal extraction conditions commonly employed in IL-UAE are as follows: more usual frequencies applied are 20–60 kHz using sonication powers in the range 150–500 W and sonication times ranging from 20 to 90 min; in some studies, in order for higher extraction yield to be achieved, several extraction cycles (3–5 cycles) are performed. The latter together with the need for preliminary soaking (2–8 h) could be considered as a drawback for IL-UAE. IL-UAE is usually performed at ambient conditions, since high temperature may result in a degradation of thermally unstable compounds. The same phenomenon was observed when increased ultrasound frequency, power, and/or sonication time is applied.

7.3.1.3 Microwave-Assisted Extraction

Microwave-assisted extraction (MAE) is widely recognized as a green technology for extracting natural products from plants [89]. MAE is based on the absorption of microwave energy by polar molecules such as water, methanol, acetone, etc. When a microwave passes through these solvents, its energy is absorbed and converted into thermal energy, thereby ensuring simultaneous heating of the whole sample (including the plant matrix) in an efficient and homogeneous way. Furthermore, the increased temperature in the matrix causes superheating and liquid vaporization within the plant cells, which might result in cell walls and/or membranes disruption. As a consequence, the penetration of extracting solvent into the plant tissues and vice versa is greatly facilitated. ILs, because of their ionic structure, tunable properties, and negligible volatility, are considered by many authors as potent candidates for VOC alternatives in MAE. Indeed, IL-assisted MAE (IL-MAE) has been successfully applied to extract variety of classes of compounds such as alkaloids, phenolic compounds, essential oils, etc. (cf. Table 7.1).

Both 1.5 M $[C_4C_1im][BF_4]$ and 1 M $[C_6C_1im][BF_4]$ have been shown to be efficient extractants for the recovery of the phenolic alkaloids liensinine, isoliensinine, and neferine from the seeds of *Nelumbo nucifera* by Lu et al. [33]. At the optimized extraction conditions, IL-MAE results in 20–50 % higher extraction efficiency compared to the conventional HRE and MAE with 80 % methanol, for a significantly reduced extraction time, from 2 h (HRE) to 1.5 min (IL-MAE). However, in case of $[C_6C_1im][BF_4]$, a strong dependence between the solid-liquid ratio and extraction efficiency is observed. It was found that the efficiency increases when the ratio is changed from 1:5 to 1:10 and then dramatically decreases with further alteration to 1:20. An increase in viscosity due to increased chain length of the imidazolium cation has been proposed as an explanation of this phenomenon, but more reasonable explanation seems to be that the increased volume of the solvent lowers the extraction efficiency due to ineffective heating for the short irradiation time applied. Moreover, a slight decrease in yields was observed when a microwave power higher than 280 W was used. Interestingly, such dependence has not been observed by Ma et al. [34] when the aporphine alkaloids *N*-nornuciferine, *O*-nornuciferine, and nuciferine were extracted from the leaves of *Nelumbo nucifera* with 1 M $[C_6C_1im]Br$, despite that an increased efficiency up to 47 % was achieved for 2 min. The latter suggests that the matrix morphology and the IL type should always be considered during the extraction optimization. Other important factors, namely, preliminary soaking and number of extraction cycles, have been studied by Wang et al. [35]. They showed that for the effective extraction of the quinoline alkaloids camptothecin and 10-hydroxycamptothecin from samara of *Camptotheca acuminata*, at least 2 h preliminary soaking and 2 extraction cycles at 8 min irradiation are necessary for IL-MAE to give ca. 17 % higher extraction outcome compared to MAE with 80 % ethanol at the same conditions.

In the first article regarding IL-MAE of phenolic compounds [44], Du et al. showed that 2.5 M [C_4C_1im]Br performs better than methanol in the recovery of *trans*-resveratrol from the rhizome of *Fallopia japonica*. Under the optimized extraction conditions (60 °C, s/l ratio 1:20) ca. 93 % extraction efficiency was achieved in 10 min. The authors stressed that the sample size is of a great importance for better extraction of *trans*-resveratrol, since particles larger than 0.45 mm hinder the solvent penetration into them, which lowers the extraction yields. Similar dependence was also observed by Du et al. [45] in the IL-MAE of resveratrol and quercetin from the tubers of *Smilax china* and by Liu et al. [48] in the extraction of quercetin and kaempferol from *Toona sinensis* and *Rosa chinensis*. Moreover, a strong temperature dependence on the extraction, leading to a decrease in yields when the temperature exceeded 60–70 °C, was documented in those studies. On the contrary, such phenomenon was not observed in the extraction of quercetin and other phenols from the leave samples of *Psidium guajava* [45] and *Myrica rubra* [46]. As in the case of alkaloids, the extraction of paeonol from the roots of *Cynanchum paniculatum* was found to be strongly dependent both on the solid-liquid ratio and particle size by Jin et al. [49]. In this case pure [C_4C_1im]Cl was used as an extractant, and the aforementioned dependence was attributed to the ability of this IL to dissolve cellulose from the matrix, which might affect negatively the mass transfer due to increase in viscosity. IL-MAE was used as a sample preparation technique for the extraction and determination of nine bioactive flavones, isoflavones, stilbenes, and coumarins from the leaves of *Cajanus cajan* by Wei et al. [51]. It was found that 0.75 M [C_4C_1im]Br is able to extract quantitatively the active compounds for 7 min at the optimized conditions (60 °C, 300 W, s/l ratio 1:12). Similar optimal extraction conditions (50 °C, 600 W, s/l ratio 1:12) were reported appropriate for the extraction of bioactive phenolic compounds named phloroglucinols from aerial parts of *Dryopteris fragrans* [50]. It was found that the relative extraction efficiency of IL-MAE with 0.75 M [C_8C_1im]Br is higher than that achieved by IL-UAE and IL-NPCE, as well as by HRE and MAE with 80 % ethanol. IL-MAE also proved an appropriate method for the effective extraction of water-insoluble phenolic compounds named depsidones from crustose lichen *Pertusaria pseudocorallina* by Bonny et al. [52]. It was found that [C_1C_1im][C_1OSO_3] performed 1.5-fold better in IL-MAE than in IL-HRE. Considering these results, IL-MAE with [C_1C_1im][C_1OSO_3] was further successfully employed in the extraction of depsides, depsones, and depsidones from other lichen, e.g., *Pertusaria amara* and *Ochrolechia parella*, respectively.

IL-MAE appears to be the preferred technique for the extraction of essential oils (EO). Zhai et al. [73] used pure [C_4C_1im][PF_6] as water substituent in IL-MAE to extract essential oils from two commonly used spices in cooking, namely *Illicium verum* (star anise) and *Cuminum cyminum* (cumin). It was found in this study that the IL is able to absorb microwave energy more readily than water, which allows the appropriate extraction temperature to be reached nearly three times faster. Thus, at the optimal conditions, IL-MAE ensures considerably shortened extraction time (15 min) in comparison with the conventional hydrodistillation, which needs 180 min to extract the essential oils completely. Moreover, it was found that the use of IL-MAE reduces the oxidation and hydrolyzation of the essential oil constituents. A similar approach has been used by Jiao et al. [77] to extract EO from the seeds of *Forsythia suspensa*, but in this case, the hydrodistillation was performed after subsequent addition of extra water into the IL when IL-MAE is completed. The authors studied the extraction kinetics to show that the use of 76 % $[C_2C_1im][C_1CO_2]$ results in *ca.* 64 % enhanced yield compared to 0.91 M $[C_2C_1im][C_1CO_2]$ aqueous solution. IL-MAE has been also applied for the simultaneous extraction and distillation of EO from the cortex of *Cinnamomum* spp. [76] and leaves of *Rosmarinus officinalis* [74]. It is noteworthy that microwave irradiation applied in the above studies strongly accelerates the extraction process, but without causing significant changes in the composition of EO. Similarly, 0.25 M $[C_{12}C_1im]Br$ was found to be the most suitable IL for the simultaneous extraction of EO and biphenyl cyclooctene lignans from the fruits of *Schisandra chinensis* [75]. In this study, the preliminary soaking for 4 h and subsequent irradiation for 40 min at 385 W were found necessary.

Summarizing, IL-MAE has been widely applied to extract active ingredients from plants, especially in case of thermally unstable analytes. Compared to the conventional extraction methods, IL-MAE provides higher extraction rates and yields and is less solvent and energy consuming. The optimal extraction conditions employed in IL-MAE depends on the plant material morphology and IL structure and could be systematized according to Table 7.1 as follows: microwave power, 280–700 W; temperature, 50–105 °C; irradiation times, 1.5–40 min; and solid-liquid ratio, 1:7–1:40. In some cases, a higher solvent volume results in lower extraction efficiency and preliminary soaking followed by several extraction cycles have to be performed.

7.3.1.4 Other Extraction Techniques

Considering the benefits of IL-UAE and IL-MAE in the extraction of secondary metabolites from plant materials, a combined ionic liquid-based ultrasonic/microwave-assisted extraction (IL-UMAE) methodology was developed by Lu et al. [54] to extract bioactive anthraquinones from *Rheum* spp. (rhubarb). A comparative analysis toward other methods such as HRE, UAE, and MAE showed that at the optimized conditions (500 W, solid liquid ratio 1:15), IL-UMAE exhibits higher extraction efficiency (19–24 % enhanced) for the remarkably short extraction time of 2 min.

An alternative method that provides higher extraction yields at reduced processing time, energy, and solvent consumption, namely, IL-based ultrahigh pressure extraction (IL-UPE), has been employed by Liu et al. [72] to extract hydrophobic bioactive compounds named tanshinones from *Salvia miltiorrhiza* with 0.5 M $[C_8C_1im][PF_6]$ in ethanol. In this procedure, extractant and solid material, with or without packing, are subjected to pressures between 100 and 800 MPa for a given period. During the optimization of the process parameters, it was found that the extraction outcome is dependent on the pressure applied, reaching a maximum at 300 mPa and then decreasing *ca.* 25 % when 500 MPa is reached. However, at the optimized conditions, IL-UPE was found to give similar extraction yield compared to IL-UAE, but for a significantly reduced extraction time of 2 min. The effectiveness of IL-UPE is attributed to the cell tissue disruption, which allows the compounds of interest to be easily washed away from the plant matrix.

In recent years, a new extraction method called negative-pressure cavitation extraction (NPCE) has gained increased popularity. Compared to classical extraction methods such as maceration, HRE, and UAE, it possesses enhanced extraction ability and has been successfully applied to recover diverse bioactive compounds from plant materials [90]. NPCE is a cheap and energy-efficient method that can keep a constant low temperature. It also ensures inert atmosphere during the extraction, since nitrogen is continuously introduced into the liquid-solid system, in order to increase the turbulence, collision, and mass transfer between the solvent and matrix. The latter suggests NCPE as an appropriate method for recovery of thermosensitive and susceptible to oxidation compounds.

IL-assisted NPCE (IL-NPCE) with [C_8C_1im]Br was recently reported by Duan et al. [60] as an efficient method for the extraction of three main flavonoids – genistin, genistein, and apigenin – from the roots of *Cajanus cajan* (pigeon pea). The process was initially performed in a lab-scale device and after optimization of the extraction parameters was shown to give higher extraction yields than IL-UAE and, similar with these, obtained by IL-HRE. In addition, IL-NPCE proved to be less solvent and time consuming and was further transferred to scale-up experiments to extract 500 g of the plant material. A comparison between the lab-scale and pilot-scale experiments showed that same extraction yields were obtained in 15 min in both cases, thus suggesting IL-NPCE as an appropriate method for industrial application.

7.3.2 Effect of Ionic Liquids on the Extraction Efficiency

Generally speaking, ILs proved to be better extractants than common molecular solvents (e.g., water, methanol, ethanol, dichloromethane, chloroform, toluene, etc.), and this was clearly demonstrated by many comparative experiments. It is a well-known fact that properties such as polarity, viscosity, density, and surface tension for a series of ILs based on the same cation are strongly dependent on the anion type, and therefore, at the very beginning of many studies, a comparative analysis of the extraction ability of different anions, coupled mainly with [C_4C_1im]$^+$ as cation, had been performed (*cf.* Table 7.1). A wide range of anions differing in their complexity and possibility for non-covalent interactions have been assessed. The results obtained from these experiments show that, with some exceptions, Br$^-$ anion appears to be the most preferred among the others. Nevertheless, anions such as Cl$^-$, [BF_4]$^-$, [PF_6]$^-$, [OTs]$^-$, and [C_1CO_2]$^-$ were also proved successful.

In particular cases, $[C_1CO_2]^-$ and $[OTs]^-$ anions showed comparable or better extraction ability than Cl^- and Br^-, but the latter were chosen due to better total extraction efficiency, in case of similar compounds are to be extracted, or due to economical reasons. Anions with well-established toxicological profiles, namely, saccharinate $\{[Sac]^-\}$ and acesulfamate $\{[Ace]^-\}$, were also employed, showing better performance than Cl^- and Br^-. Considering the properties of the anions that had been mainly selected, it could be concluded that the hydrogen-bonding ability [91] is the main factor that influences the extraction outcome, but a synergistic effect with π-π and n-π interactions, offered by some aromatic ring containing anions, could also be assumed. Exceptions are $[BF_4]^-$ and $[PF_6]^-$ anions, but the results obtained with ILs containing these ions should be accepted with attention, since, as was discussed in Section 2, they tend to hydrolyze, especially at higher temperature, and to form hydrogen fluoride [31], which for sure decreases the initial pH value of the extractive system.

Having an appropriate anion selected, the next step is the effect of the cation to be assessed, this being achieved by performing comparative extractions with ILs based on the same anion. The great importance of the "organic" nature of the cation could be rationalized by a comparison of results obtained at same conditions for solutions of both organic and inorganic salts based on the same anion. Furthermore, it is noteworthy that the extraction outcomes were found sometimes highly dependent on the cation type and sometimes not, and this clearly shows that the specific interactions between the IL species and solutes of interest have to be taken into account. Particularly, $[C_4C_1im]^+$ was proved superior to the rest toward extraction of phenolic acids [45], and considering the efficiency order found in this case $\{[C_4C_1im]^+ \geq [C_4pyr]^+ > [(C_1)_4N]\}$, the author concluded that the two electron-rich aromatic cations solvate the phenolics in a more efficient manner via π-π and n-π interactions, thus ensuring better solubilization. In contrast, a comparison between the extraction yields of caffeine obtained with $[C_4C_1im]Cl$ and $[C_4C_1pyrr]Cl$ did not show significant differences, and so the influence of the aromatic cation in this case seems negligible [42]. The length of the side alkyl chain in the cation was also found of a significant importance, since it influences the IL properties. Considering the results summarized in Table 7.1, it could be concluded that the $[C_4C_1im]^+$ cation manifested as the most promising candidate when an extended set of cations were evaluated, but nevertheless, in some cases cations with longer alkyl substituents were found more appropriate.

Another influential factor, namely, the IL concentration, is of an immense importance from an applied standpoint. For instance, the use of ILs in an insufficient concentration could result in an incomplete extraction, whereas the use of extra quantities of the ILs will increase in a meaningless manner the overall process cost. Generally speaking, the higher the IL concentration, the higher extraction outcomes obtained, but nevertheless, many studies reported an initial increase in extraction yields with the IL concentration, followed by a decrease after reaching a maximum. The latter was attributed to the increase in viscosity, caused either by the IL itself or by dissolution of additional compounds, particularly carbohydrates, from the plant matrix. Nevertheless, because IL physicochemical properties are greatly affected by the presence of cosolvents [20], variations in IL concentration could be used for fine-tuning of the extractive systems, thus allowing a selective extraction to be achieved.

Summarizing, ILs have proved to possess a strong dissolving power due to the distinct multiple interactions provided by the ions. The anion influence on the extraction efficiency was found pronounced in comparison with that of the cation, and the relative order of hydrogen-bonding ability could be employed to explain these results. For the same anion, the cation type was found significant, giving advantage to the electron-rich aromatic cations. The length of the side alkyl chain and the IL concentration are also important, since their variation could change the extractive systems properties. Based on the above, it could be concluded that both cation and anion influence the extraction efficiency mainly due to the enhanced non-covalent interactions, e.g., hydrogen-bonding and π-π and n-π interactions, with the solutes of interest.

7.3.3 Extraction Mechanism

Du et al. [45] were the first who tried to put some light onto the mechanism of solid-liquid IL-MAE by measuring the kinetics of extraction of phenolic compounds by means of $[C_4C_1im]Br$. The solutes studied were gallic acid, ellagic acid, and quercetin extracted from the leaves of *Psidium guajava* and also *trans*-resveratrol and quercetin from the tubers of *Smilax china*. The kinetic curves obtained were found similar for all samples, showing that the extraction yields of the extracted compounds increase rapidly at first and then reach an equilibrium. Further, it was observed that the equilibrium concentration of the solutes extracted from *Smilax china* tubers is attained in 6 min, which differs from the results obtained for the *Psidium guajava* leave sample, where the three solutes were extracted in 11, 10, and 6.5 min, respectively. On the one hand, this indicates that the matrix morphology has influence on the extraction and, on the other, suggests that the solute structure is also important. Additional elucidation of the extraction mechanism was performed by scanning electron microscopy (SEM) and infrared spectroscopy (FTIR). The microstructure of the leave and tuber samples studied by SEM was found obviously modified after IL-MAE and not in HRE, which suggests that the extraction efficiency in IL-MAE could be attributed to the microwave ability to cause cell explosion and thus to facilitate the solute release into the extract. Further, the FTIR analysis performed showed no changes in the sample chemical structure before and after IL-MAE, suggesting that the plant tissues were not destroyed by the IL. Based on the results obtained, the authors concluded that the IL-MAE mechanism is not related to the sample characteristics.

Similar influence of microwave irradiation on the sample microstructure was observed by Du et al. [46] during IL-MAE of myricetin and quercetin from the leaves of *Myrica rubra* by means of the acidic IL $[C_4C_1im][HSO_4]$ or acidified $[C_4C_1im]Br$ and not in case of MAE with acidified ethanol. The latter allowed a conclusion to be made that the solvent could influence the extraction mechanism, but again without causing changes of the matrix chemical structure. The same conclusion, based on SEM and FTIR analyses, has been also drawn independently by Liu et al. [74] and Ma et al. [75] after examination of the simultaneous extraction of phenolic compounds and essential oils from the leaves of *Rosmarinus officinalis* and fruits of *Schisandra chinensis*, respectively.

In other studies, Liu et al. [47, 48] demonstrated that the surface and cell wall structures of *Toona sinensis* and *Rosa chinensis* samples were visibly destroyed both after IL-MAE and IL-UAE of quercetin and kaempferol with $[C_4C_1im]Br$ or $[C_8C_1im]Br$ aqueous solutions and concluded that the mechanism of the two methods employed is based on cell destruction caused either by sudden rise in temperature by the microwave irradiation or by mechanical vibrations by ultrasound waves. The role of the temperature and microwaves as the main factors affecting the extraction process was also denoted by Yansheng et al. [86] in the protic IL-assisted MAE of biologically active lactones from *Ligusticum chuanxiong*. It was found in this study that the extraction efficiency of IL-MAE is much higher than that of the standard extraction under the same temperature, thus suggesting a synergistic reaction of breaking and heating effect of the microwaves. This was further proved by means of SEM analysis of plant samples after IL-assisted and conventional solvent MAE. The results obtained showed the same plant tissue disruptions in both cases, so a conclusion that the mechanism of IL-MAE is similar to the conventional organic solvent MAE has been made.

A more in-depth study on the mechanism of solid-liquid extraction with ILs has been published recently by Bogdanov and Svinyarov [41]. Based on a measured kinetic data for batch extraction of the aporphine alkaloid glaucine from the leaves of *Glaucium flavum* by means of 1 M $[C_4C_1im][Ace]$, the authors assessed the temperature dependence of the extraction and performed a detailed comparative analysis when IL-aqueous solution and methanol were used as extractants. The rate of extraction was found to increase sharply at the beginning of the operation and to become slower and slower approaching the saturation level for each temperature studied, which is a typical behavior for second-order processes [92–94]. It is noteworthy that this phenomenon was found more pronounced for the molecular solvent than for the IL solution, which reached the saturation level faster. The results obtained showed a great advantage of the IL-assisted extraction over the molecular solvent, since the glaucine was not only extracted quantitatively with the IL, but this was achieved for a considerably shortened soaking time, regardless of the temperature applied. To rationalize these results, the authors proposed a detailed mechanism describing the whole process in terms of solute-solvent, solute-matrix, and matrix-solvent interactions at every stage of the extraction process, and as a result, the apparent asset of the IL-assisted extraction was attributed to the ability of the IL, despite being in an aqueous solution, not only to solubilize better the solute of interest but to disrupt the cell tissues to some extent and to modify the matrix permeability by H bonding with the carbohydrate building blocks forming the cell walls. As in the case of IL-MAE, the aforementioned deduction was further

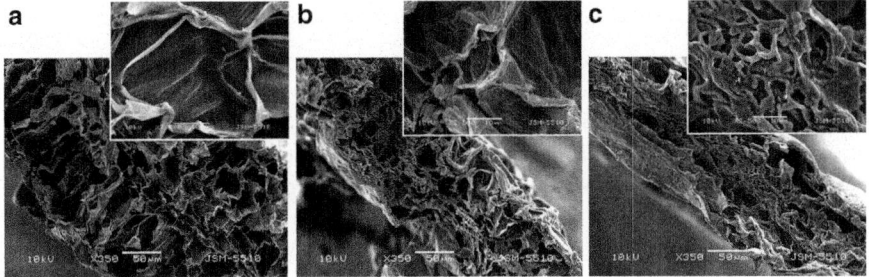

Fig. 7.3 Representative scanning electron micrographs (350× and 2,500×) of *G. flavum* leaf samples extracted with (**a**) water, (**b**) methanol, and (**c**) 1 M [C_4C_1im][Ace]

proved by SEM examination of the cross section of leave samples extracted with different solvents. The results obtained (*cf.* Fig. 7.3) clearly showed significant physical changes of both internal and bulk structure of the plant matrix after IL-assisted extraction, and not in case of water or methanol, thus proving that the ILs are able to interact with the cellulose even being dissolved in water. This was further proved by Cláudio et al. [42] who studied the batch extraction of caffeine from the seeds of *Paullinia cupana* in IL-aqueous media. The performed SEM analysis after 30 min extraction showed again that compared to pure water, the seeds' cells are broken in a higher extent in the presence of 2.34 M [C_4C_1im]Cl-aqueous solution, which significantly improves the extraction yield of caffeine in the latter. Moreover, based on FTIR and TGA analysis of the biomass prior to and after extraction, the authors concluded that although the pure ILs were shown to be good candidates to dissolve biopolymers, particularly carbohydrates, the presence of water prevents this in a great extent, thus allowing selective extraction of secondary metabolites from plant materials. Similar effect of the IL-aqueous solution on the plant material microstructure was also reported by Rasmmann et al. [43] in the solid-liquid extraction of piperine from *Piper nigrum* by means of surface-active ILs. Noticeably, the changes in the biomass morphology in this case, which could not be observed when pure water is used as extractant, were caused by ILs in 50 mM concentration.

Based on the above reasoning, a conclusion could be made that the role of ILs in the solid-liquid extraction processes from plants is not limited only to the enhanced solute-solvent interactions, which means an increased solubility of the compounds of interest, but it could also be attributed to the pronounced solvent-matrix interactions leading to a plant matrix permeability modification. In sum, due to the unique property of ILs to interact with carbohydrates via hydrogen bonding, they could be considered as worthy substituents of molecular solvents in the extraction of value-added chemicals by classical extraction methods, while the use of MAE, UAE, UPE, or other techniques that are known to cause cell tissue disruption will contribute additionally to the facilitation of the process as a whole.

7.3.4 Ionic Liquid Regeneration and Solute Recovery

Despite ILs being proved as excellent extractants, they are still expensive in comparison to the conventional molecular solvents commonly used in the production of natural products. Therefore, the efficient recycling of ILs is an important issue that addresses the economics of their use. Taking this into account, it is noteworthy that the purification of ILs after solid-liquid extraction is studied to a very limited extent. The following section considers the recent achievements regarding this issue, as well as the possible ways for the solute recovery.

In the pioneering work on the solid-liquid extraction by means of ILs, Lapkin et al. [84] demonstrated that artemisinin – a compound of pharmaceutical interest – could be recovered successfully from protic IL by addition of an anti-solvent. Artemisinin could be considered as a neutral compound, so the addition of water to the crude IL extract, in a ratio of 3:1 (v/v) with respect to the IL, causes simultaneous separation of an oil fraction and precipitation of the artemisinin at ambient conditions. The latter allowed the target compound to be isolated for 10 min in 82 % yield and in a high purity (>95 %, NMR), and the loss of 18 % artemisinin in this case was attributed to its partial solubility into the oil phase formed. It is noteworthy that the examination of the possible removal of the accumulated nonvolatile impurities from the ILs has not been performed in this particular study, which is of a great value if the long-term stability of the solvent and the impact on the process economics have to be assessed. The importance of the latter could be rationalized by considering the results obtained by Yansheng et al. [86], who studied the extraction of bioactive lactones from *Ligusticum chuanxiong* by means of the protic IL $[(HO^2)C_2OC_2(C_1)_2HN][C_2CO_2]$. It was found that the extraction efficiency toward the target compounds decreases in a significant manner after two successive extractions, and this was attributed to the ability of the pure IL to dissolve and accumulate other compounds from the plant matrix. The latter was proved by the observation that the IL viscosity increases gradually with the extraction time, which hinders the mass transfer and thus poses difficulties if successive extractions are intended to be performed. Consequently, the use of pure ILs increases the process demands due to the need of IL purification after every single extraction step. Considering this, the authors further tried to purify the IL and to recover the compounds of interest by re-extraction with *n*-hexane. The results obtained were not satisfactory, so a conclusion has been made that in order for successful back-extraction to be performed, a more specific solvent should be selected.

A success in the direction of neutral solute back-extraction and IL purification was recently reported by Ressmann et al. [43], who achieved *ca.* 95 % recovery of piperine from 50 mM [C_{12}betaine]Cl-aqueous extract by performing a single back-extraction with butyl acetate in a ratio of 1:4 (v/v) with respect to the former. After the removal of the organic solvent, the neutral alkaloid has been isolated quantitatively in an excellent purity, confirmed by ^{1}H NMR analysis. The above procedure allows not only the quick and clean solute recovery but ensures the IL-based extractant recycling at the same time. This way, the residual IL-aqueous solution could be directly used in a subsequent extraction of fresh plant material. Indeed, five successive runs without any loss of performance were further carried out. Another important factor that influences the economics and safety of the solid-liquid extraction, namely, the loss of IL due to its absorption on the processed plant material, was recently addressed by Yan et al. [71], who reported an increasing IL loss with increasing size of the particles extracted. However, after the removal of the water under vacuum, the author recycled the residual acidic IL [$(HO_3S^3C_3)C_1im$][HSO_4] from the water-soluble impurities by adding absolute ethanol and subsequent filtration. Similarly, [C_2C_1im][C_1CO_2], used for the extraction of essential oil from the fruits of *Forsythia suspensa* by Jiao et al. [77], was successfully recovered by the addition of ethanol to the IL-aqueous solution (after essential oil removal by hydrodistillation) and subsequent azeotropic distillation of water. This approach proved successful, and the recovered IL was reused in five repetitive extraction cycles, showing the same extraction efficiency.

Unlike the neutral compounds discussed above, most alkaloids are basic in nature and exist in the form of salts in the plant matrix. Thus, they could be supposed to be highly soluble both in ILs and water and therefore to be hardly recovered by distillation or by anti-solvent-induced precipitation from the IL-aqueous solutions. A possible way of alkaloids recovery, and so IL purification, was proposed by Bogdanov et al. [40], who recovered the aporphine alkaloids glaucine and cataline from [C_4C_1im][Ace]-aqueous solution by means of back-extraction with ethyl acetate. After this procedure, the IL was recovered quantitatively and in a high purity (HPLC and ^{1}H NMR), and the subsequent water removal under reduced pressure was shown to allow a successful repetitive extraction of fresh plant material to be conducted with the same IL. Similarly, Cláudio et al. [42] found that dichloromethane, chloroform, and *n*-butanol are good candidates for back-extraction of caffeine from crude [C_4C_1im]Cl-aqueous extract. Despite that the two chlorinated solvents were found to ensure better partitioning, and thus higher re-extraction yields than the alcohol used, the latter was selected for the caffeine recovery due to its lower volatility and toxicity. Furthermore, the authors showed that the same IL-based extractant could be reused at least in three successive extractions without loss of efficiency, which allows the solute of interest to be accumulated into the extract. The latter is of a great importance from a practical standpoint, since every successive extraction cycle will decrease the overall cost of the IL employed.

A combined strategy for the simultaneous solute recovery and IL recycling was developed by Bonny et al. [52] for the isolation of phenolic compounds, such as variolaric acid, alpha-alectoronic acid, and lecanoric acid from [C_1C_1im][C_1OSO_3]-aqueous solution. The approach consists of a preliminary extraction with diethyl ether of the filtered crude IL-based extract, followed by addition of aqueous acetone as anti-solvent. This way, the target phenolics were obtained as precipitates and the extraction yields were then evaluated by means of HPTLC analysis.

Another approach, based on the ability of IL to form aqueous two-phase systems (ATPSs) was also employed for phenolic compounds recovery. Tan et al. [95] studied the partition of phenolic anthraquinones in a [C_4C_1im][BF_4]/Na_2SO_4 ATPS.

The observed strong pH dependence in this case allowed the target compounds to be first preferably partitioned into the IL-rich phase at pH *ca.* 4 and then re-extracted into a fresh salt-rich phase after adjusting the pH value *ca.* 14. Following this procedure, aloe-emodin and chrysophanol were recovered in 92 and 91 % yields, respectively. Similarly, the pH dependence on the partition of natural phenolic acids in $[C_4C_1im][OTf]$-based ATPS was used as a platform for the successful recovery of gallic, vanillic, and syringic acids by Cláudio et al. [96]. It was found that the charged acidic species preferentially partition into the salt-rich phase, whereas their neutral forms tend to partition into the IL-rich phase. Thus, Na_2SO_4 was initially used as a kosmotropic salt to induce the formation of IL-based ATPS and so to concentrate the solutes of interest into the IL-rich phase. The latter had been separated and reused to form a new ATPS with Na_2CO_3 (pH *ca.* 11), and thus simultaneous back-extraction with excellent yields and IL recovery were achieved. Using similar strategy, Ribeiro et al. [69] recovered polyphenols and saponins from $[(HO)^2C_2(C_1)_3 N]Cl$-aqueous extract. In this case, K_3PO_4 was found appropriate to induce the ATPS formation, thus ensuring distribution of the solutes of interest into the IL-rich phase. The back-extraction from the latter was achieved by addition of a hydrophobic IL, so an aqueous phase concentrated of saponins and phenols has been finally obtained.

An elegant approach for the simultaneous recovery of shikimic acid and IL recycling was recently reported by Zirbs et al. [85]. After precipitation of the biopolymers dissolved into the pure IL by the addition of water as an anti-solvent, the authors employed an ion exchange resin in acetate form to separate the dissolved shikimic acid from the acetate-based IL $[C_2C_1im][C_1CO_2]$. This way, the purified $[C_2C_1im][C_1CO_2]$ is obtained as an aqueous solution, which, after water removal, could be directly used in a subsequent extraction. Shikimic acid was further isolated in a high purity via elution with 10 % acetic acid, leaving the resin ready for a consecutive separation. Similarly, Zhu et al. [67] used polyamide resin for selective trapping of the glycoside flavonoids rhodiosin and rhodionin from crude IL extract.

Summarizing, several approaches considering both the ILs and solutes of interest properties have been successfully applied for their separation and purification. The methods employed for neutral compounds recovery include anti-solvent-induced precipitation, back-extraction with organic solvents, or hydrodistillation of volatile compounds. Back-extraction with organic solvents appears to be the only successful way for recovery of basic compounds such as alkaloids and IL-based ATPS proved to be a successful method for the recovery of acidic compounds. In some cases, ion exchange resin or resin for selective trapping gave satisfactory results. Noticeably, the employment of IL-aqueous solutions should be preferred in the case when secondary metabolites are to be extracted, since the use of pure ILs increases the level of accumulated impurities into the extract, thus increasing the process demands. In conclusion, although some achievements have been done in the direction of IL recycling and solute recovery, additional work is necessary in order for more conclusive generalization to be made.

7.4 Conclusion

ILs, mainly due to the versatility of possible ion combinations and fine-tunable unique physicochemical properties, can be considered as potential substituents of the volatile, flammable, and toxic organic solvents commonly employed in the solid-liquid extraction processes. Indeed, ILs have proved to be efficient extractants in the recovery of a wide variety of value-added chemicals, such as alkaloids, polyphenolic compounds, saponins, lipids, essential oils, etc.

A range of techniques, which differ in their cost and complexity, can be employed in the IL-assisted extractions from natural sources. The most studied methods include classical batch extraction (HRE), microwave-assisted extraction (IL-MAE), ultrasound-assisted extraction (IL-UAE), IL-assisted ultrahigh pressure extraction (IL-UPE), and IL-based negative-pressure cavitation-assisted extraction (IL-NPCE). All these methods possess specific requirements and certain advantages compared to the rest in the recovery of a particular group of compounds. For instance, UAE appears to be the preferred technique in IL-assisted extraction of glycosides and IL-MAE for recovery of essential oils and phenolic compounds. Regardless of the technique employed, ILs have proved to be better extractants than common molecular solvents, and IL-assisted extractions have been shown more effective and economical in comparison to organic solvents at the same conditions.

Both pure ILs and their mixtures with water, organic solvents, other ILs, and molten salts could be successfully employed in the extraction of natural products. In all cases, IL-assisted extractions significantly reduce the extraction times and solvent consumption, and this could be attributed to the stronger dissolving power of ILs, due to the distinct multiple interactions provided by the ions. Both cation and anion influence the extraction efficiency mainly due to the enhanced non-covalent solute-solvent interactions provided, but the role of ILs in the solid-liquid extraction processes from plants is not limited only to this, but it can also be attributed to the pronounced solvent-matrix interactions leading to a plant matrix disruption and permeability modification.

Despite that ILs have proved to be excellent extractants, they are still expensive in comparison to the conventional molecular solvents commonly used in the manufacturing of natural products. Therefore, the efficient recycling of ILs is an important issue that addresses the economics of their use. It is noteworthy that the purification of ILs after solid-liquid extraction is studied to a very limited extent. Nevertheless, several approaches considering both the ILs and solutes of interest properties have been successfully applied for their separation and purification. The methods employed include anti-solvent-induced precipitation, back-extraction with organic solvents, hydrodistillation of volatile compounds, and partitioning in IL-based ATPS, and in some cases, ion exchange resin or resin for selective trapping has given satisfactory results. Noticeably, the employment of IL-aqueous solutions should be preferred in the case when secondary metabolites are to be extracted, since the use of pure ILs increases the level of accumulated impurities into the extract, thus increasing the process demands.

In conclusion, the knowledge for the IL-based extractive systems gained to date represents a promising basis for the future development of novel and improved methodologies for the recovery of useful compounds from natural sources by means of solid-liquid extraction. IL-assisted solid-liquid extractions have been successfully employed mainly on a laboratory scale, and so, the need for a transfer into an industrial scale processes could be put forward. To this end, more efforts toward the development of equipment that meets the specific requirements of IL-assisted extractions and the assessment of the overall process costs are necessary. Finally, an easily available and inexpensive novel ILs with improved properties from an environmental standpoint [97] and efficient procedures for IL recovery and recycling are also crucial factors, which will definitely decide the fate of these innovative extractive systems.

Acknowledgments The author would like to dedicate this work to Professor Willi Kantlehner, on the occasion of his 70th birthday, with gratitude for his guidance into the field of ionic liquids.

References

1. Cragg GM, Newman DJ (2013) Natural products: a continuing source of novel drug leads. Biochim Biophys Acta 1830:3670–3695
2. Brusotti G, Cesari I, Dentamaro A et al (2014) Isolation and characterization of bioactive compounds from plant resources: the role of analysis in the ethnopharmacological approach. J Pharm Biomed Anal 87:218–228
3. Azmir J, Zaidul ISM, Rahman MM et al (2013) Techniques for extraction of bioactive compounds from plant materials: a review. J Food Eng 117:426–436
4. Bucar F, Wube A, Schmid M (2013) Natural product isolation – how to get from biological material to pure compounds. Nat Prod Rep 30:525–545
5. Anastas PT, Kirchhoff MM (2002) Origins, current status, and future challenges of green chemistry. Acc Chem Res 35:686–694
6. Ventura SPM, e Silva FA, Gonçalves AMM et al (2014) Ecotoxicity analysis of cholinium-based ionic liquids to *Vibrio fischeri* marine bacteria. Ecotoxicol Environ Saf 102:48–54
7. Freemantle M (2009) An introduction to ionic liquids. RSC Publishing, Cambridge
8. Bogdanov MG, Kantlehner W (2009) Simple prediction of some physical properties of ionic liquids: the residual volume approach. Z Naturforsch B 64:215–222
9. Bogdanov MG, Iliev B, Kantlehner W (2009) The residual volume approach II: simple prediction of ionic conductivity of ionic liquids. Z Naturforsch B 64:756–764
10. Bogdanov MG, Petkova D, Hristeva S et al (2010) New guanidinium-based room-temperature ionic liquids. Substituent and anion effect on density and solubility in water. Z Naturforsch B 65:37–48
11. Hallett JP, Welton T (2011) Room-temperature ionic liquids: solvents for synthesis and catalysis. 2. Chem Rev 111:3508–3576
12. Parvulescu VI, Hardacre C (2007) Catalysis in ionic liquids. Chem Rev 107:2615–2665
13. Opallo M, Lesniewski A (2011) A review on electrodes modified with ionic liquids. J Electroanal Chem 656:2–16
14. Poole CF, Poole SK (2011) Ionic liquid stationary phases for gas chromatography. J Sep Sci 34:888–900
15. Ho TD, Zhang C, Hantao LW et al (2014) Ionic liquids in analytical chemistry: fundamentals, advances, and perspectives. Anal Chem 86:262–285

16. Reichardt C, Welton T (2011) Solvents and solvent effects in organic chemistry, 4th edn. Wiley, Weinheim
17. Reichardt C (2005) Polarity of ionic liquids determined empirically by means of solvatochromic pyridinium N-phenolate betaine dyes. Green Chem 7:339–351
18. Jessop PG, Jessop DA, Fu D et al (2012) Solvatochromic parameters for solvents of interest in green chemistry. Green Chem 14:1245–1259
19. Meindersma GW, Maase M, Haan ABD (2007) Ionic liquids. In: Ullmann's encyclopedia of industrial chemistry. Wiley-VCH, GmbH & Co. KGaA, Weinheim
20. Jacquemin J, Husson P, Padua AAH et al (2006) Density and viscosity of several pure and water-saturated ionic liquids. Green Chem 8:172–180
21. Tariq M, Freire MG, Saramago B et al (2012) Surface tension of ionic liquids and ionic liquid solutions. Chem Soc Rev 41:829–868
22. Dupont J (2004) On the solid, liquid and solution structural organization of imidazolium ionic liquids. J Braz Chem Soc 15:341–350
23. Greaves TL, Drummond CJ (2013) Solvent nanostructure, the solvophobic effect and amphiphile self-assembly in ionic liquids. Chem Soc Rev 42:1096–1120
24. Cammarata L, Kazarian SG, Salter PA et al (2001) Molecular states of water in room temperature ionic liquids. Phys Chem Chem Phys 3:5192–5200
25. Gaillon L, Sirieix-Plénet J, Letellier P (2004) Volumetric study of binary solvent mixtures constituted by amphiphilic ionic liquids at room temperature (1-alkyl-3-methylimidazolium bromide) and water. J Solut Chem 33:1333–1347
26. Sirieix-Plénet J, Gaillon L, Letellier P (2004) Behaviour of a binary solvent mixture constituted by an amphiphilic ionic liquid, 1-decyl-3-methylimidazolium bromide and water: potentiometric and conductometric studies. Talanta 63:979–986
27. Inoue T, Dong B, Zheng L-Q (2007) Phase behavior of binary mixture of 1-dodecyl-3-methylimidazolium bromide and water revealed by differential scanning calorimetry and polarized optical microscopy. J Colloid Interface Sci 307:578–581
28. Bhargava BL, Klein ML (2009) Formation of micelles in aqueous solutions of a room temperature ionic liquid: a study using coarse grained molecular dynamics. Mol Phys 107:393–401
29. Bhargava BL, Yasaka Y, Klein ML (2011) Computational studies of room temperature ionic liquid–water mixtures. Chem Commun 47:6228–6241
30. Tonova K, Svinyarov I, Bogdanov MG (2014) Hydrophobic 3-alkyl-1-methylimidazolium saccharinates as extractants for L-lactic acid recovery. Sep Purif Technol 125:239–246
31. Freire MG, Neves CMSS, Marrucho IM et al (2010) Hydrolysis of tetrafluoroborate and hexafluorophosphate counter ions in imidazolium-based ionic liquids. J Phys Chem A 114:3744–3749
32. Heinrich M, Barnes J, Gibbons S et al (2012) Fundamentals of pharmacognosy and phytotherapy, 2nd edn. Churchill Livingstone, Elsevier, New York
33. Lu Y, Ma W, Hu R et al (2008) Ionic liquid-based microwave-assisted extraction of phenolic alkaloids from the medicinal plant *Nelumbo nucifera* Gaertn. J Chromatogr A 1208:42–46
34. Ma W, Lu Y, Hu R et al (2010) Application of ionic liquids based microwave-assisted extraction of three alkaloids N-nornuciferine, O-nornuciferine, and nuciferine from lotus leaf. Talanta 80:1292–1297
35. S-y W, Yang L, Y-g Z et al (2011) Design and performance evaluation of ionic-liquids-based microwave-assisted environmentally friendly extraction technique for camptothecin and 10-hydroxycamptothecin from samara of *Camptotheca acuminata*. Ind Eng Chem Res 50:13620–13627
36. Zhang L, Geng Y, Duan W et al (2009) Ionic liquid-based ultrasound-assisted extraction of fangchinoline and tetrandrine from *Stephaniae tetrandrae*. J Sep Sci 32:3550–3554
37. Cao X, Ye X, Lu Y et al (2009) Ionic liquid-based ultrasonic-assisted extraction of piperine from white pepper. Anal Chim Acta 640:47–51
38. Yang L, Wang H, Zu Y-g et al (2011) Ultrasound-assisted extraction of the three terpenoid indole alkaloids vindoline, catharanthine and vinblastine from *Catharanthus roseus* using ionic liquid aqueous solutions. Chem Eng J 172:705–712

39. Ma C-h, Wang S-y, Yang L et al (2012) Ionic liquid-aqueous solution ultrasonic-assisted extraction of camptothecin and 10-hydroxycamptothecin from *Camptotheca acuminata* samara. Chem Eng Process 57–58:59–64
40. Bogdanov MG, Svinyarov I, Keremedchieva R et al (2012) Ionic liquid-supported solid–liquid extraction of bioactive alkaloids. I. New HPLC method for quantitative determination of glaucine in *Glaucium flavum* Cr. (Papaveraceae). Sep Purif Technol 97:221–227
41. Bogdanov MG, Svinyarov I (2013) Ionic liquid-supported solid–liquid extraction of bioactive alkaloids. II. Kinetics, modeling and mechanism of glaucine extraction from *Glaucium flavum* Cr. (Papaveraceae). Sep Purif Technol 103:279–288
42. Cláudio AFM, Ferreira AM, Freire MG et al (2013) Enhanced extraction of caffeine from Guaraná seeds using aqueous solutions of ionic liquids. Green Chem 15:2002–2010
43. Ressmann AK, Zirbs R, Pressler M et al (2013) Surface-active ionic liquids for micellar extraction of piperine from black pepper. Z Naturforsch B 68:1129–1137
44. Du F-Y, Xiao X-H, Li G-K (2007) Application of ionic liquids in the microwave-assisted extraction of *trans*-resveratrol from *Rhizma Polygoni Cuspidati*. J Chromatogr A 1140:56–62
45. Du F-Y, Xiao X-H, Luo X-J et al (2009) Application of ionic liquids in the microwave-assisted extraction of polyphenolic compounds from medicinal plants. Talanta 78:1177–1184
46. Du F-Y, Xiao X-H, Li G-K (2011) Ionic liquid aqueous solvent-based microwave-assisted hydrolysis for the extraction and HPLC determination of myricetin and quercetin from *Myrica rubra* leaves. Biomed Chromatogr 25:472–478
47. Liu X, Wang Y, Kong J et al (2012) Application of ionic liquids in the microwave-assisted extraction of quercetin from Chinese herbal medicine. Anal Methods 4:1012–1018
48. Liu X, Huang X, Wang Y et al (2013) Design and performance evaluation of ionic liquid-based microwave-assisted simultaneous extraction of kaempferol and quercetin from Chinese medicinal plants. Anal Methods 5:2591–2601
49. Jin R, Fan L, An X (2011) Microwave assisted ionic liquid pretreatment of medicinal plants for fast solvent extraction of active ingredients. Sep Purif Technol 83:45–49
50. Li X-J, Yu H-M, Gao C et al (2012) Application of ionic liquid-based surfactants in the microwave-assisted extraction for the determination of four main phloroglucinols from *Dryopteris fragrans*. J Sep Sci 35:3600–3608
51. Wei Z, Zu Y, Fu Y et al (2013) Ionic liquids-based microwave-assisted extraction of active components from pigeon pea leaves for quantitative analysis. Sep Purif Technol 102:75–81
52. Bonny S, Paquin L, Carrié D et al (2011) Ionic liquids based microwave-assisted extraction of lichen compounds with quantitative spectrophotodensitometry analysis. Anal Chim Acta 707:69–75
53. Zhang L, Wang X (2010) Hydrophobic ionic liquid-based ultrasound-assisted extraction of magnolol and honokiol from cortex *Magnoliae officinalis*. J Sep Sci 33:2035–2038
54. Lu C, Wang H, Lv W et al (2011) Ionic liquid-based ultrasonic/microwave-assisted extraction combined with UPLC for the determination of anthraquinones in Rhubarb. Chromatographia 74:139–144
55. Zhang L, Liu J, Zhang P et al (2011) Ionic liquid-based ultrasound-assisted extraction of chlorogenic acid from *Lonicera japonica* Thunb. Chromatographia 73:129–133
56. Jin R, Fan L, An X (2011) Ionic liquid-assisted extraction of paeonol from *Cynanchum paniculatum*. Chromatographia 73:787–792
57. Han D, Row KH (2011) Determination of luteolin and apigenin in celery using ultrasonic-assisted extraction based on aqueous solution of ionic liquid coupled with HPLC quantification. J Sci Food Agric 91:2888–2892
58. Tang B, Lee YJ, Lee YR et al (2013) Examination of 1-methylimidazole series ionic liquids in the extraction of flavonoids from *Chamaecyparis obtuse* leaves using a response surface methodology. J Chromatogr B 933:8–14
59. Sun X, Jin Z, Yang L, et al (2013) Ultrasonic-assisted extraction of procyanidins using ionic liquid solution from *Larix gmelinii* bark. J Chem (1):1–9

60. Duan M-H, Luo M, Zhao C-J et al (2013) Ionic liquid-based negative pressure cavitation-assisted extraction of three main flavonoids from the pigeonpea roots and its pilot-scale application. Sep Purif Technol 107:26–36
61. Sun Y, Liu Z, Wang J et al (2013) Aqueous ionic liquid based ultrasonic assisted extraction of four acetophenones from the Chinese medicinal plant *Cynanchum bungei* Decne. Ultrasonics Sonochem 20:180–186
62. Wei W, Fu Y-j, Zu Y-g et al (2012) Ionic liquid-based microwave-assisted extraction for the determination of flavonoid glycosides in pigeon pea leaves by high-performance liquid chromatography-diode array detector with pentafluorophenyl column. J Sep Science 35:2875–2883
63. Cao X, Qiao J, Wang L et al (2012) Screening of glycoside isomers in *P. scrophulariiflora* using ionic liquid-based ultrasonic-assisted extraction and ultra-performance liquid chromatography/electrospray ionization quadrupole time-of-flight tandem mass spectrometry. Rapid Commun Mass Spectrom 26:740–748
64. Yang L, Liu Y, Y-g Z et al (2011) Optimize the process of ionic liquid-based ultrasonic-assisted extraction of aesculin and aesculetin from *Cortex fraxini* by response surface methodology. Chem Eng J 175:539–547
65. Yang L, Li L-l, Liu T-t et al (2013) Development of sample preparation method for isoliquiritigenin, liquiritin, and glycyrrhizic acid analysis in licorice by ionic liquids-ultrasound based extraction and high-performance liquid chromatography detection. Food Chem 138:173–179
66. Yang L, Ge H, Wang W et al (2013) Development of sample preparation method for eleutheroside B and E analysis in *Acanthopanax senticosus* by ionic liquids-ultrasound based extraction and high-performance liquid chromatography detection. Food Chem 141:2426–2433
67. Zhu S, Ma C, Fu Q et al (2013) Application of ionic liquids in an online ultrasonic assisted extraction and solid-phase trapping of rhodiosin and rhodionin from *Rhodiola rosea* for UPLC. Chromatographia 76:195–200
68. Fan J-P, Cao J, Zhang X-H et al (2012) Optimization of ionic liquid based ultrasonic assisted extraction of puerarin from *Radix Puerariae Lobatae* by response surface methodology. Food Chem 135:2299–2306
69. Ribeiro BD, Coelho MAZ, Rebelo LPN et al (2013) Ionic liquids as additives for extraction of saponins and polyphenols from mate (*Ilex paraguariensis*) and tea (*Camellia sinensis*). Ind Eng Chem Res 52:12146–12153
70. Lin H, Zhang Y, Han M et al (2013) Aqueous ionic liquid based ultrasonic assisted extraction of eight ginsenosides from ginseng root. Ultrason Sonochem 20:680–684
71. Yan W, Ji L, Hang S et al (2013) New ionic liquid-based preparative method for diosgenin from *Rhizoma dioscoreae nipponicae*. Pharmacogn Mag 9:250–254
72. Liu F, Wang D, Liu W et al (2013) Ionic liquid-based ultrahigh pressure extraction of five tanshinones from *Salvia miltiorrhiza* Bunge. Sep Purif Technol 110:86–92
73. Zhai Y, Sun S, Wang Z et al (2009) Microwave extraction of essential oils from dried fruits of *Illicium verum* Hook. f. and *Cuminum cyminum* L. using ionic liquid as the microwave absorption medium. J Sep Sci 32:3544–3549
74. Liu T, Sui X, Zhang R et al (2011) Application of ionic liquids based microwave-assisted simultaneous extraction of carnosic acid, rosmarinic acid and essential oil from *Rosmarinus officinalis*. J Chromatogr A 1218:8480–8489
75. Ma C-h, Liu T-t, Yang L et al (2011) Ionic liquid-based microwave-assisted extraction of essential oil and biphenyl cyclooctene lignans from *Schisandra chinensis* Baill fruits. J Chromatogr A 1218:8573–8580
76. Liu Y, Yang L, Zu Y et al (2012) Development of an ionic liquid-based microwave-assisted method for simultaneous extraction and distillation for determination of proanthocyanidins and essential oil in *Cortex cinnamomi*. Food Chem 135:2514–2521
77. Jiao J, Gai Q-Y, Fu Y-J et al (2013) Microwave-assisted ionic liquids treatment followed by hydro-distillation for the efficient isolation of essential oil from *Fructus forsythiae* seed. Sep Purif Technol 107:228–237

78. Severa G, Kumar G, Cooney MJ (2013) Corecovery of bio-oil and fermentable sugars from oil-bearing biomass. Int J Chem Eng, Article ID 617274
79. Severa G, Kumar G, Troung M et al (2013) Simultaneous extraction and separation of phorbol esters and bio-oil from *Jatropha* biomass using ionic liquid–methanol co-solvents. Sep Purif Technol 116:265–270
80. Choi S-A, Oh Y-K, Jeong M-J et al (2014) Effects of ionic liquid mixtures on lipid extraction from *Chlorella vulgaris*. Renew Energ 65:169–174
81. Choi S-A, Lee J-S, Oh Y-K et al (2014) Lipid extraction from *Chlorella vulgaris* by molten-salt/ionic-liquid mixtures. Algal Res 3:44–48
82. Kim Y-H, Choi Y-K, Park J et al (2012) Ionic liquid-mediated extraction of lipids from algal biomass. Bioresour Technol 109:312–315
83. Young G, Nippgen F, Titterbrandt S et al (2010) Lipid extraction from biomass using co-solvent mixtures of ionic liquids and polar covalent molecules. Sep Purif Technol 72:118–121
84. Lapkin AA, Plucinski PK, Cutler M (2006) Comparative assessment of technologies for extraction of artemisinin. J Nat Prod 69:1653–1664
85. Zirbs R, Strassl K, Gaertner P et al (2013) Exploring ionic liquid–biomass interactions: towards the improved isolation of shikimic acid from star anise pods. RSC Adv 3:26010–26016
86. Yansheng C, Zhida Z, Changping L et al (2011) Microwave-assisted extraction of lactones from *Ligusticum chuanxiong* Hort. using protic ionic liquids. Green Chem 13:666–670
87. Ma C-h, Liu T-t, Yang L et al (2011) Study on ionic liquid-based ultrasonic-assisted extraction of biphenyl cyclooctene lignans from the fruit of *Schisandra chinensis* Baill. Anal Chim Acta 689:110–116
88. Vilkhu K, Mawson R, Simons L et al (2008) Applications and opportunities for ultrasound assisted extraction in the food industry – a review. Innov Food Sci Emerg Technol 9:161–169
89. Zhang H-F, Yang X-H, Wang Y (2011) Microwave assisted extraction of secondary metabolites from plants: current status and future directions. Trends Food Sci Technol 22:672–688
90. Dong L-L, Fu Y-J, Zu Y-G et al (2011) Negative pressure cavitation accelerated processing for extraction of main bioactive flavonoids from *Radix Scutellariae*. Chem Eng Process 50:780–789
91. Cláudio AFM, Swift L, Hallett JP et al (2014) Extended scale for the hydrogen-bond basicity of ionic liquids. Phys Chem Chem Phys 16(14):6593–6601. doi:10.1039/C3CP55285C
92. Rakotondramasy-Rabesiaka L, Havet J-L, Porte C et al (2007) Solid–liquid extraction of protopine from *Fumaria officinalis* L. Analysis determination, kinetic reaction and model building. Sep Purif Technol 54:253–261
93. Harouna-Oumarou HA, Fauduet H, Porte C et al (2007) Comparison of kinetic models for the aqueous solid–liquid extraction of *Tilia sapwood* in a continuous stirred tank reactor. Chem Eng Commun 194:537–552
94. Su C-H, Liu C-S, Yang P-C et al (2014) Solid–liquid extraction of phycocyanin from *Spirulina platensis*: kinetic modeling of influential factors. Sep Purif Technol 123:64–68
95. Tan Z, Li F, Xu X (2012) Isolation and purification of aloe anthraquinones based on an ionic liquid/salt aqueous two-phase system. Sep Purif Technol 98:150–157
96. Cláudio AFM, Marques CFC, Boal-Palheiros I et al (2014) Development of back-extraction and recyclability routes for ionic-liquid-based aqueous two-phase systems. Green Chem 16:259–268
97. Bubalo MC, Radošević K, Redovniković IR et al (2014) A brief overview of the potential environmental hazards of ionic liquids. Ecotoxicol Environ Saf 99:1–12

Chapter 8
Enzymatic Aqueous Extraction (EAE)

Lionel Muniglia, Nathalie Claisse, Paul-Hubert Baudelet, and Guillaume Ricochon

Abstract Aqueous enzymatic extraction is employed for fractionation of plant raw material and for extraction of molecules of interest in a safe manner. For many years, the improvement of industrial enzymes lead to new potentialities and new products and implies today an entire rethinking of green extraction and its economic prospects.

This chapter deals with enzymatic aqueous extraction as an alternative method for green extraction. The interests of the use of enzymatic mixtures during green extraction processes of natural molecules are detailed through successful and recent improvements. A focus is done on vegetable products. Advantages and drawbacks of enzymatic-based technologies are described: implementation, availability of enzymes, diversity of activities, development of new enzymatic activities, cost, safety, efficiency, etc.

From lab to industrial scale, examples illustrate the state of the art in enzymatic aqueous extraction.

These technologies are also considered through economical and environmental considerations dealing with actual knowledge. This allows us to envisage future industrial development of enzymatic aqueous extraction processes and to position them as green processes.

L. Muniglia (✉) • P.-H. Baudelet
Laboratoire Ingénierie des Biomolécules, ENSAIA, 2 Avenue de la Forêt de Haye TSA40602, 54518 Vandoeuvre Cedex, France
e-mail: lionel.muniglia@univ-lorraine.fr; paul-hubert.baudelet@univ-lorraine.fr

N. Claisse • G. Ricochon
Biolie SAS, 24, 30 Rue de Lionnois, BP60120, 54003 Nancy Cedex, France
e-mail: nathalie.claisse@biolie.fr; guillaume.ricochon@biolie.fr

Finally, some elements are taken into account in assessing the potential benefits of combining various green technologies to promote synergies with green extraction technologies and improve efficiency, improve the economic balance, and reduce environmental impact.

8.1 Introduction

Enzyme-assisted aqueous extraction (EAE) processes were developed for about 40 years with more and more interesting outlooks and an industrial feasibility now about of success. Indeed, many pitfalls have been raised with the progress of technology, and besides a whole system needs to be reinvented with the use of alternative extraction technologies. Today, advances in enzyme catalysis, the availability and diversity of enzymes plus increasingly strong environmental constraints pave the way for numerous credible industrial developments. A recent report (published in 2011) by the French Environment and Energy Management Agency (called ADEME, a French public agency under the joint authority of the Ministry for Ecology, Sustainable Development and Energy and the Ministry for Higher Education and Research) entitled "Barriers to the replacement of volatile organic compounds in industrial processes" listed the "solutions to overcome technical and economic bottlenecks to the substitution of solvent-based products...." Among the different solutions, two were devoted to enzyme technology and cited as follows:

- Mechanical solution with enzyme preparation
- Water extraction with enzyme preparation

Considering the six principles of eco-extraction defined in the book entitled *Plant Eco-Extraction, Innovative Processes and Alternative Solvents* [1], enzyme-assisted aqueous extraction enables compliance with established criteria. First of all, EAE is based on the valuation of renewable substances as a whole, the raw material is derived from plants, and renewable and waste generation is minimized. All parts of the raw material are valued and the economical value of each products is increased. Unit processing operations are minimized and often simplified. They are also energy efficient and comparable to other alternative technologies. Finally, and this is a fundamental point, products are better preserved and their intrinsic properties less distorted.

Thus, this chapter illustrates the potential of aqueous processes assisted by enzymes. The principles of enzyme technology will be described with the features of the method and of the enzymes that are key to its efficiency. Various examples will be presented to illustrate the advances in this field. Initially, enzyme-assisted aqueous extractions were applied mainly to oil commodities. Today's applications are varied, and fractionation of non-oil plant material is of great interest with respect to the products formed.

8.2 Detailed Understanding of EAE

8.2.1 Mechanisms Observed: The Role of Water as Solvent

Aqueous extraction processes have water as unique solvent. The principle of extraction is thus at the opposite of the process based on the use of organic solvent. Those principles were initially developed by Johnson [2] and then followed by Rosenthal [3].

The processes of extraction by organic solvent are based on their capacity to dissolve and extract oils. In EAE processes, oil does not have the same affinity with water and thus cannot be dragged away as in solvent extraction. The spontaneous dissolution of a species in a solvent is always followed by a decrease of the free energy of the system. The Gibbs equation associates the free energy (ΔG), the enthalpy (ΔH), the absolute temperature (T), and the entropy (ΔS) as

$$\Delta G = \Delta H - T\Delta S$$

During the dissolution, the thermic energy of the system is consumed to split the molecules of the solute and the molecules of the solvent. The system's energy is gained back, while the molecules of the solute interact with the molecule of the solvent. The global variation of enthalpy is exothermic (negative) if the energy lost by the system in the disruption of interactions solute/solvent and solvent/solute is higher than the energy absorbed by the creation of interaction solute/solvent. In that case, the dissolution is spontaneous. In the case of aqueous extraction processes, the interactions between triglyceride and water are particularly weak and do not allow the split of the molecules of water gathered by hydrogen bonds. Oils are not soluble in water in normal conditions of dissolution.

Extraction of oil in aqueous medium is thus based on the opposite phenomenon of dissolution: the insolubility of oil in water. The goal is to dissolve the water-soluble compounds, which are responsible for the retaining of oil in a closed structure, triggering then its natural release. In theory, yields of extraction of water-soluble compounds are favored by all the techniques (enzyme, milling, etc.) permitting to increase their dissolution in water. Hence, for enzymatic aqueous extractions, it is a priority to have a complete knowledge of the cell wall composition, of the enzymatic activities, and of the influence of the pretreatments and working conditions.

8.2.2 Enzyme as the Main Tool

An enzyme is a biological molecule, which has the capacity to catalyze a chemical reaction. Except for ribozymes (RNA), all the enzymes known up to now have a proteic nature. Enzymes are the biological tools of nature for life. Every organism possesses its own toolbox to run the chemical reactions necessary for its good functions. Enzymes are catalysts: they contribute to increase the kinetic of chemical reactions (10^6–10^{12} faster) and make them feasible more rapidly. They differ from chemical catalysts on three points:

- Softer conditions of reaction (pH, temperature, pressure, etc.)
- Greater specificity of reactions
- Regulation of the catalysis possible *via* external parameters (concentration in products and substrates, pH, temperature, metabolism, etc.)

There is a large number of available enzymes: up to now more than 6,000 different chemical reactions are biocatalyzed. More than 90,000 names of enzymes have been listed, and the proteic structure of 27,700 of them is identified (BRENDA database: http://www.brenda-enzymes.org).

Nomenclature of enzymes was recently homogenized. Historically, one enzyme could have several different names, or one name could correspond to several enzymes. Today, enzymes are referenced under an E.C. number composed by a category (x), a subcategory (y), a last category under that one (z), and a number of order (w). Each enzyme is thus associated with a number defined as E.C. x.y.z.w. E.C. means "Enzyme Commission"; x defines the type of chemical reaction catalyzed; y and z, respectively, designate the main substrate and cosubstrate of the enzyme; and w is the serial number. In this classification, enzymes with a number starting with 1.y.z.w catalyze reduction-oxidation reaction; enzymes with number starting with 2.y.z.w. catalyze the transfer of chemical functions. The generic name of enzyme is attributed by adding the suffix "ase" to the name of the reaction catalyzed, for example, oxidoreductase, transferase, etc.

Since enzymes catalyze specific reactions, in the context of EAE, it is important to know the composition of the raw material, in order to determine the most appropriate enzymes to run the extraction.

8.2.3 Superior Vegetal Cell Wall

In order to reach the compounds of interest stocked in vegetal cells, several barriers have to be crossed: extracellular cell walls, cell walls, and oleosomes. Each of those cell walls is composed of its own constituents organized in a complex structure. Those cell walls are synthesized and hydrolyzed naturally by specific enzymes. The main constituents of the vegetal cell wall are illustrated in Fig. 8.1.

8.2.4 Extracellular Cell Walls

In plants, the extracellular matrix is mainly composed of sugars and proteins and confers rigidity and protection to the protoplasts. There are two types of cell walls:

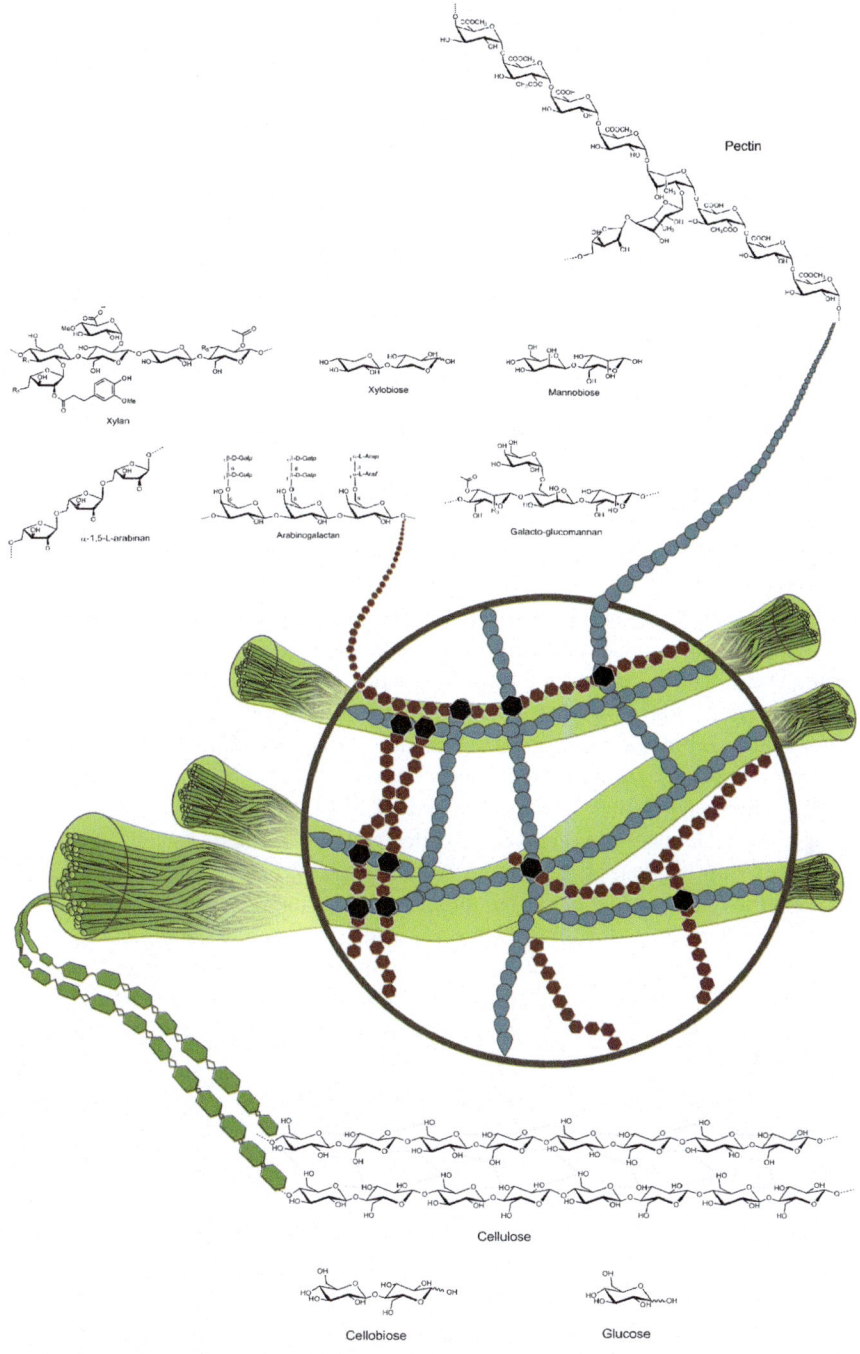

Fig. 8.1 General diagram of a plant cell wall (Frédéric Paulien(c))

- The primary cell wall surrounding growing cells
- The secondary cell wall surrounding mature cells

8.2.4.1 Primary Cell Wall

It is constituted of four types of polymers arranged between them.

Cellulose is a linear glucan with units linked *via* β-1,4 bonds. A molecule of cellulose can contain more than 3,000 units of glucose, and even though it is hydrophilic, cellulose is not soluble in water. In the cell wall, chains of cellulose are assembled in parallel to form microfibrils. Those microfibrils have a diameter comprising between 5 and 12 nm, and they are composed of 36–1,200 molecules of cellulose linked together with the hydrogen bonds of the neighboring glucose hydroxyl groups. Microfibrils are measuring between 5 and 15 nm, and they are separated from each other from 20 to 40 nm. Microfibrils are constituted of amorphous domains as well as crystalline domains, which are very organized and thus not easily accessible for enzymes. Microfibrils are entangled in an amorphous matrix composed essentially of hemicellulose and pectic compounds.

The hemicelluloses are a complex mixture of heterogeneous carbohydrates and carbohydrate-derived compounds forming a branched network. It contains hexoses (glucose, mannose, galactose), pentoses (arabinose, xylose), methylated derivatives (rhamnose, fucose), and acids (uronic, acetic). Hemicelluloses are constituted of an osidic backbone (linear structure being often homologs of cellulose like xyloglucan, arabinogalactan, galactomannan, etc.). Hemicellulose and cellulose are linked together *via* hydrogen bonds. This osidic backbone is branched with short segments positioned in a defined order (xylogalactose fucose in the case of xyloglucans). Hemicellulose can be extracted with weak alkali (NaOH 7.5 %).

Considering pectins, pectic compounds are a set of heterogeneous polysaccharides rich in galacturonic acid. The main pectin chains are constituted of uronic acid units linked in α-1,4 (polygalacturonic acid). Those units are sometimes found with rhamnose linked in 1,4 or in 1,2 creating thus branches called "pectic elbows." Those formed structures are called zigzag rhamnogalacturonan. Units of rhamnoses or galacturonic acids can have lateral chains of galactans, arabinans, or mixtures of arabinogalactans.

Glycoproteins are polypeptides containing hydroxyproline named HRGP (hydroxyproline-rich glycoprotein). Those molecules have several particularities:

- Numerous amino acids bearing hydroxyl groups (hydroxyproline, serine, tyrosine)
- Repeating peptidic sequences (ser hyp hyp hyp hyp)
- Peptidic sequences including tyrosine, allowing the formation of diphenol bonds with hemicelluloses or pectins
- Presence of galactose and arabinose involved in the formation of bonds with polypeptides and polyholosides for the formation of the constitutive cell walls

8.2.4.2 Secondary Cell Wall

The secondary cell wall is composed of around 45 % (w/w of dry weight) of cellulose, hemicellulose, and a small part of pectic compounds. Most of the secondary cell walls are also composed of lignin (up to 35 % of dry weight). The lignin is a complex system of cross-linked bonds between simple phenolic alcohols (*p*-coumaryl alcohol, coniferyl alcohol, and sinapyl alcohol). It is recalcitrant to any extraction process, which would not trigger its degradation.

The formation of lignin is possible thanks to free radicals, which react spontaneously and randomly, issued from the action of peroxidases. Moreover, those free radicals can form covalent bonds with cellulose and hemicellulose. Only a few bacteria and fungi are capable of biodegrading lignin (lignolysis). This degradation is realized thanks to powerful exo-cellular peroxidases that are capable of opening aromatic cycles, allowing then a complete lignolysis.

8.2.4.3 Oleosomes

Oil is stocked as droplets surrounded by a non-membranous layer inside the cells. Those systems are called oleosomes. Oleosomes are formed when a particular proteic structure constituted of oleosins is bonded at the surface of droplets to form a limit layer. Oleosomes are thus constituted of a core of triglycerides surrounded by a monolayer of phospholipids in which are connected the oleosins. Those proteins are formed of three amphiphilic domains N- and C-terminal and of a central hydrophobic segment comprising 72 amino acids. The goal of this limit layer is to provide oleosomes from coalescence.

8.2.5 Algal Cell Wall

The cell walls of algae are composed of most of the regular constituents of terrestrial plants: cellulose, hemicellulose, pectin, etc.

However, there is a wide diversity of structures and of cell wall compositions between family, species, and even within one species [4]. Particular and original compounds are regularly discovered. The cyanobacteria are prokaryote microalgae being gram negative bacteria and therefore possess cell walls constituted of two membranes (inner and outer) containing a layer of peptidoglycans [5]. Green algae can contain chitin and chitosan which are also constitutive compounds of the exoskeleton of insects and cell wall of fungi [6]. Cell walls can be composed of algaenans, which is a family of molecules comprising long aliphatic C-chains (C30 to C120) conferring a great resistance to algae. Some of them have a structure close to sporopollenins which are constituents of the cell wall of pollen grains [7, 8]. Green algae can also contain ulvans which are sulfated polysaccharides from Ulvophyceae [9].

Red algae can be composed of lignin (abundant polymer in the secondary vascularized cell walls), carrageenans, porphyrans, and agarose (conferring the gelatinous aspect of their cell walls) [10]. Brown algae are composed of alginates or alginic acids, polysaccharide formed of mannuronic acid and guluronic acid, fucoidans (family of sulfated polysaccharides constituted of sulfated esters of fucose (F-fucoidan) or glucuronic acids (U-fucoidan)), homofucans, and collagen (fibrillary proteins) [11]. Some of the microalgae have an organic structure composed of silica called frustule (diatoms, etc.), a mineral structure made of calcium carbonate scales as cell walls [12] or even no cell wall (e.g., *Dunaliella tertiolecta*) [13].

8.2.6 Enzymes Able of Hydrolyzing Vegetal Cell Walls

The use of enzymes in the food and biotechnologies industries is now widespread. Four families of enzymes are commonly used to hydrolyze raw materials: cellulases, hemicellulases, pectinases, and proteases. Among those enzymes, all of them have the capacity to hydrolyze part of the constituents of vegetal cell walls.

- Cellulases are hydrolyzing the cellulose to release molecules of cellobiose and glucose when the hydrolysis is complete.
- Hemicellulases are hydrolyzing hemicelluloses producing simple sugars or oligosaccharides.
- Pectinases are hydrolyzing the different types of pectins, releasing uronic acids.
- Proteases are capable of hydrolyzing proteins releasing peptides, amino acids, etc.

The first three cited enzymes can be used alone or in a mixture; however, proteases have to act separately for three main reasons. They are capable of hydrolyzing the enzymatic proteins and thus of deeply decreasing the specific activities of hydrolytic mixture of enzymes. Moreover, if the extraction is used in order to produce oil, the peptides released by the action of protease will highly stabilize the resulting emulsions. Finally, if the goal is the extraction of proteins, proteases must be prohibited in order to preserve their native structure.

8.2.6.1 Cellulases

The insoluble form of cellulose (crystalline microfibrils) is very recalcitrant to enzymatic hydrolysis. However, there are microorganisms capable of efficiently hydrolyzing cellulose producing a variety of enzymes called cellulases.

Mechanisms of action of cellulases are not entirely clearly identified. Zhang et al. consider that the chemical mechanisms the most widely accepted for enzymatic hydrolysis result from the synergic action of three main types of enzymes (Fig. 8.2) [14]:

8 Enzymatic Aqueous Extraction (EAE)

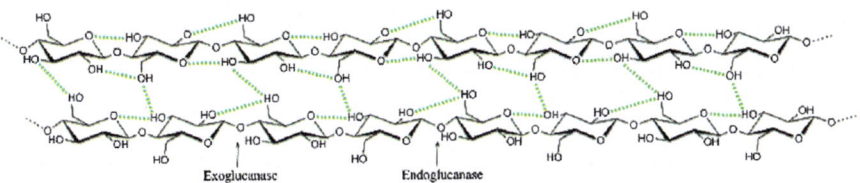

Fig. 8.2 Cellulase action on cellulose

- Endoglucanases are capable of hydrolyzing the β-1,4 glucosidic intramolecular bonds.
- Exoglucanases hydrolyze cellulose from the ends of glucosidic chains, releasing then glucose or cellobiose.
- Finally, β-glucosidases are hydrolyzing molecules of cellobiose into two molecules of glucose in order to eliminate that dimer.

Two steps are important in the hydrolysis: the hydrolysis in the solid phase and the one in the liquid phase. During the primary hydrolysis (solid phase), endo- and exoglucanases depolymerize the substrate until it becomes soluble (degree of polymerization <6). The limiting factor of that hydrolysis is thus the concentration of endo- and exoglucanases. The secondary hydrolysis, which takes place in the liquid phase, implies essentially the hydrolysis of cellobiose into glucose by β-glucosidases, even though β-glucosidases have also the capacity to hydrolyze the soluble cyclodextrins. The fixation of cellulases is a surface phenomenon, and it might sometimes be difficult for enzymes to rapidly degrade cellulose. The chains of glucan of microfibrils are very close to each other in the crystalline part: the access of enzymes is limited. In order to penetrate the microfibrils and give access to enzymes, it is necessary to dissociate the molecules of cellulose by breaking the intermolecular hydrogen bonds.

8.2.6.2 Hemicellulases

Hemicelluloses are linear polysaccharides, branched to microfibrils of cellulose by hydrogen bonds or to pectin by covalent bonds. This structure is forming a complex solid network around vegetal cells.

The principle of the degradation of hemicellulose by enzymes is the same as the one for cellulose. This structure is however more complex and requires a greater diversity of enzymatic activities due to the large variety of hemicelluloses. There are numerous oligosaccharides forming hemicelluloses, and consequently a large number of different linkages are formed. Enzymes are in general specific to one type of linkage; as a consequence, a lot of them are involved in the degradation of hemicelluloses. The mode of action of the main hemicellulases is presented in Fig. 8.3 according to reviews published in 2003 by Howard and Shallom [15, 16]. Table 8.1 resumes the main hemicellulases activities.

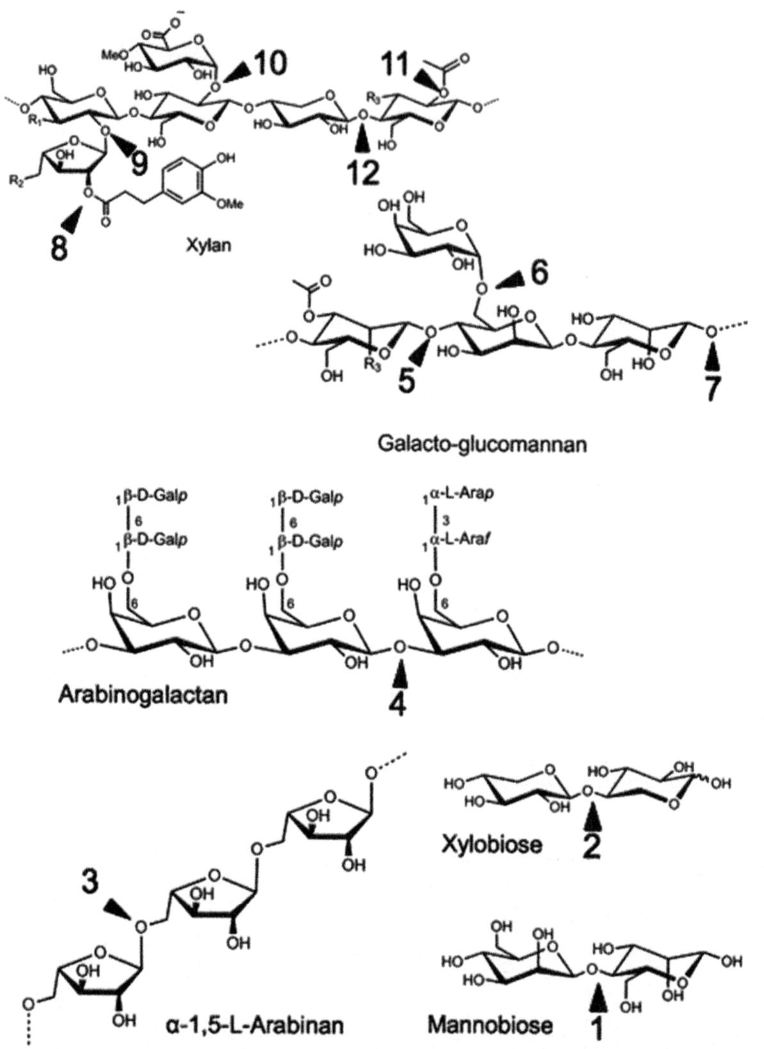

Fig. 8.3 Hemicellulase action on main hemicelluloses components. *1* Exo- β-1,4-mannosidase; *2* β-1,4-xylosidase; *3* endo- α -1,5-arabinanase; *4* β -galactosidase; *5* endo- β-1,4-mannanase; *6* α-galactosidase; *7* β-glucosidase; *8* feruloyl esterase; *9* α-L-arabinofuranosidase; *10* α-glucuronidase; *11* acetylxylan esterase; *12* endo- β-1,4-xylanase

8.2.6.3 Pectinases

The biodegradation of pectins is realized thanks to pectinases and specifically endopolygalacturonases, which only act on acid units (non-esterified). The esterified units are degraded by methyl pectin esterases, which permit to adapt the substrate

Table 8.1 Main hemicellulases and esterases necessary to hydrolyze hemicelluloses in vegetal cell walls

Enzymes	Substrate	EC number
β-1,4-xylosidase	β-1,4-xylo-oligomers (xylobiose)	3.2.1.37
Endo-β-1,4-xylanase	β-1,4-xylan	3.2.1.8
Exo-β-1,4-mannosidase	β-1,4-manno-oligomers (mannobiose)	3.2.1.25
Endo-β-1,4-mannanase	β-1,4-mannan	3.2.1.78
Endo-α-1,5-arabinanase	α-1,5-arabinan	3.2.1.99
β-Galactosidase	Terminal nonreducing β-D-galactose residues	3.2.1.23
β-Glucosidase	Nonreducing β-D-glucosyl residues	3.2.1.21
α-Glucuronidase	α-D-glucuronoside	3.2.1.139
α-L-arabinofuranosidase	α-Arabinofuranosyl (1 → 2) ou (1 → 3) xylo-oligomères α-1,5-arabinan	3.2.1.55
α-Galactosidase	α-galactopyranose(1 → 6) manno-oligomers	3.2.1.22
Endo-galactanase	β-1,4-galactan	3.2.1.89
Acetylxylan esterase	2- or 3-O-acetyl xylan	3.1.1.72
Feruloyl esterase	4-hydroxy-3-methoxycinnamoyl group from an esterified sugar	3.1.1.73

to react with the previous enzyme. In the case of highly methylated pectins, only the pectin lyases from bacterial origin (non-detected in plants) are capable of hydrolyzing them.

Pectinases were the first enzymes, which were commercially available in the 1930s for their use in the production of wine and fruit juices. While the structure of vegetal cell walls was identified in the 1960s, scientists were capable of formulating more appropriate mixtures of enzymes to the industrials which were interested in enzymatic reactions.

Nowadays, pectinases are at the heart of the processes developed in the industries of fruit juices, as well as in the textile industries. More recently, pectinases have also found applications in the biotechnologies industries. As for hemicelluloses and celluloses, the complete hydrolysis of pectins requires the synergic action of several pectinases (Table 8.2, Fig. 8.4).

8.2.6.4 Algae Hydrolytic Enzymes

Some algae have in their cell walls or extracellular matrices original homo- or heteropolysaccharides such as alginates and fucoidans of brown algae or carrageenans of red algae. Enzymes from various resources including viruses, bacteria, fungi, marine mollusks, and algae have been reported to have specific hydrolysis activities on those algal cell wall polysaccharides.

Table 8.2 Main pectinases and their classification

Enzymes	Substrate	EC number
Pectin-esterase	Methyl ester of galacturonates	3.1.1.11
Polygalacturonase	Random hydrolysis of (1 → 4)-α-D-galactosiduronic linkages in pectate and other galacturonans	3.2.1.15
Exopolygalacturonase	Hydrolysis of (1 → 4)-α-D-galactosiduronic linkages in pectate from their nonreducing end	3.2.1.67
Pectin lyase	Eliminative cleavage of (1 → 4)-α-D-galacturonan methyl ester to give oligosaccharides with 4-deoxy-6-O-methyl α-D-galact-4-enuronosyl groups at their nonreducing ends	4.2.2.10

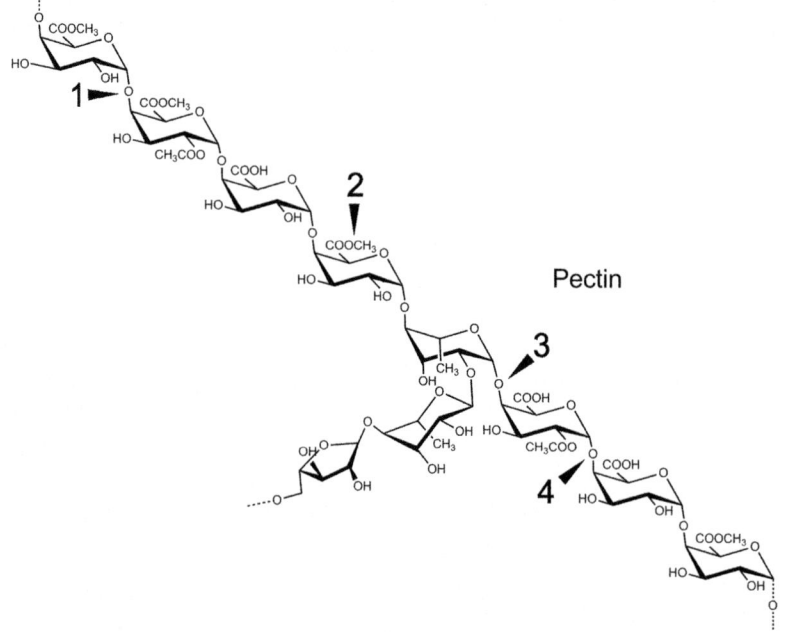

Fig. 8.4 Pectinase. *1* EndoPectin lyase; *2* Pectinesterase; *3* Rhamnogalacturonase; *4* EndoPolygalacturonase

An enzyme from the marine bacteria *Formosa algae* shows specific hydrolysis activity on fucoidan of brown algae [17]. A fucoidan endohydrolase enzyme (EC 3.2.1.44) from a bacterium of the *Flavobacteriaceae* family hydrolyzes fucoidan into tetra- or hexasaccharide forms from fucopyranose disaccharidic units [18].

Kim et al. reported recently an alginate lyase activity from *Microbacterium oxydans* [19], and the hydrolytic activity on β-1,4 linkages of alginate has also been shown from an enzyme of the chlorovirus CVN1 infecting the green algae *Chlorella NC64A* [20] (EC 4.2.2.11 and EC 4.2.2.3).

The sea cucumber intestine bacteria identified as *Pseudoalteromonas* sp. and the marine bacteria *Cellulophaga* sp. show carrageenase activities with enzymes able to hydrolyze, respectively, kappa- and iota-carrageenans (EC 3.2.1.83 and EC 3.2.1.157) [21, 22]. Agarose is hydrolyzed by an exo-β-agarase (EC 3.2.1.83) from endophytic marine bacterium *Pseudomonas sp.* of the red algae *Gracilaria dura* [23] and an α-agarase from *Thalassomonas* sp. which also hydrolyzes β-1,3-linkages of porphyran [24].

Enzymes hydrolyzing green algae polysaccharides found in cell walls such as ulvan have also been reported from a marine bacterium [25]. The enzyme named "Val-1," from the chlorovirus CVK2 has been shown to cleave chains of β- and α-1,4-glucuronic acids found in an original polysaccharide of the *Chlorella strain NC64A* cell wall [26].

8.3 Enzymatic Aqueous Extraction

8.3.1 Press or Solvent Process Using Enzymes

Enzymes can be used as additional tools to regular processes in order to increase yields of extraction by weakening the vegetal cell walls.

8.3.1.1 Oil Extraction

The first studies on enzyme-assisted aqueous extraction of oil were performed in 1972. At that time, Sherba published their research work on the fractionation of soybean with proteases [27]. In 1975, Lanzani published a preliminary report on the utilization of enzymes for extraction of vegetal oil [28]. In 1977, Fantozzi et al. [29] used enzymes as a pretreatment step on olive fruit before pressing by conventional techniques. Fullbrook pursued this topic of research in 1983 [30]. His results were encouraging since he demonstrated that solvent penetrates more easily inside the seeds when the vegetal cell walls are partially hydrolyzed by enzymes. The yields were significantly improved since he obtained 50 % more oil from rapeseed and 90 % more from soybean. Sosulski et al. formulated a mixture of β-glucanases, pectinases, cellulases, and hemicellulases and showed that an enzymatic pretreatment step before the Soxhlet extraction increased the yield in oil by 45 %. He also demonstrated in 1993 that an enzymatic pretreatment step before pressing improved the yield in oil up to 30–50 % on rapeseed, depending on the variety [31]. Further research works on soybean and sunflower using commercial enzymatic mixture showed improved yields in oil by 8–10 % and 4 %, respectively [32].

This yield improvement is due to the breakdown of vegetal cells: enzymes are efficient tools for combination with other techniques. It will be shown in a later

Table 8.3 Yields and process conditions of enzyme-assisted aqueous extraction of rapeseed

Enzyme	T (°C)	t (h)	yld extrac	References
Protease	40–50–65	3	74 % total oil	[28]
Protease pectinase a-amylase	40–50–60	3	78 % total oil	[28]
β-Glucanase protease	50–63	3	72 % total oil	[30]
Hemicellulase	50–53	3	75 % total oil	[30]
Multi-activities	50	4	Nd	[36]
Multi-activities	45–50	6	Nd	[37]
Multi-activities	Nd	Nd	80 % total oil	[38]
Pectinase cellulase β-glucanase	48	5	92.5	

T temperature, t time, *yld extrac* yield of extraction

section that enzymes provide significant improvements combined with other eco-friendly technologies.

8.3.1.2 Other Compounds Extraction

Cerda et al. (2013) studied the phenolic content of thyme extract [33]. This study shows that the total phenolic content extracted with a mixture of water and ethanol was improved by 70 % using enzymes as pretreatment. Similar results were obtained on the phenolic extraction of unripe apples. Total phenolic content and caffeic acid content were about 2- and 13-folds higher than those of the control with different enzymatic mixtures [34].

In the study of [35], bioactive compounds were extracted from ginger (*Zingiber officinale Roscoe*) with organic solvents and a combination of solvents after a pretreatment with different types of enzymes. Extraction yields of oleoresin and 6-gingerol content were higher than values obtained without the enzymatic pretreatment. Similarly, the largest total polyphenol content is obtained by ethanol extraction of ginger pretreated with cellulase.

8.3.2 Enzyme-Assisted Aqueous Extraction

Enzymatic aqueous extractions have been studied for several decades, and the yields reached with that technique have constantly improved. Table 8.3 displays some examples representing the evolution of yields of rapeseed oils from 1975 to today.

In Table 8.3, enzymatic mixtures are first composed of proteases and then progressively replaced by cellulosic activities. Enzymes seem to be more efficient in mixture than separately. Yields of extraction increased regularly from 1975 up to now. This is partly due to the evolution of equipments (better stirring, better regulation of parameters, etc.) but also to the better quality, selection, efficiency, and concentration of available enzymes.

Numerous other substrates have been investigated. It has been proved that the use of enzymes capable of hydrolyzing cellulose, hemicelluloses, pectins, and proteins enhances the yield of oil extraction on most substrates, such as soybean [39, 40], corn [41], olive [29, 42, 43], coconut, [44, 45], avocado [46, 47], sunflower [28, 48], palm [49], etc.

8.3.3 Pretreatment of EAE

Pretreatments consist in preparing the seeds before the extraction step itself. In the case of enzyme-assisted aqueous extraction, the role of pretreatment is to prepare the seeds for the action of enzymes. Pretreatments aim to reduce or suppress the structures of compounds that could limit the enzymatic hydrolysis. Pretreatments and conditions of their applications depend on the type of biomass to process. There are two categories of pretreatment methods: physical or chemical (sometimes both). On one side, physical treatments include grinding, vapor-phase cracking, and hydrothermolysis [50]. On the other side, chemical treatments consist in acidic or basic processes (using generally H_2SO_4 or NaOH). Both types of treatments are designed to remove lignin or hemicelluloses, which are limiting in further steps of the process. For example, pretreatments are frequently used for the production of ethanol since it is necessary to strongly increase the yield in molecules of glucose.

8.3.3.1 Inactivation of Endogenous Enzymes

It is necessary to inhibit the action of endogenous enzymes such as myrosinases or lipases during aqueous extraction, in order to preserve the quality of oil. Eylen et al. studied the kinetic of inactivation of myrosinases by a thermic treatment of whole broccoli [51]. They demonstrated that this enzyme is not anymore detected after an exposure time of 30 min at 72.5 °C or after 10 min at 75 °C. It was also shown that the time required for inactivation is decreased if the seed is not intact.

In 2006, Zhang showed that the totality of myrosinases were inactivated while dipping rapeseeds in boiling water for 5 min, which decrease the glucosinolate content of 11.28 % [52]. They demonstrated that the inactivation is facilitated in the presence of an aqueous medium (90 °C as reference temperature for the study). The thermic treatment is more efficient using microwave heating than vapor heating. The percentage of free fatty acids increases with the period of treatment similarly with both heating techniques.

8.3.3.2 Grinding

The grinding of seeds aims to reduce the size of particles in order to increase the accessible surface of cell walls for enzymatic hydrolysis, that is, the enzyme-substrate interactions. Supposing that the particles of seeds are spherical, it is necessary to reduce drastically the volume to increase the accessible surface. Sphere is the geometrical structure that has the largest volume for the smallest surface. This property implies to minimize the volume of seed particles to favor the access of enzymes.

The ratio surface over volume is 3 times the opposite of the radius of the sphere. This means that the volume of the sphere is reduced 3 times faster than the surface when the radius is shortened. From this point of view, it is of great importance to grind seeds as finely as possible.

The grinding of seeds in a process of enzyme-assisted aqueous extraction is thus an essential step. Everyone agrees to affirm that the finer the ground is, the better is the extraction process. By opposition, the grinding is not much studied as a parameter for the entire process. The main publications on the topic briefly described the parameters of grinding and eventually the particle size data issued from the seeds processed with an aqueous treatment. The size of particles is however very detailed for animal feeding applications. Many research studies were performed on particle sizes to best fit with animal digestion in order that digestive enzymes efficiently hydrolyze the totality of ingested food. Unfortunately it is not possible to extrapolate those results to the vegetal enzymatic extraction process because the period in the intestinal tract and thus the period of hydrolysis changes with the sizes of particles. A ground material will be easily but rapidly digested, whereas a material roughly ground will be less digestible but will stay longer in contact with enzymes [53].

Concerning the study of grinding in aqueous extraction, Rosenthal et al. demonstrated in 1998 that the damages caused by a grinder to vegetal cell walls were sufficient to influence the yields of extraction in oil and proteins [54]. A finer grinding will affect deeper the cell walls and will thus yield to better results. Their study allows concluding that the effect of enzymes on soybean is not significant compared to the role of the size of particles. The yields of oil after an enzyme-assisted aqueous extraction vary from 28 to 66 %, while the average size of soybean particles ranges between 850 and 150 μm. The same observation was done on corn germs [55]. In 2007, Evon et al. tested the impact of coupling a double screw extruder with an enzymatic process of extraction on sunflower meal, and 55 % of oil were recovered [56], which is not better than the yields reached by Rosenthal with a simple grinding step [54].

8.3.3.3 Vapor-Phase Cracking

The material is disposed in a reactor under a very high pressure, then the vapor is quickly removed, and the pressure decreases accordingly. The main physical or chemical changes of biomass processes with this technique are often attributed to the removal of hemicelluloses. This hydrolysis is realized *via* the release of acetic acid produced by the hydrolysis of acetyl groups linked to the hemicelluloses. Moreover, water acts as an acid at high temperatures. The vapor allows the transfer of large amounts of heat inside the seeds, limiting thus the dilution of biomass in large volume of water (ratio water/biomass of 6 %).

8 Enzymatic Aqueous Extraction (EAE)

Weil enlightened the efficiency of this treatment in 1997. The treatment of the seeds at around 250 °C for one hour produces after cooling yields of 80–90 % of cellulose converted into glucose, whereas the yields only reach 50–60 % with no pretreatment. Weil also noticed that the initial pH of 5 drops to 3 at the end of the reaction [57].

The improvement of enzymatic treatments depends mainly on the removal of hemicelluloses, on the reduction of the size of particles, and on the raising of the volume of pores.

Research studies of Shankar et al. in 1997 showed that the flaking of soybean coupled with a vapor treatment facilitates the action of enzymes [58].

8.3.3.4 Hydrothermolysis

This technique is based on the same principle as vapor-phase cracking. The main difference of this pretreatment is that the pressure is high enough for the water to remain in a liquid state at high temperatures (at 200 °C, the pressure must be greater than 15.5 bars). This implies significantly shorter time of reaction (a few minutes). As for vapor-phase cracking, hemicelluloses are rapidly hydrolyzed by the liberation of acetic acid. In particular conditions (220 °C for 2 min [59]), it is possible to partially solubilize lignin (50–66 %).

8.3.3.5 Acidic Treatments

Diluted sulfuric acid is mixed with biomass to hydrolyze hemicelluloses into simple carbohydrates. In order to solubilize biomass, Leea et al. tested different acid concentrations from 0.0735 %, 0.4015 %, to 0.735 % by weight for 10, 15, and 20 min, as well as different temperatures from 140 to 204 °C [60]. They succeeded in solubilizing from 83 to almost 100 % of hemicelluloses, 26.3–52.5 % of lignin; 95.2–79.6 % of hemicelluloses were hydrolyzed into simple carbohydrates, and the residual fraction was composed of oligomers.

Cellulose processed with this treatment is highly accessible to enzymes (more than 90 % of cellulose is concerned).

8.3.3.6 Alkali Treatments

In comparison with the acidic pretreatments, alkali treatments are realized at lower temperatures and pressures. They can be performed at room temperature; however, the time of reaction reaches several hours or days instead of minutes. Moreover, on the contrary of acidic pretreatments, salts are formed within the biomass during the reaction. Those salts are then not easy to remove.

The treatment with lime was used, for example, on wheat straw or on poplar [61, 62].

The main result of an alkali treatment is the removal of lignin from biomass, which increases the reactivity of resulting polysaccharides. Moreover, this type of technique removes the acetyl substitutes and the uronic acids linked to hemicelluloses which can provide enzymes from accessing the glycosidic linkages [63].

It is however hardly considerable to use such a technique for a process in food or cosmetic industries.

However, Jiang et al. improved yields of oil extraction and processing of protein hydrolysates from peanut in a three-step process involving chronologically alkaline extraction and two steps of enzymatic hydrolysis [64].

8.4 Influence of Extraction Parameters

8.4.1 Enzymatic Mixture

The composition of the enzymes mixture is a key factor; the nature of its activities and their relative proportions determine all or part of process performances. As explained, in order to destroy completely the cell wall, it is necessary to use different and complementary enzyme activities.

When enzymatic activities are numerous, extraction yields are better: the effects of an enzyme will promote the release of a favorable substrate to another family of enzymes. This has been demonstrated on many substrates using xylanases and β-glucanases [65], which confirms that the efficiency of oil extraction is dependent of the level of degradation of plant cell walls. When the reaction medium contains a high amount of a large number of enzymes, the oil yield will be high and rapidly achieved. However, the optimal enzymatic mixture must be adapted to each kind of seeds, each substrate, due to the variation of the parietal compositions between species and varieties. It is impossible to use a specific enzymatic mixture for all plant species.

The study of Ramadan on seeds and skin of goldenberry pomace illustrates this effect [66]. A lot of different industrial enzyme mixtures have been tested: CellulaseEC (Extrakt Chemie, Stadthagen, Germany) with cellulase, cellobiase, glucosidase, pectinase, xylanase, and amylase activities; Cellulbrix (Novozymes A/S, Bagsvaerd, Denmark) containing cellulase and cellobiase activities; Rohapect VR-C supplied by AB Enzymes (Darmstadt, Germany, pectinase, protease, and hemicellulase); Pektinase L-40 (pectinase) from ASA Spezialenzyme GmbH; Rapidase citrus oil (pectinase and proteases) produced DSM Food specialties; and Gammazym ANP Z1143 produced by Gamma Chemie GmbH (proteases). All these enzymes have an optimum pH between 3.5 and 5 and optimal temperatures between 40 and 60 °C. Enzymatic hydrolysis has been realized at 50 °C under stirring (150 rpm) and at pH 4.3.

It appears that the four best yields are obtained with enzymatic mixtures composed with a large range of activities, and not with a single enzyme. Among these four mixtures, three appear to be equivalent, and only the mixture with pectinases and protease appears to be less efficient. The best three yields are obtained with mixtures of cellulases and pectinases, with or without proteases.

In a recent study, the extractions of turmeric oil with an enzymatic pretreatment or not were compared [67]. The results show that the extraction yields are better with enzymes than without, either for oleoresin, for curcumin, or for the volatile oil. However, the choice of enzymes is of prime importance since some mixtures are of benefit to the extraction (α-amylase and glucoamylase), while others are inefficient (xylanase and cellulase).

Enzymatic extraction operates in this study because the enzymes have a selected pH and optimum temperature close to each other.

Rosenthal et al. have demonstrated that combining protease activity with cellulases significantly improves extraction yields of oil and protein from soybean [68]. However, proteins are hydrolyzed during extraction process, and a hydrolyzate is obtained with oil. Multistep enzymatic extraction is now in development at the laboratory scale. Either because the activities that have been identified as necessary for the hydrolysis of the walls do not work in the same ranges of pH or temperature or because it seeks to extract several products that are sensitive to enzymes (extract protein at first, and then add protease to extract oil in a second time) [69].

8.4.2 pH

pH is an important parameter for two main reasons: its action on vegetal walls and its effect on enzyme efficacy.

We can notice about the pH action on the cell wall polysaccharides that:

- Hydrogen bonds' stability depends on the interval of pH (breaking in acid environment).
- Furthermore for breaking hydrogen bonds, an acidic pH allows mobility of Ca^{++}, which is involved in the polyuronic chain support and makes thus easier the falling off of pectin cohesion.

So, acidic pH provides an increase in wall plasticity and intermolecular sliding and facilitates the extension. But at this moment, molecules are not lysed.

The second important effect of an acidic pH is the enzyme activity dependence. Indeed, enzymes are proteins so they will have different ionization depending on the pH (as well as some of their substrates).

So pH changes the enzymatic protein structure and thus its bonding capacity to the substrate. As the enzyme catalytic capacity is strongly depending on this bonding capacity between protein and substrate, an inappropriate pH might strongly decrease the enzymatic activity. pH-related denaturation may be irreversible to some enzymes.

Moreover, it is surely likely that with an enzyme mixture, the optimal pH of each enzyme will not be the same.

It is necessary to define an average pH, which is suitable for each enzyme activity in the mixture.

8.4.3 Temperature

Like the pH, temperature will have an effect on raw materials, for example, on environment viscosity and also on enzymes efficacy.

The raw material that we have to hydrolyze is most of the time composed of a mix of polysaccharides. The rheological behavior of these polymers is depending on temperature while they are in water. The viscosity of a mixture of cellulose and water might slightly decrease when the temperature increases. For example, starch is a gelling agent that increases strongly the environment viscosity with temperature increasing. The pectin, which is soluble at low temperature, has the capacity to form a gel by cooling down after a heating step: this is the principle of making jelly and jam. The medium viscosity could thus change a lot with temperature variation. These modifications might be problematic during the process for the heating or stuff transfers.

Enzymes are also really temperature sensitive. Like for most of the chemical reactions, catalytic activity of the enzyme is higher with increasing temperature. However, since they are also proteins, thermal denaturing is possible with a heating step. A 3D structure modification, even a little one, could stop the substrate from bonding and thus avoid the chemical reaction wanted. Denaturing temperature is globally related to the producer organism. Generally, an animal enzyme has a denaturing temperature of 40–45 °C, whereas this temperature is more than 60–65 °C for enzymes produced by microorganisms. Nevertheless, some enzymes are thermo-resistant and could tolerate more than 100 °C during several minutes. The temperature reaction is an important factor that should be kept in mind for enzymatic extraction. Up to a certain point, the temperature could improve the enzymes efficiency whose result is a decreasing viscosity of the environment by hydrolyzing polysaccharides. With temperature effect, these same compounds might increase the environment viscosity and thus disturb the enzyme activity by stopping heating and stuff transfer.

8.4.4 Stirring

Stirring does not directly affect enzymes or raw material. It is however important to consider this parameter for several reasons:

- It must ensure a good heat transfer to support the action of enzymes.
- It must allow a good mass transfer. The enzymes should be present in the entire reaction volume and interact with their substrate at every moment.
- It must limit the shear to prevent emulsification if the raw material processed contains oil.

8.4.5 Seed/Water Ratio

Hagenmaier studied the extraction of sunflower oil at a seed/water ratio of 1:10 [70]. On the same seeds, Dominguez et al. coupled enzymatic hydrolysis followed by pressing [48]. The seed/water ratio was studied to 2:5 (m:v) that promotes the activity of enzymes; for 1:5 it would provide a better digestibility of the cake. The same team fixed 2 years later the optimum ratio to 1:8. In 1992, Badr and Sitohy determined an optimum for 1:3 [71].

The seed/water ratio is very often discussed in the literature; it is always different depending on the raw material studied. The ratio is very often different even on the same raw material as shown in previous example. Agitation is a critical factor, and the viscosity of the medium is dependent of the seed/water ratio. The experimenter will define this ratio according to equipments available in the lab for stirring. Results shown in articles that will optimize this ratio do it accordingly to the equipement available. It is normal that these results vary from one study to another.

8.4.6 Enzymes/Seeds Ratio and Hydrolysis Time

Likely to the seed/water ratio, seed/enzyme ratio and hydrolysis time are very discussed. In theory, hydrolysis time and enzyme concentration are very correlated: for example, time of hydrolysis can be divided by two, by doubling the concentration of enzymes for similar results (if not saturated). On the other hand, by decreasing the amount of enzyme and increasing the hydrolysis time in a proportional manner, comparative results can be achieved. In practice the effectiveness of the enzymes decreases more or less quickly when the hydrolysis time is prolonged. Different inhibition factors may occur: from the raw material, from products of reactions, from the process factors, etc. Besides, enzymes effectiveness decreases over time.

To find a good ratio, it is necessary to take different factors into consideration:

- Enzymes have a cost, as the energy required for the process. It is necessary to find a compromise between the amount of enzyme and the hydrolysis time.
- Yield: another economic factor. The reaction must be sufficiently effective to give good yields and quality of molecules of interest.
- Microbiology: a mixture of water and organic matter is a very favorable environment for the development of microorganism. The overall hydrolysis time must take into account this factor.
- Product quality: some products may be sensitive to oxidation or temperature. An extraction for too long can affect the quality of the oil or the concentration of antioxidant that is sought to be extracted.

8.5 A Wide Variety of Products

8.5.1 EAE Impacts on Products Quality

Most of the time, oils extracted by enzymatic aqueous process are of better quality than other oils considering the general quality attributes such as fatty acid profiles, free fatty acid contents, fatty acids composition, iodine value, saponification number, unsaponifiable matter, peroxide value, phosphorus amount, nutrient content (amount of tocopherols, etc.), refractive index, density, color, etc. [69, 72].

Considering other fractions obtained from the plant from enzyme-assisted aqueous extraction, coproducts with a higher added value than those obtained by methods using organic solvents are produced. Thus, proteins, polysaccharides, or active compounds such as polyphenols can be extracted with good yields while preserving their intrinsic qualities.

The main difference of EAE with other methods is that the enzymes can selectively cut covalent bonds in the raw material. Only enzymes enable this directed and selective hydrolysis of covalent bonds within the cell wall. As a result, molecules of interest such as phenolic acids present in vegetable cell walls can be released, and extracted yields are sometimes clearly superior to those obtained even with organic solvents. Thus, the choice of the composition of enzyme mixtures for the enzymatic modification of the plant is a crucial point. For example, if the objective of splitting the raw material is the production of proteins in their native form, proteases should be excluded from the process, and the extraction conditions will remain soft (temperature, pH, etc.). Most of the time, these soft physicochemical conditions are required to preserve the quality of enzymes and the efficiency of EAE.

Today the products obtained by EAE are more numerous (proteins, polysaccharides, oligosaccharides, oil, phenolic compounds, etc.) and abundantly described in the literature. This increase is very significant in recent years, and this phenomenon is reinforced by effective and growing consideration of environmental issues and the sustainable development of processes.

Among different molecules sought by enzymatic aqueous extraction, the oil is the product that has undergone the greatest number of program search for many years. Today, the enzymatic aqueous extraction projects grow much more widely, and all of the compounds of interest from a plant raw material are desired.

The oil quality is most often estimated by the authors through the amount of free fatty acids contained in the oil after extraction. Enzymatic aqueous extraction processes could potentially affect the oxidation of oils; free fatty acids and peroxide value are therefore the first factors considered. Historically, oils extracted in the early years had problems of oxidation stability. Thus, the oil obtained by pressing assisted by enzymes showed a lower quality than that obtained by cold pressing, according to Sosulski et al., but the oil quality was better than the oils treated by solvent [31]. Free fatty acids were present in larger quantities in the study of Bocevska (1993) with 1.5 % free fatty acid against 1.1 % for the control [41].

Ramadan et al. (2009) concluded on higher peroxide value in the case of aqueous extraction of goldenberry pomace oil [66]. They have noticed a decrease in radical scavenging activity and oxidative stability for this oil extracted with enzymes probably due to a relative decrease in the levels of unsaponifiables and polar fractions (polar lipids and phenolics). This decrease can be attributed to the extraction in aqueous phase. In contrast, other authors, such as Najafian et al. (2009), saw no significant differences in acidity and peroxide or iodine values, by comparing the two types of processes [73]. Tano-Debrah did not notice loss of quality in extraction assisted by enzymes (using proteases, cellulases, and pectinases) of coconut oil [74]. According to Towa (2010) and Latif and Anwar (2009), enzymatically extracted soybean oil has less free fatty acid than its oil control [75, 76]. This level of quality, equivalent to that obtained by conventional methods, was also reached on rapeseed by Zhang in 2007 [77]. Compared to a method of solvent extraction, rapeseed oil extracted enzymatically in this study shows more free fatty acids but has a lower peroxide value, and the color is darker than that obtained from solvent. Iodine value, saponification index, and fatty acid composition are similar in both cases. In the same way, no significant differences were observed in the fatty acid profile, the density, the refractometry index, the amount of free fatty acids, the iodine value, the color index, the saponification, and the amount of unsaponifiable material of soybean oils [76]. The same study reports an improvement on the tocopherol content and on the oxidative stability. The same authors have already seen an increase in the tocopherol content of 10 % in 2009 on a sunflower oil [78]. Ramadan et al. (2009) also show that the levels of minor compounds values, such as phenols and tocopherols, may be higher in the oils extracted with water [66]. However, the opposite effect was observed for cottonseed oil, as demonstrated by Latif and Anwar (2011) [76], which may indicate that these results are dependent on the raw material, the extraction technique, and the enzymatic treatment conditions including the nature of the enzymes employed. Soto et al. (2007) confirmed this hypothesis [79].

Thus, according to the studies examined, oils from enzymatic processes have sometimes better quality and sometimes lower quality than oils extracted by simple cold pressing or solvent. These results are linked to the conditions of aqueous extraction. However, some general trends are confirmed. The fatty acid composition does not vary and therefore also the iodine value. The color of the oil extracted and the unsaponifiable matter is the same or higher than the oils witnesses but is never lower. The levels of free fatty acids and peroxide indices are dependent on the raw material, the conditions of implementation of the method, the oil separation, and the storage of oil and seeds. The tocopherols and phenolic compounds (both antioxidants) are generally higher in oils extracted enzymatically conferring better oxidation resistance to oils [76, 78], contrary to what has been shown by Bocevska et al. in 1993 [41] but confirmed in recent studies. Thus, the quality of virgin coconut oil extracted thanks to enzymes is enhanced and better than commercial virgin oil for free fatty acids and peroxides [72]. Moreover, contents in lauric acid and vitamin E were found to be higher in virgin oil extracted thanks to enzymes than in others oils (commercial virgin coconut oil and commercial standard coconut oil). Considering the sensory analysis, coconut oil extracted with enzymes gives better scores than both other oils for desirable attributes like freshness and nutty odor. One last advantage to quote but not least, overall quality of oils extract enzymatically facilitates refining stages [80].

Generally, enzymatically extracted oils prove to be better than other oils. It seems that the oxidation problems sometimes observed would be more related to storage conditions rather than process.

Vegetable proteins are the second category of sought molecules thanks to enzymatic extraction. An increasing interest is focused on proteins which are described as a valuable coproduct of oil extraction. Proteins can be a benefit to enhance nutritive value and functional properties of food [81]. However, the process used to fractionate oilseed to valuable oil and proteins among all has a direct impact on end-product performances. This is particularly true for the functional properties of proteins that can be affected by the operational conditions of extraction associated with denaturation (organic solvent, pH, temperature, salts, or ionic strength). Ramadan et al. (2009) [66] have demonstrated that the enzymatic attack of the cell wall during enzymatic extraction of oil from goldenberry pomace helps to improve the extraction yields but also the quality of the meal with reduced fiber content and enhanced digestibility. Many authors confirmed the improvement of the quality of the meal obtained after enzymatic extraction of borage oil [31, 32, 79]. A recent study from Bagnasco et al. (2013) presents an enzyme-assisted aqueous extraction method performed on rice to value all different fractions obtained after conversion of raw rice into white rice like rice hull and bran mainly [82]. A promising way to value proteins is to combine an enzymatic hydrolysis of proteins during extraction to prepare mixtures composed of proteins and peptides with sensory properties from rice middlings as raw material.

The other constituents of the plant raw material are not to overlook as they represent a potential. Thus, polysaccharides and their hydrolysis products, phenolic compounds, and other nutrients are some of interesting molecules.

According to Aliakbarian et al. (2008) who quoted several authors, the addition of enzymes helps reducing the complexation of hydrophilic phenolic compounds with polysaccharides and thus increases the amount of free phenol present in the oil and aqueous phase [83]. Moreover, the authors indicate a correlation between the concentration of enzymatic mixtures and highest levels of total polyphenols and antiradical power. Fan et al. (2011) compared the traditional hot water extraction of polysaccharides from *Grifola frondosa* with enzyme-assisted aqueous extractions [84]. Several enzyme mixtures were tested with activities used alone or in mixtures. The results are clear: extraction with enzyme mixtures contributes, on the one hand, to improve the extraction yields and, on the other hand, to improve the antioxidant efficacy of extracted polysaccharides. The main explanation is that the hydrolysis due to enzymes helps to reduce the molecular weight of polysaccharides, physical characteristic that is in favor of the antioxidant activity. While not exhaustive, mention may be made to Fu et al. [85] who first proposed the extraction of luteolin and apigenin from pigeon pea (*Cajanus cajan* (L.) Millsp.) leaves. They compared efficiency of pectinase, β-glucosidase, and cellulase to optimize extraction of these active flavonoids.

Among raw materials, which will be to consider in the near future, algae represent a new wide field of investigation to produce new valuable molecules. Enzymatic aqueous extraction is one of technologies that will split the algal material in an eco-designed manner [85].

8.5.2 Reducing Emulsion Formed During EAE

During extractions performed on oil commodities, an emulsion composed of water and oil may occur and block the recovery of free oil. Often this major pitfall was the reason for the abandonment of this method for the extraction of oil materials. Strategies can be implemented to prevent the formation of these emulsions (e.g., by limiting the production of emulsifying molecules) or then to destabilize them at the end of the process. During the extraction of oil from soybean flour, Chabrand and Glatz (2009) described the formation of a stable emulsion and the near absence of free oil [86]. They proposed to combine the action of protease and phospholipase in a mixture to hydrolyze interfacial proteins and phospholipids to foster the breakdown of oil droplets and promote coalescence. In order to be cost-effective, they describe very high percentages of protease activity recycling (90 %). Similarly, enzymes (aspartic endoprotease) are employed to successfully destabilize oil-in-water emulsion found in coconut milk to produce virgin coconut oil [72]. The protease hydrolyzes peptide bonds in the protein chain and helps to reduce the emulsifying properties, which leads to coalescence of oil droplets.

Although it is more advantageous to reduce the formation of emulsions, certain raw materials such as soybeans result in significant amounts of emulsified molecules because of seed composition. Jung et al. (2009) proposed using several enzyme mixtures, accurately selected, to help to deconstruct the emulsion formed during EAE of soybeans and to improve yields of free oil [80]. Analyses showed that the produced oil is of high quality and requires less refining steps than hexane-extracted oil (degumming). Recently, Li (2014) proposed treating the cream obtained after enzymatic-assisted extraction of soybean oil by an ethanol ultrasound-assisted destabilization [87]. Almost 95 % of the oil contained in the cream was recovered after optimization of operating conditions of the ultrasound system.

One other valuable option would be to promote these natural emulsions in areas that need and request them such as cosmetics or detergents industry.

8.5.3 Bifunctionality of Enzymes

Chen et al. (2011) present an original approach where enzyme-assisted aqueous extraction was combined with enzymatic modification using a unique activity of cellulase from *Penicillium decumbens*, both hydrolyzing plant cell walls and leading to transglycosylation of flavonols extracts [88]. In this study, flavonoids are extracted thanks to aqueous and enzymatic process from *Ginkgo biloba* leaves. That cellulase is known to possess a strong transglycosylation activity that can change the flavonol aglycones into polar glycosides whose higher solubility in ethanol-water extractant leads to improved extraction without affecting the bioactivity of the compounds. This example rather rarely described in the literature highlights another strong asset of enzymatic processes in which extraction and functionalization are combined in one step using a single catalyst. A similar approach is presented by Xu et al. (2013) to enhance extraction of bioactives from *Glycyrrhizae radix* [89]. They confirm the synergistic action of bifunctional enzymes contributing to both improve extraction yields degrading the cell walls but also deglycosylate flavonoids extracts helping to make them more active. The use of enzymes able to perform both transformations in the same batch contributes significantly to the improvement of technical and economic performances of EAE processes.

8.6 Combination of Different Alternative Methods

Additional processing could very likely improve the efficiency of enzymatic treatments or contribute to improved yields. These complementary treatments may occur before, after, or in parallel with enzymatic treatment of biomass. Among the alternatives that can be coupled to the enzyme extraction, mention may be made to microwave- and ultrasound-assisted extractions and also hydrodistillation.

In a recent study of Li et al. (2013), the combination of the effect of microwave-assisted extraction to the enzymatic hydrolysis seemed advantageous for the oil and protein recovery from yellowhorn seed kernels [90]. Jiao et al. (2014) confirmed it by obtaining the best extraction yields of pumpkin seed oil performing an enzymatic pretreatment (cellulase, pectinase, protease) on the seeds before treatment with microwaves [91]. Furthermore, the oil obtained by this combination of methods is comparable in respect to physicochemical aspects with oil extracted with hexane (refractive index, gravity, acid, and saponification values) and is more stable to oxidation due to higher contents of phenolics and tocopherols among others. Gai et al. (2013) have combined the enzymatic extraction with microwaves treatments to improve yields of oil extraction of *Isatis indigotica* seeds and *Forsythia suspense* seeds [92]. In both cases, they confirm the better oxidative stability of oils with comparable physicochemical properties than solvent extracted oils.

Ultrasound pretreatments have also been associated with enzymatic aqueous extraction for extracting watermelon seed oil [93]. Yields obtained are 20 % higher than those obtained by the enzymatic extraction only. The same effect are measured by Konwarh et al. (2012) who reported improvement of the extraction of lycopene combining enzymatic extraction employed with ultrasounds [94]. Chen et al. (2012) present a study combining ultrasounds and enzymes to improve the extraction of crude polysaccharides from Epimedium leaves to produce active ingredients from plant materials [95]. Xu et al. (2013) have compared the extraction of flavonoids from *Glycyrrhizae radix* by enzyme aqueous extraction or sonication-assisted extraction [89]. The results show an equal efficiency of both processes carried out separately. Da Porto et al. (2013) assessed the impact of an enzymatic pretreatment on the extraction of grape-seed oil by conventional solvent extraction and ultrasound one [96]. The results show that the enzymes significantly improved the extraction with solvent, but have no significant effect on the ultrasound extraction, which yields remain unchanged. These last studies demonstrate the need to optimize the sequence of methods and their settings.

Adulkar et al. (2014) propose combining the action of hydrolytic enzymes and ultrasound to remove grease wastewater output of dairy [97]. Although this method does not involve vegetable raw materials, it illustrates the synergistic action of these two methods. The optimization technique takes into account several parameters including enzyme concentration, temperature, ultrasound power, frequency, duty cycle, and agitation speed. The conclusion confirms the benefit of ultrasonication combined with enzymatic catalysis.

Hosni et al. (2013) perform an enzymatic pretreatment (with cellulase, hemicellulase alone, and a mixture of them) prior to hydrodistillation for recovery of essential oils from thyme (*Thymus capitatus L.*) and rosemary (*Rosmarinus officinalis* L.) [98]. As well as for yields of extraction than for antimicrobial activities of essential oils, results are better when an enzymatic pretreatment is done.

Nevertheless, when different extraction methods are used together to achieve better treatment efficacy and better quality products, it is sometimes necessary to consider the negative impact of certain methods on the activity of enzymes and their subsequent denaturation. Thus, as shown by Kapturowska et al. (2013), the ultrasound treatment may be deleterious for enzymatic activity of lipases extracted from *Yarrowia lipolytica* [99]. In these conditions, unless they have ensured the preservation of enzymatic activity, protocols involving different successive steps should be favored.

In conclusion, most of the time, the combination of different alternative extraction methods is favorable for yields and product quality. These studies, varied but not exhaustive, demonstrate the synergies that can be beneficial to the performance of clean extraction processes. However, the mixed results of some attempts show the importance of optimizing the combinations defining whether their actions are simultaneous or successive and in what order. The number of articles with such studies is growing and probably foreshadows a key sector in the coming years.

8.7 Applications of EAE

EAE encounters many applications from reactions at the laboratory scale up to industrial processes. This technique is considered as an alternative method to produce valuable products *via* smooth conditions. Due to the large variety of available enzymes, possibilities to find the most appropriate enzymatic tools for the extraction are almost infinite. One of the main advantages of EAE, besides its eco-friendly character, is the specificity of enzymes. Depending on the starting raw material, one can choose the most suitable enzyme or mixture of enzymes to drive the desired reaction. EAE is widely used at the laboratory scale, and many research studies are conducted on this topic. Emerging niche markets encourage the development of such techniques. For example, the cosmetics market is increasingly demanding of natural extracts produced in a "greener" way. EAE is then gaining more importance and interest in this field.

8.7.1 Laboratory Scale

EAE is easily applicable at the laboratory scale as the process does not request very sophisticated equipment. The enzyme-assisted aqueous extraction is mostly studied to produce extracts containing valuable molecules but also to extract-specific compounds difficult to isolate with conventional techniques. At a laboratory scale, more enzymes are available which makes the development of such elaborated processes easier to develop. Research works are also driven in parallel of extraction processes to broaden the databank of available enzymes in order to find more appropriate tools. The main objective is to define the best biological catalyst to run the extraction specifically on a plant or on parts of a plant.

The enzyme-assisted aqueous extraction of molecules from plants is getting more and more studied to develop alternative solutions than the use of hexane and other harmful solvents. EAE is a process used for the extraction of oils or valuable extracts with high contents in molecules of interest. Vegetal substrates such as soybean, sunflower, rapeseed, and peanut are widely studied because they imply large quantities at an industrial scale and thus imply large volumes of solvents for their process [100, 101].

8.7.2 Pilot/Industrial Scale

8.7.2.1 Biofuel Applications

Some industrials are already using EAE at large scale. The most noteworthy example is the extraction of cellulose-based materials for the production of fermentable sugars for the biofuel industries. As mentioned previously, this chapter mainly focuses on the EAE of non-lignocellulosic vegetal. However, as a brief illustration, one can cite the pilot plant FUTUROL located in Pomacle-Bazancourt in the north-east region of France. In this project launched in 2008 (Euros 76.4 million), a consortium of several R&D collaborators (IFP, ARD, INRA, Lesaffre) is developing new energy solutions including EAE of biomass for the production of biofuels. The main objective is to produce efficient alternative solutions for energy production valorizing agriculture by-products. The first results from the pilot plant (capacity 180,000 L/year of bioethanol) are promising since biofuels with competitive prices can be produced. The third phase of the project concerning the transition to commercial-scale production, and the increase of the capacity up to 3.5 million liters per year is scheduled to be launched in 2016.

8.7.2.2 Industrial Applications (Food, Health, and Beauty Care Products)

Enzymes are widely used at an industrial scale but mainly as biocatalysts for synthesis and modifications of molecules today. Schmid et al. and Rolle have published reviews establishing non-exhaustive lists of enzymatic reactions run at an industrial scale [102–104]. Industrial biocatalysts users count BASF, Schering, Lonza, Shimizu, Kyowa Hakko Kogyo Co. Ltd, Tosoh, DSM, and many more. There is a broad range of applications mainly concerning the synthesis of pharmaceuticals, compounds for chemistry of specialty, polymers, etc.

Roquette and Cargill are using enzymatic processes to produce polysaccharide-derived ingredients (starch, oligosaccharides). Enzymatic extraction is also used for the production of vegetal proteins: protein-rich plant materials are digested by specifically designed enzymes, which are able to hydrolyze the constitutive plant cell walls. This digestion results thus in the release of the inner content of vegetal cells (proteins, phenolic compounds, sugars, etc.). In this context, EAE is a potential powerful technique for the food industry for the production of vegetal proteins in replacement of animal origin proteins. In this field, the RuBisCO protein (ribulose-1,5-bisphosphate carboxylase/oxygenase), which main function is to catalyze the first step of CO_2 fixation in most autotrophic organisms, is considered as the most abundant vegetal protein on Earth [105]. It is at the heart of concerns for meat protein substitutes. The RuBisCO products taste very much like meat, and its bite has almost the same consistence. Some research groups have developed interesting processes for the extraction and purification of RuBisCO protein for food applications from Lucerne [106–108].

Enzyme-assisted extraction allows the production of a broad variety of bioactive compounds. In their review published in 2012, Puri et al. have established a short list of bioactive compounds available by enzymatic extraction processes [109]. Those compounds (oils and carotenoids, glycosides, phenolic compounds, proteins, flavonoids, fibers, etc.) are produced choosing the most appropriate enzymatic tools and have usually interesting yields compared with conventional techniques. EAE already have several industrial applications. For example, pectinases are used since more than 30 years in the juice industry for de-pulping and for clarification [110]. Proteases and lipases are used in the leather-making industry as well as in the protein-enriched food industry and in the pharmaceutical industry. Lipases find applications in the production of skin care products while phospholipases are mainly used as degumming agent for oil-rich seeds (soybean, canola, flax seed, hemp, etc.) [111, 112]. As early as the 1960s, Roquette launched the first enzymatic industrial reaction for the bioconversion of starch. Rapidly after, they were pioneering with a formulated dual-enzymes hydrolysis range for the production of sugar syrups. The company ARD is also a leader in the field and produces ingredients *via* enzymatic fractionation of vegetal through its daughter company Soliance cosmetics. NIZO research group, located in the Netherlands, are using proteases to produce and extract valuable peptides from proteins for the food industry. Enzymatic tools also find applications in synthesis and modification of molecules. For example, Cargill is using enzyme technology to modify molecules to improve food application products performance (transesterification) (Transcend TM enzymatic alternatives).

EAE is also interesting as a pretreatment step to predigest plant materials to increase the bio-disponibility of the molecules of interest and allows the reduction of reaction time [32, 113]. As an example, a predigestion step of tomato with pancreatin can increase the extraction yield of lycopene up to 2.5-fold compared with the solvent extraction only [114]. As mentioned previously, the enzymatic extraction can be used successively or in combination with other technology processes.

8.7.2.3 Biorefinery

New companies are developing EAE following the principle of biorefinery [115]. This concept established in 1997 defined the biorefinery as an overall concept of a processing plant where biomass feedstocks are converted and extracted into a spectrum of valuable products (E3 handbook, US Dept of Energy). This principle leads to the reduction of production costs since every phase issued from the extraction is valuable (aqueous extract, oily phase, residual undigested material, emulsion).

A few existing industrial applications of enzymatic biorefinery already exist, but it does not represent the main part of EAE applications. A young French biotechnology company, BIOLIE, located in Nancy (France) has developed a new technology based on this concept. The extraction of vegetal is realized using suitable formulated cocktails of enzymes and leads to the production of oils and active ingredients among others. The valorization of each phase of the extraction (lipidic phase, emulsion, aqueous phase, insoluble un-hydrolyzed residual material) is then possible (Fig. 8.5).

Further work on those phases leads in several steps to the separation, purification, and enrichment to obtain compounds of interest. The eventual additional costs of production for one ingredient is thus globally balanced with the exploitation of the additional phases, which can be further refined depending on the needs. The process is currently barely exploited, but its potential is very promising.

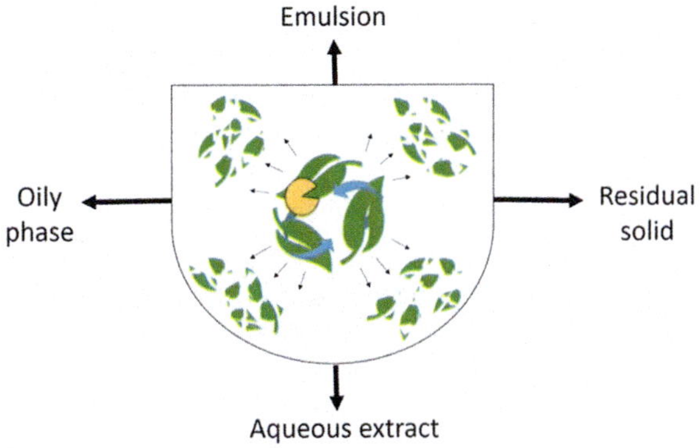

Fig. 8.5 Principle of enzymatic aqueous extraction

8.7.3 Pros and Cons of EAE at an Industrial Scale

At the industrial scale, EAE has two main advantages being firstly to avoid the risk and safety issues surrounding the use of solvent and particularly the use of hexane. EAE is considered as an environmentally friendly alternative to the solvent extractions of oils and as a sustainable process for the limitation of VOC emissions. Secondly, even though yields could be lower with EAE than those obtained with conventional solvent extractions, the quality of the products produced *via* this smooth process is usually higher. It is well known that products issued from enzymatic extraction like oils and aqueous extracts are richer in valuable actives and have higher contents in nutriments. The "smooth" reaction conditions are preserving the nutritional and organoleptic properties of ingredients. The process preserves the integrity of the benefit molecules of plants. This specificity of the process is a major sale argument in the food and cosmetic industries. At an industrial scale, this point is a serious advantage concerning the economic part since, in a single step, products have a better quality and are significantly enriched in valuable compounds.

One of the drawbacks of EAE is the cost of production. For large volumes of production, the price of enzymes which are not always commercially available represents the bottleneck of this technology [116]. For food applications, concerns are given to the presence of eventual residual traces of proteins in the products as a source of known allergens. Legislation is not completely defined on this topic, and no cases have been described yet.

Besides this point, EAE may lead to the reduction of production costs (savings in energy, time, and equipment facilities but also savings in costs of waste disposal, etc.) combined with improved product quality. Most of the available enzymes request fairly low reaction temperatures (usually less than 50 °C), resulting in a reduction of energy needs. Moreover, the efficiency of enzymes and the specificity of their chemical activity allow shorter extraction times.

Purification of extracts issued from EAE is usually more complex than those produced from solvent extractions. This is due to the global release of a higher diversity of molecules in the phases [109].

EAE has potential applications in new product generation for food and cosmetic markets [117]. Molecules of interest such as phenolic compounds can be extracted by physicochemical or mechanical methods; however, those compounds can be anchored in the plant cell walls *via* covalent bonds. In that case, extraction by conventional techniques is less efficient than an enzymatic extraction where enzymes are able to act specifically.

It is possible to build a theoretical economic model for an EAE process, even with expensive biocatalysts since as detailed previously the costs of production can be globally reduced. However, very few commercial biocatalysts are available from off the catalog, and this new technique might be in some cases overconsidered as a solution for the extraction of plant material necessitating specific enzymes. Moreover, enzymatic reactions are usually more difficult to scale up for large volumes since enzymes behave differently depending on parameters. Some alternative methods are using supported enzymes, which allows the recycling of the biocatalyst several times with no loss of activity [118, 119]. In that case, the life cycle of enzyme is directly related to its resistance properties and to its condition of use. The impact of the cost of enzymes is thus smoothed.

8.8 Conclusion

Observing the increasing number of articles on the topic of eco-extraction of the plant associated with studies demonstrating the relevance of the fractionation of plant raw materials by enzymatic aqueous extraction (EAE), that sector will show significant growth in the coming years. Due to environmental constraints increasingly strong and awareness of the need to consume healthy products, techniques to promote all fractions of the plant while preserving the intrinsic qualities of natural products are being valued. In this context, the success of biocatalysis depends ultimately on the economics of this process and thus on the exploitation of the maximum of products. The extraction process following the principles of biorefinery appears to be the best solution. At the same time, prices related to the implementation of enzymatic processes will decrease while the value-added products will grow. Economic balance may then equilibrate. The enzyme-assisted aqueous extraction provides rich phases with high potential in valuable molecule contents. As the process is raising increasing interests from academics and industrials, experience and knowledge on this technique will be rapidly broaden, and its application will become easier. The design of a broader range of enzymatic tools and the availability of those enzymes at an industrial scale should encourage the development of applications and the uses of this technique.

A new industry is to build with new raw materials leading to new quality products to sell on the market.

References

1. Chemat F (2011) Les six principes de l'éco-extraction. In: Eco-extraction du végétal. Procédés innovants et solvants alternatifs. Technique et ingénierie. Dunod Ed, pp 3–24
2. Johnson L, Lusas E (1983) Comparison of alternative solvents for oils extraction. J Am Oil Chem Soc 60(2):229–242
3. Rosenthal A, Pyle DL, Niranjan K (1996) Aqueous and enzymatic processes for edible oil extraction Enzyme. Microb Technol 19(6):402–420
4. Yamada T, Sakaguchi K (1982) Comparative studies on chlorella cell walls: induction of protoplast formation. Arch Microbiol 132(1):10–13
5. Vladimirescu A (2010) Isolation of permeaplasts and spheroplasts from *spirulina platensis*. Rom Biotechnol Lett 15(3):5361–5368
6. Blanc G, Duncan G, Agarkova I, Borodovsky M, Gurnon J, Kuo A, Lindquist E, Lucas S, Pangilinan J, Polle J, Salamov A, Terry A, Yamada T, Dunigan D, Grigoriev I, Claverie JM, Van Etten J (2010) The chlorella variabilis NC64A genome reveals adaptation to photosymbiosis, coevolution with viruses, and cryptic sex. Plant Cell 22(9):2943–2955
7. Kodner R, Summons R, Knoll A (2009) Phylogenetic investigation of the aliphatic, non-hydrolyzable biopolymer algaenan, with a focus on green algae. Org Geochem 40(8):854–862
8. Allard B, Templier J (2001) High molecular weight lipids from the trilaminar outer wall (tls)-containing microalgae Chlorella emersonii, Scenedesmus communis and Tetraedron minimum. Phytochemistry 57(3):459–467
9. Domozych D, Ciancia M, Fangel J, Mikkelsen M, Ulvskov P, Willats W (2012) The cell walls of green algae: a journey through evolution and diversity. Front Plant Sci 3(82):1–7
10. Popper Z, Michel G, Herve C, Domozych D, Willats W, Tuohy M, Kloareg B, Stengel D (2011) Evolution and diversity of plant cell walls: from algae to flowering plants. Annu Rev Plant Biol 62:567–588
11. Michel G, Tonon T, Scornet D, Cock J, Kloareg B (2010) The cell wall polysaccharide metabolism of the brown alga Ectocarpus siliculosus. Insights into the evolution of extracellular matrix polysaccharides in Eukaryotes. New Phytol 188(1):82–97
12. Pasquet V, Cherouvrier JR, Farhat F, Thiery V, Piot JM, Berard JB, Kaas R, Serive B, Patrice T, Cadoret JP, Picot L (2011) Study on the microalgal pigments extraction process: performance of microwave assisted extraction. Process Biochem 46(1):59–67
13. Liu C, Lin L (2001) Ultrastructural study and lipid formation of *isochrysis sp ccmp1324*. Bot Bull Acad Sin 42:207–214
14. Zhang Y-HP, Lynd LR (2006) A functionally based model for hydrolysis of cellulose by fungal cellulase. Biotechnol Bioeng 94:888–898
15. Howard R, Abotsi E, Jansen van Rensburg EL, Howard S (2003) Lignocellulose biotechnology: issues of bioconversion and enzyme production. Afr J Biotechnol 2(12):602–619
16. Shallom D, Shoham Y (2003) Microbial hemicellulases. Curr Opin Microbiol 6(3):219–228
17. Silchenko AS, Kusaykin MI, Kurilenko VV, Zakharenko AM, Isakov VV, Zaporozhets TS, Gazha AK, Zvyagintseva TN (2013) Hydrolysis of fucoidan by fucoidanase isolated from the marine bacterium, Formosa algae. Mar Drugs 11(7):2413–2430
18. Descamps V, Colin S, Lahaye M, Jam M, Richard C, Potin P, Barbeyron T, Yvin JC, Kloareg B (2006) Isolation and culture of a marine bacterium degrading the sulfated fucans from marine brown algae. Marine Biotechnol 8(1):27–39
19. Kim EJ, Fathoni A, Jeong GT, Jeong HD, Nam TJ, Kong IS, Kim JK (2013) *Microbacterium oxydans*, a novel alginate- and laminarin-degrading bacterium for the reutilization of brown-seaweed waste. J Environ Manage 130:153–159
20. Suda K, Tanji Y, Hori K, Unno H (1999) Evidence for a novel Chlorella virus- encoded alginate lyase. FEMS Microbiol Lett 180(1):45–53
21. Ma S, Tan YL, Yu WG, Han F (2013) Cloning, expression and characterization of a new ι-carrageenase from marine bacterium, Cellulophaga sp. Biotechnol Lett 35(10):1617–1622

22. Ma YX, Dong SL, Jiang XL, Li J, Mou HJ (2010) Purification and characterization of carrageenase from marine bacterium mutant strain Pseudoalteromonas Sp. Aj5-13 and its degraded products. J Food Biochem 34:661–678
23. Gupta V, Trivedi N, Kumar M, Reddy C, Jha B (2013) Purification and characterization of exo-β-agarase from an endophytic marine bacterium and its catalytic potential in bioconversion of red algal cell wall polysaccharides into galactans. Biomass Bioenerg 49:290–298
24. Hatada Y, Ohta Y, Horikoshi K (2006) Hyperproduction and application of alpha-agarase to enzymatic enhancement of antioxidant activity of porphyran. J Agric Food Chem 54(26):9895–9900
25. Lahaye M, Brunel M, Bonnin E (1997) Fine chemical structure analysis of oligosaccharides produced by an ulvan-lyase degradation of the water-soluble cell-wall polysaccharides from Ulva sp, (Ulvales, Chlorophyta). Carbohydr Res 304(3–4):325–333
26. Sugimoto I, Onimatsu H, Fujie M, Usami S, Yamada T (2004) vAL-1, a novel polysaccharide lyase encoded by chlorovirus CVK2. FEBS Lett 559(1–3):51–56
27. Sherba SE, Faith WTJ, Smythe CV, Steigerwalt RB (1972) Soybean fractionation employing protease. Patent US3640725A
28. Lanzani A, Petrini M, Cozzoli O, Gallavresi P, Carola C, Jacini G (1975) On the use of enzymes for vegetable-oil extraction. A preliminary report. Riv Ital Sostanze Grasse 11:226–229
29. Fantozzi P, Petruccioli G, Montedoro G (1977) Enzymatic treatment of olive pastes after single pressing extraction. Effect of cultivar, harvesting time, and storage. Rils Ital Sosfanze Grusse 54:381–388
30. Fullbrook P (1983) The use of enzymes in the processing of oilseeds. J Am Oil Chem Soc 60(2):476–478
31. Sosulski K, Sosulski F (1993) Enzyme-aided vs. two-stage processing of canola: technology, product quality and cost evaluation. J Am Oil Chem Soc 70(9):825–829
32. Dominguez H, Nunez M, Lema J (1994) Enzymatic pretreatment to enhance oil extraction from fruits and oilseeds: a review. Food Chem 49(3):271–286
33. Cerda A, Martinez ME, Soto C, Poirrier P, Perez-Correa JR, Vergara-Salinas JR, Zuniga ME (2013) The enhancement of antioxidant compounds extracted from *thymus vulgaris* using enzymes and effect of extracting effect. Food Chem 139:138–143
34. Zheng H, Hwang I, Kim S, Lee S, Chung S (2010) Optimization of carbohydrate-hydrolyzing enzyme aided polyphenol extraction from unripe apples. J Korean Soc Appl Biol 53(3): 342–350
35. Nagendra Chari KL, Mannasa D, Srinivas P, Sowbhagya HB (2013) Enzyme-assisted extraction of bioactive compounds from ginger (*Zingiber officinale* Roscoe). Food Chem 139:509–514
36. Olsen H (1988) Aqueous extraction of oil from seeds. In: Asian food conference, Bangkok, Thailande, 24–26 Oct 1988
37. Sosulski K, Sosulski FW (1990) Enzyme pre-treatment to enhance oil extractability in canola. In: Shahidi F (ed) Canola and rapeseed: production, chemistry, nutrition and processing technology. Van Nostrand Reinhold Publishers, New York, pp 277–289
38. Deng Y, Pyle D, Niranjan K (1992) Studies of aqueous enzymatic extraction of oil from rapeseed. Agric Eng Rural Dev 1 Conf Proc 1
39. Kim IH, Yoon SH (1990) Effect of extraction solvents on oxidative stability of crude soybean oil. J Am Oil Chem Soc 67(3):165–167
40. Bargale PC, Sosulski K, Sosulski FW (2000) Enzymatic hydrolysis of soybean for solvent and mechanical oil extraction. J Food Process Eng 23(4):321–327
41. Bocevska M, Karlovic D, Turkulov J, Pericin D (1993) Quality of corn germ oil obtained by aqueous enzymatic extraction. J Am Oil Chem Soc 70(12):1273–1277
42. Ranalli A, Martinelli N (1994) Extraction of the oil from the olive pastes by biological and not conventional industrial technics. Ind Alimentari 33(331):1073–1083
43. Ranalli A, Martinelli N (1995) Integral centrifuges for olive oil extraction, at the third millennium threshold. Transformation yields. Grasas Y Aceites 46(4–5):255–263

44. McGlone O, Lopez-Munguia C, Cater J (1986) Coconut oil extraction by a new enzymatic process. J Food Sci 51(3):695–697
45. Christensen F (1989) Enzyme technology versus engineering technology in the food industry. Biotechnol Appl Biochem 11:249–265
46. Buenrostro M, Lopez-Munguia CA (1986) Enzymatic extraction of avocado oil. Biotechnol Lett 8(7):505–506
47. Freitas SP, Lago RCA, Jablonka FH, Hartman L (1993) Aqueous enzymatic extraction of avocado oil from fresh pulp. Revue Francaise des Corps Gras 40:365–371
48. Dominguez H, Sineiro J, Nunez M, Lema J (1995) Enzymatic treatment of sunflower kernels before oil extraction. Food Res Int 28(6):537–545
49. Cheah SC, Augustin MA, Ooi L (1990) Enzymatic extraction of palm oil. Palm Oil Res Malaysia Bull 20:30–36
50. Mosier N, Wyman C, Dale B, Elander R, Lee Y, Holtzapple M, LaDisch M (2005) Features of promising technologies for pretreatment of lignocellulosic biomass. Bioresour Technol 96(6):673–686
51. Van Eylen D, Indrawati M, Hendrickx M, Van Loey A (2006) Temperature and pressure stability of mustard seed (Sinapis alba L.) myrosinase. Food Chem 97:263–267
52. Zhang Z, Ober JA, Kliebenstein DJ (2006) The gene controlling the quantitative trait locus EPITHIOSPECIFIER MODIFIER1 alters glucosinolate hydrolysis and insect resistance in Arabidopsis. Plant Cell 18(6):1524–1536
53. Carre B (2000) Effets de la taille des particules alimentaires sur les processus digestifs chez les oiseaux d'élevage. INRA Prod Anim 13(2):131–136
54. Rosenthal A, Pyle D, Niranjan K (1998) Mechanisms in the simultaneous aqueous extraction of oil and protein from soybean. Food Bioprod Process 76:224–230
55. Dickey L, Kurantz M, Parris N (2008) Oil separation from wet-milled corn germ dispersions by aqueous oil extraction and aqueous enzymatic oil extraction. Ind Crops Prod 27(3): 303–307
56. Evon P, Vandenbossche V, Pontalier PY, Rigal L (2007) Direct extraction of oil from sunflower seeds by twin-screw extruder according to an aqueous extraction process: feasibility study and influence of operating conditions. Ind Crops Prod 2:351–359
57. Weil J, Sarikaya A, Rau SL, Goetz J, Ladisch C, Brewer M, Hendrickson R, Ladisch M (1997) Pretreatment of yellow poplar sawdust by pressure cooking in water. Appl Biochem Biotechnol 68(1–2):21–40
58. Shankar D, Agrawal Y, Sarkar B, Singh B (1997) Enzymatic hydrolysis in conjunction with conventional pretreatments to soybean for enhanced oil availability and recovery. J Am Oil Chem Soc 74(12):1543–1547
59. Weil J, Brewer M, Hendrickson R, Sarikaya A, Ladisch M (1998) Continuous pH monitoring during pretreatment of yellow poplar wood sawdust by pressure cooking in water. Appl Biochem Biotechnol 70(2):99–111
60. Lee Y, Wu Z, Torget R (2000) Modeling of countercurrent shrinking-bed reactor in dilute-acid total-hydrolysis of lignocellulosic biomass. Bioresour Technol 71(1):29–39
61. Chang V, Holtzapple M (2000) Fundamental factors affecting biomass enzymatic reactivity. Appl Biochem Biotechnol 84(6):5–37
62. Chang V, Nagwani M, Holtzapple M (1998) Lime pretreatment of crop residues bagasse and wheat straw. Appl Biochem Biotechnol 74(3):135–159
63. Chang V, Nagwani M, Kim C, Holtzapple M (2001) Oxidative lime pretreatment of high-lignin biomass – poplar wood and newspaper. Appl Biochem Biotechnol 94(1):1–28
64. Jiang L, Hua D, Wang Z, Shiying X (2010) Aqueous enzymatic extraction of peanut oil and proteins hydrolysates. Food Bioprod Process 88(C2–3):233–238
65. Mathlouthi N, Saulnier L, Quemener B, Larbier M (2002) Xylanase, b-glucanase, and other side enzymatic activities have greater effects on the viscosity of several feedstuffs than xylanase and b-glucanase used alone or in combination. J Agric Food Chem 50(18): 5121–5127

66. Ramadan MF, Morsel J-T (2009) Oil extractability from enzymatically-treated goldenberry (*Physalis peruviana* L.) pomace: range of operational variables. Int J Food Sci Technol 44:435–444
67. Kurmudle N, Kagliwal LD, Bankar SB, Singhal RS (2013) Enzyme-assisted extraction for enhanced yields of turmeric oleoresin and its constituents. Food Biosci 3:36–41
68. Rosenthal A, Pyle DL, Niranjan K, Gilmour S, Trinca L (2001) Combined effect of operational variables and enzyme activity on aqueous enzymatic extraction of oil and protein from soybean. Enzyme Microb Technol 28(6):499–509
69. Tabtabaei S, Diosady LL (2013) Aqueous and enzymatic extraction processes for the production of food-grade proteins and industrial oil from dehulled yellow mustard flour. Food Res Int 52(2):547–556
70. Hagenmaier FD (1974) Aqueous processing of full-fat sunflower seeds: yields of oil and protein. J Am Oil Chem Soc 51(10):470–471
71. Badr FH, Sitohy MZ (1992) Optimizing conditions for enzymatic extraction of sunflower oil. Grasas y Aceites 43(5):281–283
72. Raghavendra SN, Raghavarao KSMS (2011) Aqueous extraction and enzymatic destabilization of coconut milk emulsions. J Am Oil Chem Soc 88:481–487
73. Najafian L, Ghodsvali A, Haddad Khodaparast M, Diosady L (2009) Aqueous extraction of virgin olive oil using industrial enzymes. Food Res Int 42(1):171–175
74. Tano-Debrah K, Ohta Y (1997) Aqueous extraction of coconut oil by an enzyme-assisted process. J Sci Food Agric 74(4):497–502
75. Towa LT, Kapchie VN, Hauck C, Murphy PA (2010) Enzyme assisted aqueous-extraction of oil from isolated oleosomes of soybean flour. J Am Oil Chem Soc 87:347–354
76. Latif S, Anwar F (2011) Aqueous enzymatic sesame oil and protein extraction. Food Chem 125:679–684
77. Zhang S, Wang Z, Xu SY (2007) Optimization of the aqueous enzymatic extraction of rapeseed oil and protein hydrolysates. J Am Oil Chem Soc 84:97–105
78. Latif S, Anwar F (2009) Effect of aqueous enzymatic processes on sunflower oil quality. J Am Oil Chem Soc 86(4):393–400
79. Soto C, Chamy R, Zuniga ME (2007) Enzymatic hydrolysis and pressing conditions effect on borage oil extraction by cold pressing. Food Chem 102:834–840
80. Jung S, Maurer D, Johnson LA (2009) Factors affecting emulsion stability and quality of oil recovered from enzyme-assisted aqueous extraction of soybeans. Bioresour Technol 100(21):5340–5347
81. Moure A, Sineiro J, Dominguez H, Parajo JC (2006) Functionnality of oilseed protein products: a review. Food Res Int 39:945–963
82. Bagnasco L, Pappalardo VM, Meregaglia A, Kaewmanee T, Ubiali D, Speranza G, Cosulich ME (2013) Use of food grade proteases to recover protein-peptide mixtures from rice middlings. Food Res Int 50:420–427
83. Aliakbarian B, De Faveri D, Converti A, Perego P (2008) Optimization of olive oil extraction by means of enzyme processing aids using response surface methodology. Biochem Eng J 42(1):34–40
84. Fan Y, Wu X, Zhang M, Zhao T, Zhou Y, Han L, Yang L (2011) Physical characteristics and antioxidant effect of polysaccharides extracted by boiling water and enzymolysis from Grifola frondosa. Int J Biol Macromol 48(5):798–803
85. Fu YJ, Liu W, Zu YG, Tong MH, Li SM, Yan MM, Efferth T, Luo H (2008) Enzyme assisted extraction of luteolin and apigenin from pigeonpea [Cajanus cajan (L.) Millsp.] leaves. Food Chem 111(2):508–512
86. Chabrand RM, Glatz CE (2009) Destabilization of the emulsion formed during the enzyme-assisted aqueous extraction of oil from soybean flour. Enzym Microb 45:28–35
87. Li Y, Sui X, Qi B, Zhang Y, Feng H, Zhang Y, Jiang L, Wang T (2014) Optimization of ethanol-ultrasound-assisted destabilization of a cream recovered from enzymatic extraction of soybean oil. J Am Oil Chem Soc 91(1):159–168

88. Chen S, Xing XH, Huang JJ, Xu MS (2011) Enzyme-assisted extraction of flavonoids from *Ginkgo biloba* leaves: improvement effect of flavonol transglycosylation catalyzed by Penicillium decumbens cellulase. Enzyme Microb Technol 48(1):100–105
89. Xu MS, Chen S, Wang WQ, Liu SQ (2013) Employing bifunctional enzymes for enhanced extraction of bioactives from plants: flavonoids as an example. J Agric Food Chem 61(33):7941–7948
90. Li J, Zu YG, Luo M, Gu CB, Zhao CJ, Efferth T, Fu YJ (2013) Aqueous enzymatic process assisted by microwave extraction of oil from yellow horn (*Xanthoceras sorbifolia* Bunge.) seed kernels and its quality evaluation. Food Chem 138(4):2152–2158
91. Jiao J, Li ZG, Gai QY, Li XJ, Wei FY, Fu YJ, Ma W (2014) Microwave-assisted aqueous enzymatic extraction of oil from pumpkin seeds and evaluation of its physicochemical properties, fatty acid compositions and antioxidant activities. Food Chem 147:17–24
92. Gai QY, Jiao J, Wei FY, Luo M, Wang W, Zu YG, Fu YJ (2013) Enzyme-assisted aqueous extraction of oil from *Forsythia suspense* seed and its physicochemical property and antioxidant activity. Ind Crops Prod 51:274–278
93. Liu S, Jiang L, Li Y (2011) Research of aqueous enzymatic extraction of watermelon seed oil of ultrasonic pretreatment assisted. Proced Eng 15:4949–4955
94. Konwarh R, Pramanik S, Kalita D, Mahanta CL, Karak N (2012) Ultrasonication, a complementary "green chemistry" tool to biocatalysis: a laboratory-scale study of lycopene extraction. Ultrason Sonochem 19(2):292–299
95. Chen R, Li S, Liu C, Yang S, Li X (2012) Ultrasound complex enzymes assisted extraction and biochemical activities of polysaccharides from Epimedium leaves. Process Biochem 47(12):2040–2050
96. Da Porto C, Decorti D, Natolino A (2013) Effect of commercial enzymatic preparation with pectolytic activities on conventional extraction and ultrasound-assisted extraction of oil from grape seed (*Vitis vinifera* L.). Int J Food Sci Technol 48:2127–2132
97. Adulkar TV, Rathod VK (2014) Ultrasound assisted enzymatic pre-treatment of high fat content dairy wastewater. Ultrason Sonochem 21(3):1083–1089
98. Hosni K, Hassen I, Chaâbane H, Jemli M, Dallali S, Sebei H, Casabianca H (2013) Enzyme-assisted extraction of essential oils from thyme (Thymus capitatus L.) and rosemary (Rosmarinus officinalis L.): impact on yield, chemical composition and antimicrobial activity. Ind Crops Prod 47:291–299
99. Kapturowska AU, Stolarzewicz IA, Krzyczkowska J, Białecka-Florjańczyk E (2012) Studies on the lipolytic activity of sonicated enzymes from *Yarrowia lipolytica*. Ultrason Sonochem 19(1):186–191
100. Kapchie VN, Wei D, Hauck C, Murphy PA (2008) Enzyme-assisted aqueous extraction of oleosomes from soybeans (Glycine max). J Agric Food Chem 56(5):1766–1771
101. Sharma A, Khare SK, Gupta MN (2002) Enzyme-assisted aqueous extraction of peanut oil. J Am Oil Chem Soc 79(3):215–218
102. Schmid A, Dordick JS, Hauer B, Kiener A, Wubbolts M, Witholt B (2001) Industrial biocatalysis today and tomorrow. Nature 409(6817):258–268
103. Schmid A, Hollmann F, Park JB, Bühler B (2002) The use of enzymes in the chemical industry in Europe. Curr Opin Biotechnol 13(4):359–366
104. Rolle RS (1998) Review: enzyme applications for agro-processing in developing countries: an inventory of current and potential applications. World J Microbiol Biotechnol 14:611–619
105. Ellis RJ (1979) Reviews the most abundant protein in the world. Trends Biochem Sci 4:241–244
106. Kapel R, Chabeau A, Lesage J, Riviere G, Ravallecple R, Lecouturier D, Wartelle M, Guillochon D, Dhulster P (2006) Production, in continuous enzymatic membrane reactor, of an anti-hypertensive hydrolysate from an industrial alfalfa white protein concentrate exhibiting ACE inhibitory and opioid activities. Food Chem 98(1):120–126
107. Boschetti E, Righetti PG (2012) Breakfast at Tiffany's? Only with a low-abundance proteomic signature! Electrophoresis 33(15):2228–2239

108. Andersson I, Backlund A (2008) Structure and function of Rubisco. Plant Physiol Biochem 46(3):275–291
109. Puri M, Sharma D, Barrow CJ (2012) Enzyme-assisted extraction of bioactives from plants. Trends Biotechnol 30(1):37–44
110. Alkorta I, Garb C (1998) Industrial applications of pectic enzymes: a review. Process Biochem 33(I):21–28
111. De Maria L, Vind J, Oxenbøll KM, Svendsen A, Patkar S (2007) Phospholipases and their industrial applications. Appl Microbiol Biotechnol 74(2):290–300
112. Houde A, Kademi A, Leblanc D (2004) Lipases and their industrial applications: an overview. Appl Biochem Biotechnol 118(1–3):155–170
113. Zuniga M, Soto C, Mora A, Chamy R, Lema J (2003) Enzymic pre-treatment of Guevina avellana mol oil extraction by pressing. Process Biochem 39:51–57
114. Dehghan-Shoar Z, Hardacre AK, Meerdink G, Brennan CS (2011) Lycopene extraction from extruded products containing tomato skin. Int J Food Sci Technol 46(2):365–371
115. Kamm B, Kamm M (2004) Principles of biorefineries. Appl Microbiol Biotechnol 64(2):137–145
116. Klein-Marcuschamer D, Oleskowicz-Popiel P, Simmons BA, Blanch HW (2012) The challenge of enzyme cost in the production of lignocellulosic biofuels. Biotechnol Bioeng 109(4):1083–1087
117. Veit T (2004) Biocatalysis for the production of cosmetic ingredients. Eng Life Sci 4(6):508–511
118. Bornscheuer UT (2003) Immobilizing enzymes: how to create more suitable biocatalysts. Angewandte Chemie (International ed. in English) 42(29):3336–3337
119. Sheldon R (2007) Enzyme immobilization: the quest for optimum performance. Adv Synth Catal 349(8–9):1289–1307

Chapter 9
Terpenes as Green Solvents for Natural Products Extraction

Chahrazed Boutekedjiret, Maryline Abert Vian, and Farid Chemat

Abstract This chapter presents a complete picture of current knowledge on useful and green bio-solvent "terpenes" obtained from aromatic plants and spices through a steam distillation procedure followed by a deterpenation process. Terpenes could be a successful substitute for petroleum solvents, such as dichloromethane, toluene, or hexane, for the extraction of natural products. This chapter provides the necessary theoretical background and some details about extraction using terpenes, the techniques, the mechanism, some applications, and environmental impacts. The main benefits are decreases in extraction times, the amount of energy used, solvents recycled, and CO_2 emissions.

9.1 Essential Oils as Sources of Terpenes: Recovery and Composition

Essential oils are a natural complex mixture of volatile compounds synthesized by aromatic plants. Known for their medicinal properties and their fragrance, they have been used since ancient times for various purposes including medical treatments, food preservatives, and flavoring of food. According to ISO and AFNOR standards, essential oils are defined as products obtained from raw plant materials that must be isolated by physical methods such as steam distillation, water distillation, water-steam distillation, or cold pressing for citrus peel oils. Following distillation, the

---

C. Boutekedjiret (✉)
Laboratoire des Sciences et Techniques de l'Environnement (LSTE), École Nationale Polytechnique, BP 182, El Harrach, 16200 Alger, Algérie
e-mail: chahrazed.boutekedjiret@g.enp.edu.dz

M. Abert Vian • F. Chemat
Green Extraction Team, Université d'Avignon et des Pays de Vaucluse, INRA, UMR 408, F-84000 Avignon, France

essential oil is physically separated from the water phase [1, 2]. They can also undergo a secondary treatment such as deterpenation or rectification intended to eliminate partially or completely a component or a group of components [3–5].

Essential oils exist almost only at the higher plants. The kinds able to synthesize the components which compose them are distributed in about 50 families, of which much is of *Lamiaceae, Asteraceae, Rutaceae, Lauraceae*, and *Magnoliaceae*. They can be obtained from various parts of an aromatic plant such as flowers, fruits, leaves, buds, seeds, twigs, bark, herbs, wood, and roots.

Essential oils are stored in specialized histological structures (glandular trichomes, secreting cells, epidermic cells, cavities, channels), often localized on or near the surface of the plant. If all the parts of the same species can contain an essential oil, the composition of this one can vary according to its localization. Thus, in the case of the bitter orange tree, the peel of the fruit provides the essential oil of bitter orange or Curaçao essence; the flower provides the neroli essence and the water distillation of sheet, branchless, and small fruits leads to the small grain bigaradier essence. The chemical compositions of these three essential oils are different.

Essential oils are liquid, volatile, soluble in usual organic solvents, liposoluble, and generally lighter than water in which it is insoluble. The quantity (yield which is generally between 0.005 and 10 %; often lower than 1 %) and quality (chemical composition) of an essential oil depend on several parameters: extraction process (steam distillation, water distillation, or water-steam distillation), operating conditions (length of distillation time, temperature, pressure, etc.), and plant material (part of the plant, environmental factors, cultivation methods, existence of chemotypes, and influence of the vegetative cycle, etc.) [6].

The chemical composition of essential oils is relatively complex. The number of compounds varies from oil to another and can exceed the 100 components; the major compounds can represent more than 85 % of essential oil [6]. Two types of compounds can be found in essential oils: hydrocarbon compounds known as terpenes (monoterpenes, sesquiterpenes) and oxygenated compounds or terpenoids which are functionalized terpenes (alcohols, aldehydes, ketones, esters, phenols, etc.). These compounds are classified according to the number of isoprene units (2-methyl butadiene) which constitute them. Monoterpenes, the structure of which presents two isoprene units, are natural compounds characteristic of essential oils. They can, sometimes, represent more than 90 % of the chemical composition as it is the case of citrus essential oil with more than 95 % of limonene, or turpentine oils (85 % of α-pinene) [1].

Essential oils can be recovered using a number of isolation methods. These may include the conventional or innovative methods. Steam distillation, water distillation, and water-steam distillation are the conventional methods usually used. Although these methods are simple to use, they present disadvantages such as long durations of treatment that can deteriorate the quality of extracted oils and a very significant energy consumption. The innovative technique may include use of microwaves, liquid carbon dioxide, and mainly low- or high-pressure distillation employing boiling water or hot steam. These methods are more efficient and energy saving and give a better essential oils.

Essential oils are principally used in the perfume industry; they are also used in the pharmaceutical industry, in particular in the field of external disinfectants and more generally for aromatization of the medicaments intended to be managed by oral way or as raw materials for the synthesis of active ingredients of medicaments, vitamins, etc. Essential oils also find applications in various industries such as the food industry (soft drinks, confectionery, dairy products, soups, sauces, snack bars, bakery, and in animal nutrition).

In addition, substantial quantities of essential oils are used in the preparation of toilet soaps, perfumes, cosmetics, and other home care products [7]. These last years, new aspects concerning the use of essential oils for exploitation on the production of bio-solvent have gained increasing interest. This interest was justified by several researches undertaken on the use of terpenes as green solvents for substitutions of petrochemical solvents. Essential oils constitute a safe and economically attractive renewable source of these solvents.

9.2 Terpenes: Physicochemical and Solvation Properties

The relevant properties of terpene solvent as compared to *n*-hexane as a solvent are listed in Table 9.1. Terpenes have similar molecular weights and structures to substitute *n*-hexane. Solubility parameters of solvents have been studied by means of Hansen Solubility Parameters (HSPs) [8]. The HSPs were developed by Charles M. Hansen and provide a way to describe a solvent in terms of its nonpolar, polar, and hydrogen-bonding characteristics. The HSPs work on the idea of "like dissolves like" where one molecule is defined as being "like" another if it bonds to itself in a similar way. The overall behavior of a solvent is characterized by three HSPs: δ_d, the energy from dispersion bonds between molecules; δ_p, the energy from dipolar intermolecular force between molecules; and δ_h, the energy from hydrogen bonds between molecules. *n*-Hexane and terpenes have similar values of the three descriptive terms; they likely behave similarly in practice. From this point of view, the terpenes are as effective as hexane to dissolve oils.

Figure 9.1 shows the Hansen model by plotting the δ_p parameter against the δ_h parameter, representing the dipole and hydrogen-bonding interactions of each chemical, respectively, for lipid classes of *Nannochloropsis oculata* and *Dunaliella salina* microalgae functions of different solvents. From this figure, it is interesting to spot visually miscibility homogeneous area where it can find extraction solvents such as n-hexane, terpenes and chloroform, and microalgae lipids of interest such as triacylglycerols (TAG), diacylglycerols (DAG), monoacylglycerols (MAG), and free fatty acids (FFA).

Table 9.1 Relevant properties of n-hexane and terpenes

	n-Hexane	d-Limonene	α-Pinene	p-Cymene
N° CAS	110-54-3	5989-27-5	80-56-8	99-87-6
Chemical structure				
Molecular formula	$C_{10}H_{14}$	$C_{10}H_{16}$	$C_{10}H_{16}$	$C_{10}H_{14}$
Properties				
Molar weight (g/mol)	86.17	136.23	136.23	134.22
Molar refractivity (cm^3)	29.84	45.35	43.96	45.26
Molar volume (cm^3)	127.5	163.2	154.9	155.7
Boiling point (°C)	68.54	175	158	174
Flash point (°C)	−23	48.3	32	47.2
Viscosity 25 °C (Cp)	0.31	0.83	1.32	0.83
Index of refraction	1.384	1.467	1.479	1.492
Surface tension (dyne/cm)	20.3	25.8	25.3	28.5
Density (g/cm^3)	0.675	0.834	0.879	0.861
Dielectric constant, 20 °C	1.87	2.44	2.58	2.34
Polarizability (cm^3)	11.83	17.98	17.42	17.94
Vapor pressure, 25 °C (mmHg)	150.9	1.54	3.49	1.65
Enthalpy of vaporization (kJ/mol)	28.85	39.49	37.83	39.34
Log P	3.94	4.45	4.37	4.02
Solubility in pure water, 25 °C (mg/ml)	0.11	0.012	0.069	0.025
Rate evaporation, 25 °C	8.30	0.25	0.41	0.14
Hansen parameters				
δ_d	14.9	17.2	17	18.5
δ_p	0	1.8	1.3	2.6
δ_h	0	4.3	2	1.9

9 Terpenes as Green Solvents for Natural Products Extraction

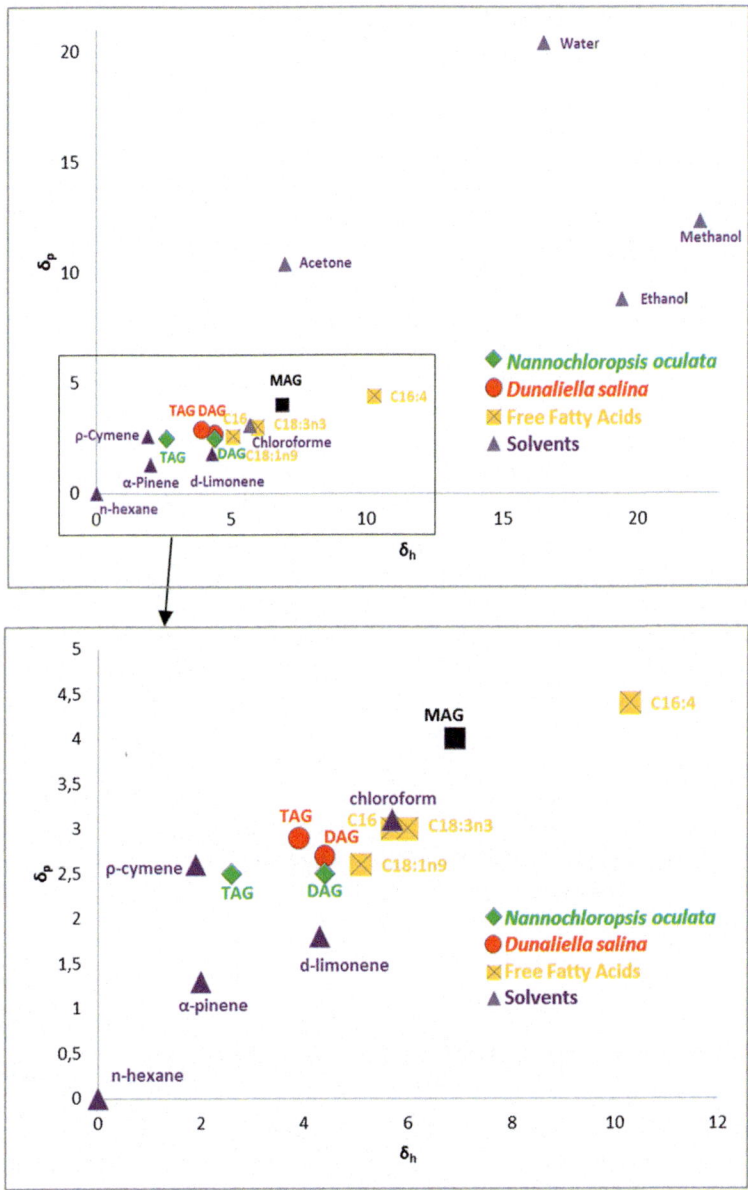

Fig. 9.1 Hansen parameters for lipid classes of *Nannochloropsis oculata* and *Dunaliella salina* functions of different solvents

9.3 Examples of Extraction Using Terpenes

Because of the consumers', and numerous regulation authorities, concerns with safety, environment, and health, which require a better control in the chemical and food industries, a new tendency to return towards the natural products is currently observed. Natural products, such as fruits and vegetables, spices, aromatic herbs, and medicinal plants, are complex mixtures of bioactive compounds such as lipids, proteins, vitamins, sugars, fibers, aromas, essential oils, pigments, antioxidants, etc. Extraction of these bio-compounds requires the use of petroleum solvents such as hexane, dichloromethane toluene, acetone, chloroform, etc. However, these solvents are classified as hazardous for the environment and health.

Due to these negative effects and the increasingly severe regulations aiming at the restriction of their use or their total elimination, such solvent has to be avoided as much as possible. Therefore, increasing interest was given to find alternative solvents more reliable and safer for the environment and health. In this context, several innovations towards green solvents have been developed: solvent-free technology [9, 10], use of water as alternative solvent [11], and use of ionic liquids that have low vapor pressure and less emission of COV [12, 13].

Terpenes were also investigated in this field. They are found in essential oils and oleoresins of fruit and aromatic plants and considered as renewable solvents, which have a safety impact, less hazard risks, and less environmental impact; consequently they can be a real substitution to petroleum solvents. The most commonly used terpene as solvent is probably d-limonene which represents a major by-product of the citrus fruits industry [14, 15]. Its physical properties were compared with those of hexane in order to extract fat and oil from oleaginous seeds [16, 17] or oil from rice bran [18, 19]. Limonene was also compared with toluene in the Dean-Stark procedure based on its ability to form an azeotropic mixture with water [20]. Recently, d-limonene was also used as a green solvent as a substitute of dichloromethane for carotenoid extraction especially lycopene [21].

9.3.1 Pinene: Origin, Applications, and Properties

Another monoterpene susceptible to be an interesting alternative solvent is α-pinene. It is a monoterpene hydrocarbon which represents the major constituent of turpentine oil from most conifers and a component of the wood and leaf oils obtained from leaves, bark, and wood of a wide variety of plants like rosemary, parsley, basil, yarrow, and roses [22].

Turpentine is a by-product of the wood and paper industry; its annual world production was more than 130–150,000 T/year, which makes it an abundant and cheap product. It constitutes 30 % of pine resin and is the most significant source in volume of volatile organic compounds. Its composition is generally rich in pinenes, 60 % of α-pinene and its isomer β-pinene; their respective proportions vary according to the geographical origin of the pines. Pinene was generally obtained by fractional distillation of steam-distilled wood turpentine. It is commonly used in the fragrance and flavor industry – and as an insecticide, solvent, and perfume bases as well as for camphor's synthesis. It is completely miscible with oils and insoluble in water. These last years, several researches were carried out to test the possibilities of using pinene as a substitute of petroleum solvents for the extraction of bioactive compounds.

9.3.2 Pinene as an Alternative Solvent for Soxhlet Extraction

Oils and fats constitute a significant share of food and can have various origins: animal or vegetable. Because of its very varied composition (complex mixture of glycerides, free fatty acid, squalene, sterols, tocopherols, alkaloids, etc.), the definition of lipids was not yet clearly established. However, it is this composition which confers its taste, texture, odor and it is at the same time characteristic and particular according to its source [23]. Fats constitute a subclass of lipids; they gather the whole of fatty acids isolated in a lipidic extract [24]. These compounds, provided by food, can play an essential role in all the forms of lives in order to provide daily energy. They also intervene in certain biological mechanisms like the transport of hormones and vitamins or the integrity of the membranes of the cells [25, 26]. The interest brought to the fats is today growing, in particular because of consumers and medical authorities who require a better control of quantities and a quality of these compounds potentially absorptive in food. Consequently, dietetic and nutritional properties of these compounds, as their implications on health, are more and more controlled and require fast and effective methods of analyses.

Nowadays n-hexane is the most used solvent for extraction of oils and fats using the Soxhlet extraction [27–32]. This choice is based on its properties, namely, nonpolar, a high selectivity to fats and oils, a relatively low boiling point (69 °C), a rather low latent heat of vaporization (29.74 kJ/mol) which allows an easy evaporation, an efficient extraction, and a limited energy cost. Despite these advantages, it is ranked on top of the list of hazardous solvents and classified as harmful, irritant, and dangerous for the environment and may cause disorders of the central nervous system and fertility problems. Due to these negative effects, the possibility to use α-pinene as a substitute solvent to n-hexane for extraction of oil was investigated.

9.3.2.1 Fats and Oils from Crops

In 2013, Bertouche et al. [33] proposed to use α-pinene to extract oil of some oilseed products: peanuts, soya, sunflower, and olive. Oils were recovered using Soxhlet extraction (Fig. 9.2), according to standardized procedure [34]. The comparison of the results with that obtained with n-hexane showed that yields of α-pinene

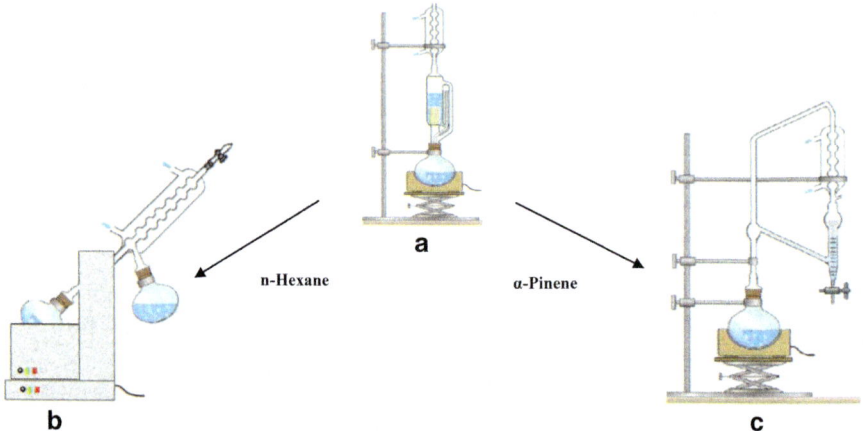

Fig. 9.2 Fat extraction and recycling procedure using *n*-hexane and α-pinene. (**a**) Soxhlet extraction. (**b**) Vacuum rotary evaporator. (**c**) Clevenger distillation

extracts were slightly higher than that of *n*-hexane. This difference is probably due to the polarity of the α-pinene slightly higher than that of hexane, which has as a consequence a more significant capacity for triglyceride dissolution. Gas chromatography coupled to mass spectrometry (GC-MS) and gas chromatography (GC) analyses of free fatty acid methyl ester (FAME) derivatives indicate that fatty acids extracted by both solvents are equivalent in terms of compounds identified and relative proportions. The data revealed a good agreement with literature data, and no significant differences ($P > 0.05$) were detected for both methods. Peanut and olive oils were characterized by strong monounsaturated fatty acid (MUFA) contents including oleic acid (C18:1) as a main component, whereas sunflower and soya oils are richer in polyunsaturated fatty acids (PUFAs) with linoleic acid (C18:2) as a principal compound.

9.3.2.2 Lipids from Microalgae

Another application using α-pinene instead of *n*-hexane was developed by Dejoye Tanzi et al. [35]. It concerns the extraction of oil from microalgae (*Chlorella vulgaris*) by means of Soxhlet extraction. In this case also, α-pinene gives better yield of oils than *n*-hexane, and the fatty acid composition is similar for both solvents. The main compounds are palmitic acid (C16:0), oleic acid (C18:1), and linoleic acid (C18:3). This composition is comparable to that observed by other authors [36, 37].

α-Pinene was also used in a simultaneous distillation and extraction process (SDEP) for extraction of lipids from wet microalgae (*Nannochloropsis oculata* and

Dunaliella salina) [38]. This procedure makes it possible to eliminate simultaneously water present in the sample followed by the extraction of oil. The innovation brought by this method is double: on one hand drying algae before extraction of oils is not anymore necessary, and on the other hand, only one green solvent (α-pinene) is used instead of drying procedure followed by petrochemical solvents – *n*-hexane to extract oil. Extracted lipids obtained using this new procedure and conventional Soxhlet with *n*-hexane have been compared in terms of total lipid content and fatty acid composition. Lipid yields for *N. oculata* and *D. salina* obtained by SDEP procedure were higher than that obtained by Soxhlet extraction. These results were in agreement with that previously reported for *Chlorella vulgaris* and oilseed products and that reported in the literature [16–19] and explained by the difference of polarity between the solvents used. On the other hand, in SDEP procedure the matrix is in direct contact with the boiling solvent which is not the case with the conventional Soxhlet. A higher dissolving ability of terpenes for lipids might also be pointed out by the higher temperature used to boil this solvent which could produce a lower viscosity of the analytes in the matrix and, accordingly, a better diffusion rate of the solute from the solid phase to the solvent. From a qualitative point of view, there is no significant difference in fatty acid composition obtained by the two methods using bio-based (pinene) and petroleum (hexane) solvent.

9.3.2.3 α-Pinene Recycling Capacity

In addition to the physical properties (polarity, selectivity, capacity of dissolution, toxicity, etc.), one of the parameters to be taken into account in the choice of a solvent is its capacity of recycling. In the case of the extraction by n-hexane, the solvent is separated from the extract in a vacuum rotary evaporator (Fig. 9.2). The boiling point of hexane is low (69 °C); this procedure is simple to realize. But for α-pinene, the boiling point is very high (156–158 °C); its elimination with the rotavapor requires a high vacuum that could degrade the recovered extracts with a more significant energy consumption. In order to resolve this problem in terms of energy and temperature, the recovery of oil was carried out using a Clevenger distillation of a mixture (oil + α-pinene), a method suggested by Virot et al. [18] for lipid extraction by *d*-limonene. This method was inspired by hydrodistillation using a Clevenger apparatus of essential oils, whose terpenes are the primary constituents (Fig. 9.2). This process, based on the principle of an azeotropic distillation with water, allows the extraction of compounds at a temperature lower than 100 °C (97–98 °C) at atmospheric pressure and even lower if reduced pressure is applied regardless of the high boiling point (150–300 °C) of terpenes. The recycling rate of α-pinene by this method, which is close to 90 %, is significantly higher than that of n-hexane (50 %), which constitutes an additional advantage for its use for the extraction of oils and fatty acids.

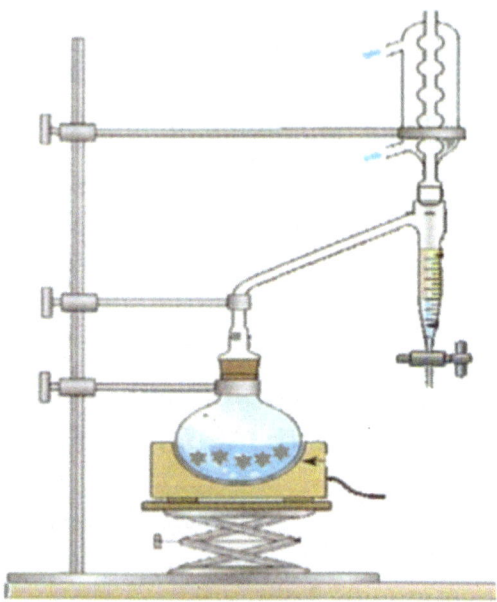

Fig. 9.3 Dean-Stark apparatus for moisture determination of vegetable matrices

9.3.3 Pinene as an Alternative Solvent for Dean-Stark Distillation of "In Situ" Water

Oven drying is the most common method used for moisture determination which represents a key step in food analysis. However, for a sample containing volatile compounds, the distillation method is the most suitable method. Several methods have been developed, and nowadays, the reference method for moisture determination in food products containing volatile compounds is the Dean-Stark distillation [39]. The principle of this method consists of an azeotropic distillation between water and petroleum solvents: toluene or xylene. However, these solvents are flammable and dangerous fire risk. They are toxic by ingestion, inhalation, and skin absorption and have detrimental health effects, especially on the nervous system, on the liver, and on the auditory function [40, 41]. Consequently, they are to be avoided as much as possible. In this context, Bertouche et al. [42] investigate the possibility to use α-pinene instead of toluene in the Dean-Stark procedure for moisture determination in food products (Fig. 9.3). The results of the moisture determination of all investigated matrices (coriander and caraway seeds, onion, garlic, carrot, leek, olive, and oregano) show that the values obtained with the two solvents are comparable and were not statistically different.

In order to confirm the effectiveness of pinene, the kinetic distillation for both toluene and pinene for moisture determination of carvi seeds was followed. As shown in Fig. 9.4, the kinetics were similar for the two solvents, and only small variations could be observed in the beginning of the water recovery with pinene which is delayed for 4 min. These variations can be explained by the difference in

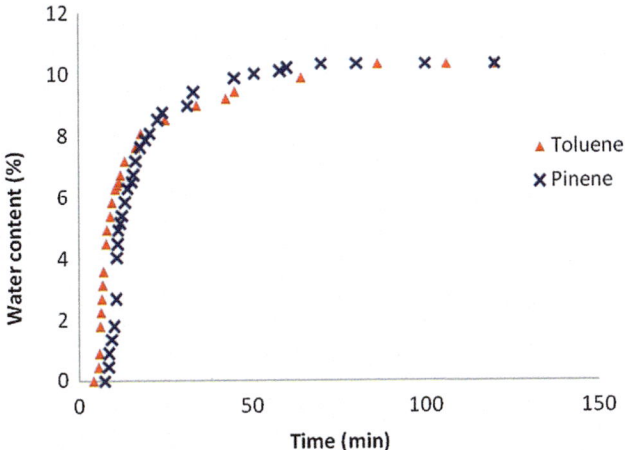

Fig. 9.4 Kinetics of water distillation of carvi seeds depending on the solvent

boiling point of the solvents. Indeed, the boiling point of α-pinene (BP = 154 °C) is higher than toluene (111 °C), as the mean time boiling point of azeotropic pinene/water (BP = 97–98 °C) is also higher than azeotropic toluene/water mixture (84 °C); consequently, the beginning of the water recovery is delayed for pinene. However, when the distillation started the water recovery was faster when using α-pinene. Indeed 40 min provides the water content comparable to those obtained after 105 min with toluene. This reduction in the processing time represents a profit of more than 60 % in terms of time and thus in consumption of energy. These results were in agreement with those cited in the literature for limonene [20], from moisture content point of view, reproducibility of results, and kinetics of distillation. Thus, α-pinene is as effective as limonene and can be used like green solvent for the determination of the water content of food products to replace toluene.

9.3.4 Pinene as an Alternative Solvent for Extraction of Carotenoids from By-Products

Carotenoids are orange-red pigments belonging to the chemical family of terpenoids. They are formed by polymerization of isoprene units to an aliphatic or alicyclic structure. This group of compounds can be synthesized by a great number of plants, algae, and bacteria and present very interesting antioxidant properties and potential beneficial health properties such as prevention of cancer [43], cardiovascular diseases [44], or macular degeneration [45]. They are used as food additives, cosmetic colorants, and antioxidants in the pharmaceutical industry. Due to these properties and an increase in demand for natural products, the interest carried to these compounds is increasing.

Fig. 9.5 Device of carotenoid extraction

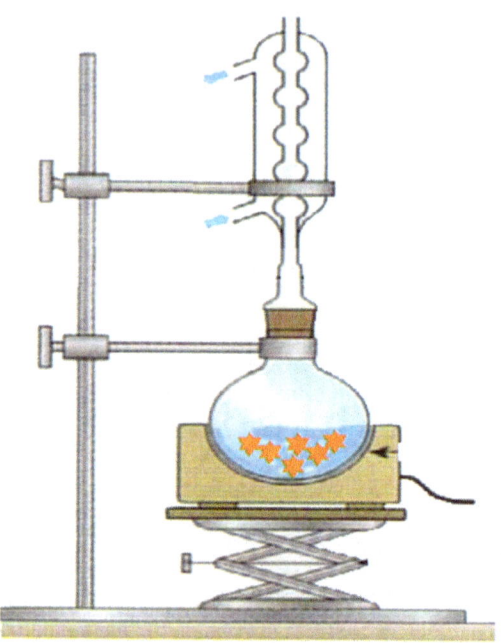

Extraction of carotenoids is usually achieved by organic solvents generating great yields of extraction. However, these solvents are harmful and generate problems of health and a great amount of waste of questionable environmental disposal. Consequently, alternative extraction methods using green solvents are under research. In this context, the use of vegetable oils for carotenoids extraction using canola, soybean, and olive oil as cosolvents has been successfully performed by supercritical fluid extraction, resulting in a yield two to four times higher [46, 47]. Carotenoids extraction was also performed using sunflower oil as extraction media in an ultrasound-assisted extraction (UAE) [48]. This original procedure was compared with conventional solvent extraction (CSE) using hexane as a solvent. The results showed that the UAE using sunflower as a solvent gives a β-carotene yield of 334.75 mg/L, in only 20 min, while CSE using hexane as a solvent gives a similar yield (321.35 mg/L) in 60 min. Limonene is another green biodegradable solvent that has been suggested as a good alternative to organic solvent for carotenoid extraction from matrices such as tomatoes [21] and microalgae [49].

Besides vegetable oils and limonene, α-pinene was also used as an alternative to hexane for carotenoid extraction. In this context, a study was performed in order to optimize the β-carotene extraction from dried ground carrot by maceration in α-pinene (Fig. 9.5). The response surface methodology using a face-centered central composite design (CCD) was carried out. The parameters chosen for optimization were temperature (ranging from 20 to 40 °C) and solid-to-solvent ratio (10-30 %). The optimal yield obtained with α-pinene was equal to 8.67 %, and the optimum conditions of β-carotene extraction obtained by statistical analysis of CCD results were 23 % for a solid-to-solvent ratio and 40 °C for temperature. For comparison, extraction of β-carotene with hexane in the optimal conditions was performed. The yield obtained (8.21 %) is similar to that obtained by α-pinene. We can assess that α-pinene may be an interesting sustainable way to replace petroleum-origin solvents for carotenoid extraction.

References

1. AFNOR (2000) Huiles essentielles, monographie relative aux huiles essentielles. Edition, Paris
2. Pharmacopée Européenne (1996) Conseil de l'Europe. Maisonneuve S.A. Editions, Sainte Ruffine
3. Diaz S, Espinosa S, Brignole EA (2005) Citrus oil deterpenation with supercritical fluids. Optimal process and solvent cycle design. J Supercrit Fluids 35:49–61
4. Arce A, Pobudkowska A, Rordíguez O, Soto A (2007) Citrus essential oil terpenless by extraction using 1-ethyl-3-methylimidazolium ethylsulfate ionic liquid: effect of the temperature. Chem Eng J 133:213–218
5. Danielski L, Brunner G, Schwänke C, Zetzl C, Hense H, Donoso JPM (2008) Deterpenation of mandarin (Citrus reticulata) peel oils by means of countercurrent multistage extraction and adsorption/desorption with supercritical CO2. J Supercrit Fluids 44:315–324
6. Naves YR (1974) Technologie et Chimie des parfums Naturels. Masson & Cie, Paris
7. Raeissi S, Diaz S, Espinosa S, Peters CJ, Brignole EA (2008) Ethane as an alternative solvent for supercritical extraction of orange peel oils. J Supercrit Fluids 45:306–313
8. Hansen CM (1967) The three dimensional solubility parameter – key to paint component affinities II. – dyes, emulsifiers, mutual solubility and compatibility, and pigments. J Paint Technol 39:505–510
9. Michel T, Destandau E, Elfakir C (2011) Evaluation of a simple and promising method for extraction of antioxidants from sea buckthorn (Hippophaërhamnoides L.) berries: pressurised solvent-free microwave assisted extraction. Food Chem 126:1380–1386
10. Khan ZH, Abert Vian M, Maingonnat JF, Chemat F (2009) Clean recovery of antioxidant flavonoids from onions: optimising solvent free microwave extraction method. J Chromatogr A 1216:5077–5085
11. Kronholm J, Hartonen K, Riekkola ML (2007) Analytical extractions with water at elevated temperatures and pressures. Trends Anal Chem 26:396–412
12. Leveque J, Cravotto G (2006) Microwave, power ultrasound and ionic liquids, a new synergy in green chemistry. Chimia 8:313–320
13. Scammells P, Scott J, Singer R (2005) Ionic liquids: the neglected issues. Aust J Chem 58:155–169
14. Njoroge SM, Koaze H, Karanja PN, Sawamura M (2005) Essential oil constituents of three varieties of Kenyan sweet oranges (Citrus sinensis). Flav Fragr J 20:80–85
15. Mira B, Blasco M, Berna A, Subirats S (1999) Supercritical CO_2 extraction of essential oil from orange peel. Effect of operation conditions on the extract composition. J Supercrit Fluids 14:95–104
16. Virot M, Tomao V, Ginies C, Visinoni F, Chemat F (2008) Green procedure with a green solvent for fats and oils' determination Microwave–integrated Soxhlet using limonene followed by microwave Clevenger distillation. J Chromatogr A 1196–1197:147–152
17. Virot M, Tomao V, Ginies C, Chemat F (2008) Total lipid extraction of food using d-limonene as an alternative to n-hexane. Chromatographia 68:311–313
18. Mamidipally PK, Liu SX (2004) First approach on rice bran oil extraction using limonene. Eur J Lipid Sci Technol 106:122–125

19. Liu SX, Mamidipally PK (2005) Quality comparison of rice bran oil extracted with d-limonene and hexane. Cereal Chem 82:209–215
20. Veillet S, Tomao V, Ruiz K, Chemat F (2010) Green procedure using limonene in the Dean-Stark apparatus for moisture determination in food products. Anal Chim Acta 674:49–52
21. Chemat-Djenni Z, Ferhat MA, Tomao V, Chemat F (2010) Carotenoid extraction from tomato using a green solvent resulting from orange processing waste. J Essent Oil Bear Plants 13:139–147
22. Fernandez X, Chemat F (2012) La chimie des huiles essentielles – Tradition et innovation. Vuibert, Paris
23. Carrasco-Pancorbo A, Navas-Iglesias N, Cuadros-Rodríguez L (2009) From lipid analysis towards lipidomics, a new challenge for the analytical chemistry of the 21st century. Part I: modern lipid analysis. Trends Anal Chem 28(3):263–278
24. Luque-García JL, Luque de Castro MD (2004) Ultrasound-assisted soxhlet extraction: an expeditive approach for solid sample treatment. Application to the extraction of total fat from oleaginous seeds. J Chromatogr A 1034:237–242
25. Desvergne B, Michalik L, Wahli W (2006) Transcriptional regulation of metabolism. Physiol Rev 86:465–514
26. Tsitouras PD, Gucciardo F, Salbe AD, Heward C, Harman SM (2008) High omega-3 fat intake improves insulin sensitivity and reduces CRP and IL6, but does not affect other endocrine axes in healthy older adults. Horm Metab Res 49:199–205
27. Luque-García JL, Luque de Castro MD (2003) Extraction of polychlorinated biphenyls from soils by automated focused microwave-assisted Soxhlet extraction. J Chromatogr A 998(1–2):21–29
28. Priego-Capote F, Luque de Castro MD (2005) Focused microwave-assisted Soxhlet extraction: a convincing alternative for total fat isolation from bakery products. Talanta 65(1):81–86
29. Virot M, Tomao V, Colnagui G, Visinoni F, Chemat F (2007) New microwave-integrated Soxhlet extraction: an advantageous tool for the extraction of lipids from food products. J Chromatogr A 1174(1–2):138–144
30. Pu W, Zhang Q, Wang Y, Wang T, Li X, Ding L, Jiang G (2010) Evaluation of Soxhlet extraction, accelerated solvent extraction and microwave-assisted extraction for the determination of polychlorinated biphenyls and polybrominated diphenyl ethers in soil and fish samples. Anal Chim Acta 663(1):43–48
31. Go-Woon J, Hee-Moon K, Byung-Soo C (2011) Characterization of wheat bran oil obtained by supercritical carbon dioxide and hexane extraction. J Ind Eng Chem 18(1):360–363
32. Eikani M, Fereshteh G, Homami S (2012) Extraction of pomegranate (*Punica granatum* L.) seed oil using superheated hexane. Food and Bioprod Process 90(1):32–36
33. Bertouche S, Tomao V, Hellal A, Boutekedjiret C, Chemat F (2013) First approach on edible oil determination in oilseeds products using α-pinene. J Essent Oil Res 25(6):439–443
34. ISO 659–1988 (E) (1988) Graines oléagineuses, détermination de la teneur en huile. International Organization for Standardization (ISO), Genève
35. Dejoye Tanzi C, Abert Vian M, Chemat F (2013) New procedure for extraction of algal lipids from wet biomass: a green clean and scalable process. Bioresour Technol 134:271–275
36. Petkov G, Garcia G (2007) Which are fatty acids of the green alga Chlorella. Biochem Syst Ecol 35:281–285
37. Plaza M, Santoyo S, Jaime L, Avalo B, Cifuentes A, Reglero G, García-Blairsy G, Señoráns FJ, Ibáñez E (2011) Comprehensive characterization of the functional activities of pressurized liquid and ultrasound-assisted extracts from Chlorella vulgaris. LWT- Food Sci Technol 46:245–253
38. Dejoye Tanzi C, Abert Vian M, Ginies C, El maataoui M, Chemat F (2012) Terpenes as green solvents for extraction of oil from microalgae. Molecules 17:8196–8205
39. AOCS Official Method Ja 2a-46 (1993) American Oil Chemist' Society, Champaign
40. Hass U, Lund SP, Hougaard KS, Simonsen L (1999) Developmental neurotoxicity after toluene inhalation exposure in rats. Neurotoxicol Teratol 21:349–357

41. Mc Williams ML, Chen GD, Fechter LD (2000) Low-level toluene alters the auditory function in guinea pigs. Toxicol Appl Pharmacol 1(167):18–29
42. Bertouche S, Tomao V, Ruiz K, Hellal A, Boutekedjiret C, Chemat F (2012) First approach on moisture determination in food products using α-pinene as an alternative solvent for dean–stark distillation. Food Chem 134:602–605
43. Andre CM, Larondelle Y, Evers D (2010) Dietary antioxidants and oxidative stress from a human and plant perspective: a review. Curr Nutr Food Sci 6:2–12
44. Riccioni G, Mancini B, Di Ilio E, Bucciarelli T, D'Orazio N (2008) Titre. Eur Rev Med Pharmacol Sci 12:183–190
45. Snodderly M (1995) Evidence for protection against age-related macular degeneration by carotenoids and antioxidant vitamins. Am J Clin Nutr 62:1448S–1461S
46. Krichnavaruk S, Shotipruk A, Goto M, Pavasant P (2008) Supercritical carbon dioxide extraction of astaxanthin from Haematococcus pluvialis with vegetable oils as co-solvent. Biores Technol 99:5556–5560
47. Sun M, Temelli F (2006) Supercritical carbon dioxide extraction of carotenoids from carrot using canola oil as a continuous co-solvent. J Supercrit Fluids 37:397–408
48. Li Y, Fabiano-Tixier AS, Tomao V, Cravotto G, Chemat F (2013) Green ultrasound-assisted extraction of carotenoids based on the bio-refinery concept using sunflower oil as an alternative solvent. Ultrason Sonochem 20:12–18
49. Castro-Puyana M, Herrero M, Urreta I, Mendiola JA, Cifuentes A, Ibáñez E, Suárez-Alvarez S (2013) Optimization of clean extraction methods to isolate carotenoids from the microalga Neochloris oleoabundans and subsequent chemical characterization using liquid chromatography tandem mass spectrometry. Anal Bioanal Chem 405:4607–4616

Chapter 10
Emulsion Extraction of Bio-products: Influence of Bio-diluents on Extraction of Gallic Acid

Ka Ho Yim, Moncef Stambouli, and Dominique Pareau

Abstract Natural products and fermentation broths are complex systems. Extraction processes such as emulsion extraction, a process derived from the industrial liquid-liquid extraction, can be used to remove molecules. This chapter presents preview studies dealing with these two processes for bio-products and notably for organic acids from biomass: different parameters were described. Then, a study on eco-conception of these processes for gallic acid with bio-diluents (one hydrogenated terpene and three ethylic fatty acid esters) was presented and compared with results using dodecane, a current petrochemical diluent. A pre-study about eco-conception of liquid-liquid extraction for gallic acid was performed. An extractant as tributyl phosphate (TBP) was necessary, and extraction yield was higher with TBP diluted in ethylic fatty acid esters than in dodecane. So it is possible with esters to reduce the TBP concentration. In extraction by emulsion with the esters as diluents, there was no need of an extractant, gallic acid being slightly soluble in these esters. However, emulsion containing these bio-diluents swelled, which do not exist with dodecane.

10.1 Introduction

Green chemistry [1] is a current concept about the environmentally friendly design of chemical products and processes. White biotechnologies using renewable resources can contribute to this issue: products are directly extracted from biomass or made from this source by microorganisms in bioreactors. Different extraction processes, like emulsion extraction which is a process derived from the classic liquid-liquid extraction, can be used in the three steps of bioprocesses: treatment

---

K. Ho Yim (✉) • M. Stambouli • D. Pareau
Laboratoire Génie des Procédés et Matériaux, Ecole Centrale Paris,
F-92290 Châtenay-Malabry, France
e-mail: ka_ho.yim@graduates.centraliens.net

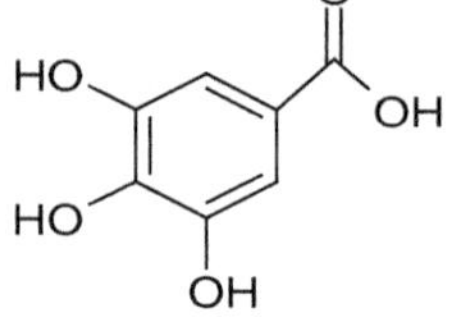

Fig. 10.1 Gallic acid (GA)

of biomass, fermentation, and purification of bio-products. Continuous extraction of bio-products during fermentation is particularly interesting because many bio-products can inhibit microorganisms.

Gallic acid (Fig. 10.1) belongs to the class of natural organic acids, the main family of bio-products. It presents in many plants and recovered by leaching the grinded plants. Gallic acid is the 3,4,5-trihydroxybenzoic acid, presenting a carboxylic acid function of pKa 4.26. Extraction of gallic acid has already been studied by different processes as aqueous two-phase extraction with a micro-channel system [2] or with ionic liquids [3] and as supported liquid membrane extraction [4]. However, gallic acid has never been extracted by emulsion, notably containing bio-reagents.

Firstly, this chapter presents a review of previous studies on liquid-liquid and emulsion extractions of organic acids from biomass, using petrochemical reagents and with eco-conception of some extraction processes. Then, as an example, a study on eco-conception of an optimized emulsion for gallic acid extraction was presented, including preliminary studies of the extraction mechanisms in liquid-liquid extraction.

10.2 Liquid-Liquid Extraction and Extraction by Emulsion of Organic Acid from Biomass

10.2.1 Principle

Liquid-liquid extraction [5] is a traditional industrial process, composed of two steps: extraction and stripping. Firstly, the molecule to be extracted, A, present in an aqueous solution, is transferred to an organic phase where it reacts with an extractant. The resulting molecule is quite soluble in the organic phase. The stripping (or back extraction) is the reverse phenomenon: the loaded organic phase is contacted with the aqueous stripping solution containing a chemical agent able to back extract A. The organic phase is then regenerated and can be recycled to the extraction step (Fig. 10.2).

Emulsion extraction (previously named extraction by emulsion liquid membrane) [6] was developed from liquid-liquid extraction for applications in petrochemistry.

First, a water-oil emulsion is formed by dispersing an aqueous solution (the stripping phase or internal phase) containing a stripping agent (T) into an organic phase (the membrane) containing a surfactant to stabilize the emulsion, an extractant (E) to improve the transfer, and a diluent. The droplets of the aqueous internal

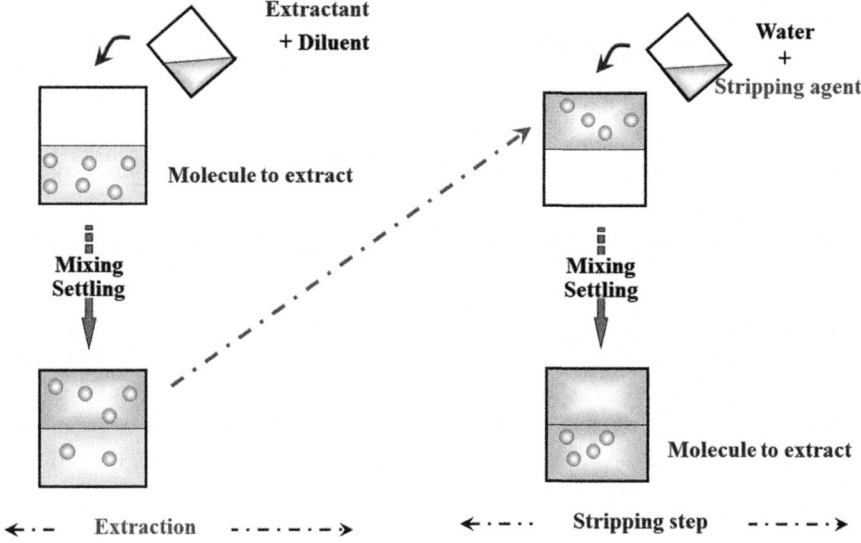

Fig. 10.2 Principle of liquid-liquid extraction

Fig. 10.3 Principle of emulsion extraction

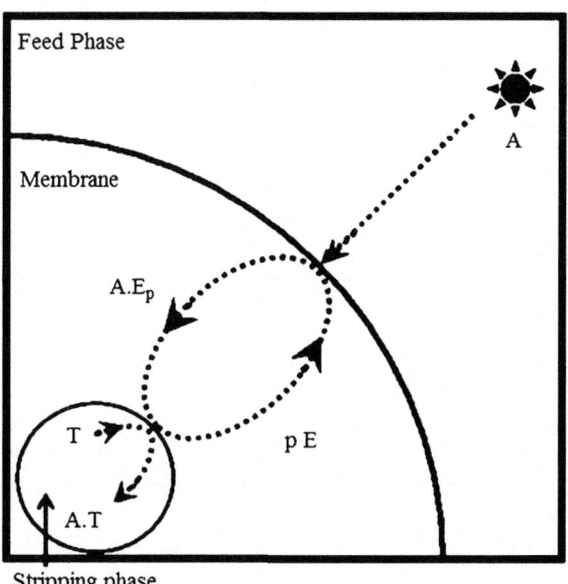

phase are very small (from 1 to 100 μm). The emulsion is then gently dispersed by agitation into the aqueous feed (the aqueous external phase) containing the solute A to be extracted; due to this agitation, a "water-oil-water" double emulsion exists. As shown in Fig. 10.3, the extractant E reacts with A and enables its extraction and its transfer across the organic membrane from the feed phase to the stripping one: it acts as a "carrier." T, the stripping agent, allows the back extraction of A in the internal phase. Emulsion extraction combines extraction and stripping in a single step. After this step, the two phases of the emulsion are separated by electro-coalescence to recover the concentrated aqueous solution and recycle the organic phase to emulsification.

In some cases, feed solutions are organic and oil-water emulsions are used; the membrane is an aqueous phase.

The advantages of emulsion extraction in comparison with liquid-liquid extraction are numerous: extraction and stripping performed in a single step, enhanced kinetics due to the huge internal interface area, good efficiency even with trace amounts of solutes, possibility of exhausting the feed solution, reduced contactor volume, and interesting concentration factors with reasonable energy consumptions. However, several drawbacks can be mentioned: no possibility of scrubbing between extraction and stripping resulting in a reduced selectivity, importance of emulsion stability to prevent back transfer of extracted solute, and sometimes difficulty in breaking the emulsion by electro-coalescence.

10.2.2 Liquid-Liquid Extraction of Bio-based Organic Acids

As said before, it is generally necessary in liquid-liquid extraction to use an extractant that is able to react with the species to be extracted. In the case of organic acids, two different types of extractants can be employed.

10.2.2.1 Extraction by Basic Extractants

Organic acids can exchange protons. So the best extractants may be proton acceptors, that is, bases as tertiary amines. There is then the formation of an ion pair, corresponding to the following equilibrium:

$AH + R_3N = \underline{A^-R_3NH^+}$ where AH is the organic acid and R_3N the tertiary amine; the underlined species are the organic ones.

Trioctylamine or TOA is one of the most used basic extractants.

Stripping agents must be bases stronger than amines, for example, soda, NaOH, or ammonia.

Extraction of organic acids from fermentation broths by TOA is one of the most studied cases: for example, lactic acid [7], glycolic acid [8], succinic acid [9], propionic acid [10], and malic acid [11].

In some cases, when the organic acid concentration is high (several g/l), a third reagent, named modificator, is added to the organic phase to avoid the formation of a third phase due to the low solubility of extracted complexes in diluents. Octanol and decanol are current modificators.

Particularly, in the case of lactic acid made from fermentation [7], different diluents (dodecane, mineral oil, and kerosene) and different modificators (hexanol, octanol, decanol, and ethyl acetate) were studied. With the same TOA (30 % v/v) and modificator (20 % v/v) concentrations in dodecane, alcohols gave better results than ester, and when comparing the alcohols, the shortest molecule (hexanol) is the best. However, due to the toxicity of alcohols for microorganisms, the less toxic (decanol) was preferred. As for the three tested diluents containing TOA 30 % v/v and decanol 20 % v/v, the same results were obtained. But dodecane being the less toxic solvent for microorganisms, it was chosen for further studies. When the TOA concentration in dodecane and decanol 20 % v/v increased, the extraction yield increased too. Using these results and the slope method, extraction mechanism was identified: the stoichiometry of the extracted complex is 1:1 (TOA: lactic acid) as expected.

Another main parameter is the pH of the feed solution. Indeed, in the case of glycolic acid extraction by TOA diluted in kerosene [8], when the feed pH decreased, the extraction yield increased: the conclusion is as expected that only the acid form is extractable.

10.2.2.2 Extraction by Solvating Extractants

In this case, the extractant is a Lewis base whose lone electron pair interacts with the hydrogen atom of the acid function. Current solvating extractants of this type are tributyl phosphate (TBP) and trioctylphosphine oxide (TOPO); in these molecules, the atom of oxygen bears two lone pairs.

The corresponding equilibrium is as follows:

$AH + S = AHS$ where S is the solvating extractant.

Stripping agents are bases able to neutralize the extracted acid giving their ionic forms which cannot be extracted by this type of extractant. Soda is a current stripping agent.

The separation of acids (acetic, propionic, valeric, and butyric acids) from fermentation broths by TOPO in kerosene [12] was studied; the pKa of these acids is between 4 and 5. As expected, extraction strongly depended on pH: with a pH (2.5) lower than pKa, extraction was better than with a higher pH (5.5). Moreover, for shorter chains in the molecule, extraction was lower due to a higher affinity with water.

Keshav et al. [13] showed that the nature of the diluent had a significant effect on extraction of propionic acid by a solvating extractant: when comparing extraction with TBP diluted in toluene, heptane, or petroleum ether, the best extraction was obtained for toluene, and as for the others, the performance is equivalent, but lower. Moreover, when TBP concentration increased, extraction yield increased too.

10.2.3 Emulsion Extraction of Bio-based Organic Acids

Emulsion extraction of bio-products from biomass and notably from fermentation broths [14] has already been studied. Most cases used "water-oil-water" emulsions because fermentation broths were generally aqueous solutions. However, extraction of bio-products from natural oils [15, 16] was directly performed with "oil-water-oil" emulsions, as extraction of polyunsaturated ethylic fatty acid esters from fish oils. The aqueous membrane contained silver ion as an extractant with concentrations between 0.5 and 1 $mol \cdot l^{-1}$ and saponin as a surfactant with concentrations between 0.5 and 1 % w/w (for O/W emulsions).

In "water-oil-water" emulsion extraction, in some cases, no extractant is needed because the solute to be extracted is slightly soluble in the organic membrane and its concentration gradient makes the driving force. A well-chosen stripping agent is added in the internal phase, to maximize the concentration gradient till the end of the process, generally by changing its chemical form (e.g., acid to ion). Extraction performance and kinetics can be improved by the addition of a suitable extractant. As an example, extraction of tyrosol [17], a natural polyphenol present in olive oil, was studied without and with extractant and modificator.

First, the organic membrane only contained a 2 % w/w nonionic surfactant (ECA 4,360 J) diluted in Shellsol, a paraffinic diluent. The stripping phase was a NaOH 1 $mol \cdot l^{-1}$ solution. The resulting extraction was slow: after 20 min, extraction yield was only 70 %. Isodecanol (2 % w/w) was then added to the organic phase as a modificator to increase polarity; extraction was quantitative (97 %) after 20 min because tyrosol, a polar molecule, is more soluble in the presence of isodecanol. Finally, when a solvating extractant as Cyanex 923 (mixture of trialkylphosphine oxides) was added in the membrane, extraction was fast, quantitative (97 %) after only 6 min. Additional experiments with Cyanex 923 in the absence of isodecanol showed the same extraction yield without the third phase. So isodecanol is useless in this case.

Viscoelastic polymers [18] can be added to the organic membrane to improve extraction and emulsion stability as in the study about extraction of penicillin G, a natural antibiotic.

These polymers increase the membrane viscosity, and the emulsion is less fragile faced with shearing forces of the agitation during the extraction process and so its breakage is reduced. A low polymer concentration – in this paper, 1 % w/w – is sufficient; a higher concentration considerably slows the solute transfer across the membrane.

In these papers and some others dealing with emulsion extraction of organic acids from biomass or synthetic fermentation media [19–21], after choosing the organic reagents, researchers perform generally parametric studies to identify the influence of chemical and operating parameters on extraction yield.

Several operating parameters of emulsification have a significant influence on extraction as well as the volume ratio (membrane/internal phase), generally higher than 1 in order to avoid emulsion detrimental inversion. Emulsification time (about

10 min) and agitation with a rotor-stator (about 10,000 rpm) might be sufficient to obtain a stable emulsion and small droplets.

Low extractant and surfactant concentrations – less than 5 % v/v – must be sufficient to get a good and fast extraction; these reagents in greater concentrations indeed increase the membrane viscosity and hence slow down solute transfer. Moreover, when there is an important concentration of surfactant, greater than its critical micelle concentration, large quantities of water can be transported by inverse micelles between both aqueous phases, resulting in an increased emulsion instability.

The initial salt concentrations of both aqueous phases have an influence on osmosis: for example, too high salt concentrations in the internal phase increase emulsion swelling due to osmosis, and its subsequent breakage; if the surfactant concentration is lower than the critical micelle concentration, osmosis occurs only if water is slightly soluble in the membrane.

pH is a very important parameter for extraction of acids as seen previously. It must be carefully controlled in both aqueous phases all along the experiment.

Increasing temperature makes the phases less viscous and can accelerate solute transfer, but the emulsion becomes more fragile.

As for the agitation during extraction, a minimal intensity is necessary to disperse emulsion in the feed phase but a strong agitation can break emulsion. In fact, agitation speed is particularly related to the volume ratio feed phase/emulsion which controls the concentration performance of the process.

So emulsion extraction efficiency depends on many parameters. Experimental design can then be very useful to obtain significant results with a minimal number of tests. Berrios et al. [22] studied emulsion extraction of gibberellic acid made by fermentation and peculiarly the effects of 12 parameters with a design of 16 experiments. First, the most important parameters were identified: these were surfactant and extractant concentrations. With these two parameters and extraction time, an optimization was performed with a second experimental design including more precise ranges of parameter values.

10.2.4 Eco-conception of Extraction with Bio-reagents

Liquid-liquid extraction was recently studied in terms of eco-conception, using bio-diluents. For example, tocopherol, a natural pharmaceutical active agent, was extracted from olive oil by ethyl lactate [23]: due to the solubility of tocopherol in ethyl lactate, there is no need of an extractant and good partition coefficients were obtained.

In hydrometallurgy [24], yttrium, which is a rare earth, was efficiently extracted by a synthetic extractant diluted in biodiesel.

Other examples about eco-conception of extraction processes from natural products and in biotechnologies were found in studies dealing with leaching of solids [25]: zein, a lipophilic protein present in corn grains, was extracted by different bio-solvents. Limonene, which is lipophilic as zein, was a better solvent than ethyl lactate which is hydrophilic. The physical properties of both extracted molecules and organic reagents are very important to adapt carefully in order to maximize extraction.

Now, as an example, an experimental study on liquid-liquid and emulsion extractions of gallic acid with some bio-diluents and its main results will be given below.

10.3 Eco-conception of Liquid-Liquid and Emulsion Extractions with Bio-diluents: Example of Gallic Acid

10.3.1 Choice of Bio-diluents

As the industrial diluents from petrochemical industry, bio-diluents must present numerous favorable characteristics. They must have a low viscosity, a very low solubility in water, and a high chemical stability. Their density must be far from water solution densities to have efficient and rapid settling. Finally, they must not be volatile and dangerous to respect green chemistry concept.

Bio-alcohols, the most famous bio-products, could not be used as bio-diluents because they are soluble in water.

Terpenes, especially monoterpenes, are dangerous, volatile, and not very stable due to their oxidation by air, according to suppliers. Hydrogenated terpenes are more stable, as commercialized squalane from olive oil (Fig. 10.4a) which was tested in our study.

About vegetal oils [26], they could not be used because they are too viscous. Fatty acids, made from these oils by acid hydrolysis, could not be used too because the unsaturated ones are not stable in the presence of air, most of saturated ones are solid in ambient conditions, and the two liquid saturated ones are soluble in water. Glycerol, a coproduct, is soluble in water too. Transesterification of natural vegetal oils with light alcohols (notably from biomass) produces less viscous fatty acid esters. Generally, the alcohol is methanol but this is the most dangerous bio-alcohol. In our case, we used ethanol, less dangerous with respect to green chemistry principles. Three ethylic fatty ester acids, ethyl caprate, ethyl laurate, and ethyl myristate, were tested: in Fig. 10.4b, n is, respectively, 8, 10, and 12.

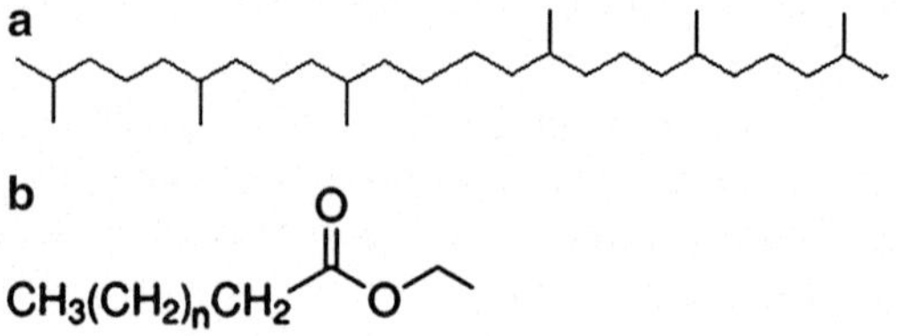

Fig. 10.4 Tested bio-diluents: (a) squalane and (b) ethylic fatty acid esters

10.3.2 Application in Liquid-Liquid Extraction

The initial aqueous solution (gallic acid 294 $\mu mol \cdot l^{-1}$) was prepared by dissolving gallic acid in distilled water.

The operating procedure was the following:

Extraction step. At ambient temperature (about 20 °C), 25 ml of the gallic acid aqueous solution and 25 ml of different organic solutions (containing TBP or not) were agitated a few minutes (equilibrium between both phases was reached as verified) and afterwards settled.

Organic phase concentration measurement. 20 ml of the loaded organic phase (light phase of the extraction experiment) was put into a new separatory funnel with 20 ml of Na_2HPO_4 (250 mmol $\cdot l^{-1}$) and agitated. Stripping is quantitative in all cases due to the high pH (9–10) of the aqueous phase, which was previously verified.

Without an extractant, squalane and ethylic fatty acid esters could not extract gallic acid, as dodecane; the extracted quantity of gallic acid was beyond the detection limit of the analysis by UV-visible spectrophotometer at 760 nm with the Folin-Ciocalteu method (15 $\mu mol \cdot l^{-1}$). So an extractant was necessary. Experiments were previously made with TOA with no extraction efficiency in all cases.

TBP was then tested. A model was developed, assuming the formation of a single organic complex between the associated form of gallic acid (AH) and TBP; the dissociated gallic acid (ionic) is not being extracted by TBP. The following equilibrium reaction can be written as follows:

AH + p TBP = $AH(TBP)_p$ where AH stands for associated gallic acid.

Due to the large amount of TBP compared to gallic acid, the equilibrium-free TBP concentration (i.e., not complexed with gallic acid) is equal to its total concentration, $[TBP]_0$.

Using the equilibrium constant of the reaction between AH and TBP and the slope method, Eq. 10.1, where q is the associated gallic acid partition coefficient (i.e., the ratio between the mass of the organic complex and the mass of AH in aqueous solution), was obtained:

$$\ln(q) = p \times \ln([TBP]_0) + \ln(K_p) \qquad (10.1)$$

Although organic phase was rather viscous, gallic acid extraction was fast: equilibrium was reached within 1 min. Figure 10.5 shows on logarithmic scale partition coefficients of gallic acid, q, in the function of TBP concentration in different diluents: results with bio-diluents were compared with dodecane, a current diluent in liquid-liquid extraction.

Partition coefficients obtained for fatty acid ethylic esters are higher than those obtained with hydrocarbons as dodecane and squalane. This difference might be attributed to a better solubility of the GA-TBP complex and maybe a small extraction of gallic acid by the esters themselves. By calculating the linear regression of $\ln(q)$ vs $\ln[TBP]$ for the five diluents, the following slopes were found, indicating the stoichiometry p of the complex (Table 10.1).

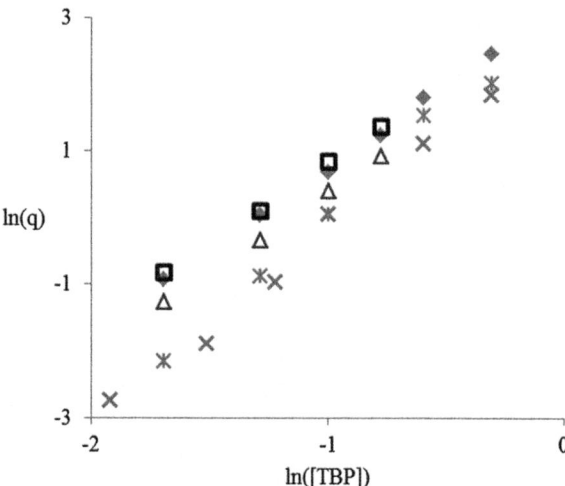

Fig. 10.5 Partition coefficients of gallic acid as a function of TBP concentration for different diluents (♦ ethyl caprate, ◻ ethyl laurate, △ ethyl myristate, ✱ squalane, and ✕ dodecane)

Table 10.1 Stoichiometry for the model with one single complex

	Ethyl caprate	Ethyl laurate	Ethyl myristate	Squalane	Dodecane
Slope	2.3	2.5	2.5	3.1	3.0
r^2	0.999	0.998	0.997	0.993	0.995

Using squalane gave the same performance and the same stoichiometry (three molecules of TBP for one molecule of gallic acid) as dodecane because they are both hydrocarbons. Contrarily, for ethylic fatty acid esters, the slopes might suggest the coexistence of two organic complexes, AH (TBP)$_2$ and AH(TBP)$_3$, whose complexation constants were, respectively, K_2 and K_3.

Assuming this hypothesis and using the two complexation constants,

$$[AH]_{org} = [AH(TBP)_2] + [AH(TBP)_3]$$

$$q/[TBP]_0^2 = K_3 \times [TBP]_0 + K_2, \qquad (10.2)$$

$q/[TBP]_0^2$ is then plotted against $[TBP]_0$; the linear regressions give K_2 and K_3 (Fig. 10.6).

The results are given in Table 10.2.

Correlation coefficients are near 1, which confirms the model. This mechanism is different from the case of dodecane and squalane. Two different complexes are formed, one with three molecules of TBP (as hydrocarbons) and the other with two molecules of TBP, suggesting a different solubilization in the diluents in relation with their physical properties. Ethyl caprate and ethyl laurate approximately exhibit

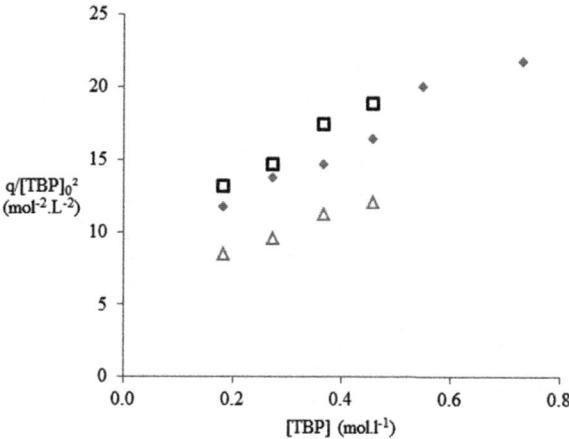

Fig. 10.6 Model of the formation of two organic complexes in ethylic fatty acid esters (ethyl caprate ♦, ethyl laurate ☐, and ethyl myristate △)

Table 10.2 Values of complexation constants, K_2 and K_3

Diluent	K_2 (M^{-2})	K_3 (M^{-3})	r^2
Ethyl caprate	8	19	0.965
Ethyl laurate	9	21	0.985
Ethyl myristate	6	13	0.985

the same constants, higher than those of ethyl myristate. In the family of esters, there are again differences in gallic acid solvation due to the physical properties of the diluents. Due to their good performances, ethyl caprate and ethyl laurate were used to study eco-conception of gallic acid extraction by emulsion.

10.4 Application in Extraction by Emulsion

As ethylic fatty acid esters could dissolve low quantities of gallic acid, the first experiments of emulsion extraction did not use any extractant; the organic membrane was an ethylic fatty acid ester containing ECA 4,360 (3 % v/v) as a surfactant (not bio-sourced). The stripping phase was a basic Na_2HPO_4 250 mmol·l^{-1} solution. The emulsification of both phases, whose volume ratio was 1, has always been performed at 13,500 rpm in 10 min with a rotor-stator (ULTRA-TURRAX T25), in an iced-water bath to limit the temperature increase. For each experiment, at ambient temperature (about 20 °C), an emulsion volume of 60 ml was mixed with 360 ml of the feed solution containing gallic acid (294 μmol·l^{-1}), with a magnetic stirrer, at 430 rpm.

Table 10.3 Extraction by emulsion of gallic acid using ethylic fatty acid esters ($[GA]_{ext,0} = 294$ μmol·l^{-1})

Diluent	t_{Ex} (min)	$[GA]_{ext}$ (μmol·l^{-1})	$[GA]_{int}$ (mmol·l^{-1})	Sw (%)	EY (%)	BI (%)
Ethyl caprate	1	125 ± 11	1.8 ± 0.1	103	57	95
	2	60 ± 19	2.5 ± 0.3	111	80	98
	5	14 ± 1	2.8 ± 0.2	119	95	100
	10	23 ± 3	2.3 ± 0.2	133	92	94
Ethyl laurate	1	193 ± 14	1.0 ± 0.2	101	34	96
	2	118 ± 16	1.9 ± 0.1	106	60	97
	5	73 ± 16	2.4 ± 0.2	111	75	99
	10	37 ± 5	2.4 ± 0.1	122	88	97

The balance index (BI) was defined by the ratio between the sum of the masses of gallic acid measured in the external, membrane, and internal phases at extraction time t and the initial one in the external phase at time 0. The extraction yield (EY) was defined by the ratio between the mass in the internal phase at extraction time t and the initial one in the external phase at time 0.

During these experiments, significant emulsion swelling was observed; this is probably due to the higher water solubility in esters than in hydrocarbons (where it is quite negligible). This solubility promotes water transfer from external to internal aqueous phase due to the difference in osmotic pressures. In the case of hydrocarbons, osmosis is negligible or at least much slower. Swelling of emulsion (Sw) was measured by the ratio of sodium concentrations in the internal phase at times 0 and t, $[Na^+]_{int,0}$ and $[Na^+]_{int}$, giving the dilution factor due to osmosis and resulting in swelling. Analysis of Na^+ was performed by spectrophotometry of atomic absorption.

Table 10.3 shows, for different extraction times (t_{Ex}), concentrations of gallic acid in external and internal phases, emulsion swelling (Sw), extraction yield (EY), and balance index (BI) corrected by emulsion swelling. Experiments were replicated three times.

No gallic acid was found in organic membranes: stripping was then totaled as expected. The balance index is close to 100 %, which is satisfactory. As Na_2HPO_4 was chosen as a stripping agent, the concentration gradient of associated gallic acid was always maximal, the concentration in the internal phase being always zero. Extraction by emulsion without TBP is efficient when ethylic fatty acid esters are used because gallic acid is probably slightly soluble in them. The extraction of gallic acid is faster for ethyl caprate than for ethyl laurate, suggesting gallic acid should be more soluble in the first ester. After 5 min of contact, the extraction yield is 95 % for ethyl caprate and only 75 % for ethyl laurate. For 10 min, the extraction remains important for ethyl caprate (92 %) and increases to 88 % for ethyl laurate.

But the internal phase concentrations are strongly influenced by emulsion swelling. Swelling is increasing with time and is more important for ethyl caprate than ethyl laurate. After 2 min of extraction, where swelling is still reduced, the concentration of gallic acid in the internal phase is higher for ethyl caprate than

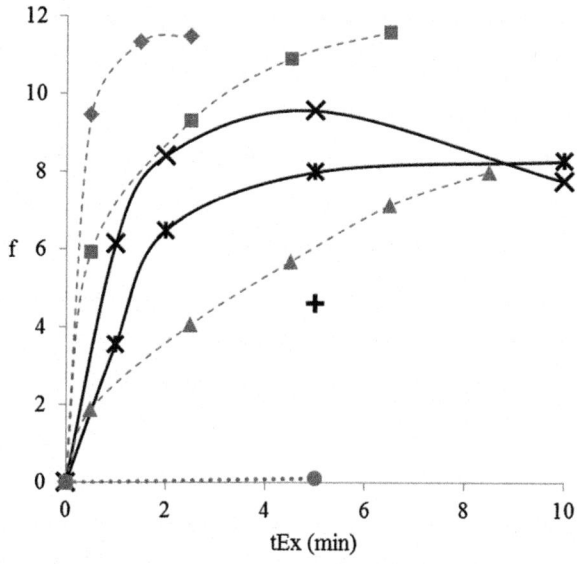

Fig. 10.7 Concentration factor as a function of time for different diluents (ethyl caprate —✕—, ethyl laurate —✻—, dodecane ··●··, dodecane-ethyl caprate (1:1) ✦, TBP 1 % v/v in dodecane --▲--, TBP 2 % v/v in dodecane --■--, and TBP 3 % v/v in dodecane --◆--); other parameters were fixed such as $V_{int}/V_m = 1$, [ECA 4,360] = 3 % v/v, agitation speed = 430 rpm, $V_{ext}/V_{em} = 6$, $[Na_2HPO_4]_{int,0} = 250$ mmol·l^{-1}, and $[GA]_{ext,0} = 294$ μmol·l^{-1}

for ethyl laurate, as expected. However, after 10 min of extraction, the same value in the internal phase was obtained for both diluents, despite the faster extraction observed for ethyl caprate. So extraction is faster than emulsion swelling, and the optimization of the contact time between the feed phase and the emulsion must be carefully optimized to get the highest concentration in the internal phase along with a good gallic acid recovery. The best conditions are obtained for ethyl caprate in 5 min with 95 % extraction and 2.8 mmol l^{-1} gallic acid in the internal phase (concentration factor of about 10).

In Fig. 10.7, these results are compared with formulations using dodecane: the concentration factor is the best parameter for comparison. The concentration factor was defined by the ratio between the concentration of total gallic acid in the internal phase at t and its concentration in the external phase at time 0.

An experiment using a mixture of dodecane-ethyl caprate was performed to check if extraction would be as efficient with lower emulsion swelling.

The maximal concentration factor of about 12 was obtained in the presence of TBP in dodecane, because TBP highly accelerates extraction and dodecane exhibits a reduced swelling, which could not be reached when ethylic fatty acid esters were used because of swelling. As said before, the best concentration factor was obtained with ethyl caprate and 5 min of extraction: $f = 9.6$. In the first minutes, the same concentration factor was obtained with ethyl caprate and dodecane containing TBP 2 % v/v, but after 5 min of extraction, swelling is diluting the internal phase for ethyl caprate.

Mixing ethyl caprate and dodecane did not show good results; after 5 min of extraction, the concentration factor was twice lower with this mixture than with pure ethyl caprate, due to the lower extraction yield (49 %) and to the same emulsion swelling (116 %).

10.5 Conclusion

Using bio-diluents in the studied extraction processes of gallic acid instead of dodecane (petrochemical product) made them more environmentally friendly. Especially, ethylic fatty acid esters were good choices because they are made by transesterification of vegetal oils with ethanol, and all the reagents of this production can be bio-sourced.

In liquid-liquid extraction, it is necessary to add TBP which does not come from green chemistry. For the same TBP concentration, higher partition coefficients of gallic acid are obtained with the esters than with dodecane. So it is possible with esters to reduce the TBP concentration. In extraction by emulsion with the esters as diluents, there is an evidence and major advantage because there is no need of an extractant, gallic acid being slightly soluble in these esters. Due to the basic medium in the internal phase, the associated gallic acid concentration gradient is always maximal, resulting in a fast and almost quantitative transfer, as for ethyl caprate in 5 min. These esters however give emulsion swelling, which does not exist with dodecane. So the internal phase is diluted, and the best obtained concentration factor is about 9.6, using ethyl caprate after 5 min of extraction, instead of 12 for dodecane, but with TBP.

In extraction by emulsion, even if the extractant is not mandatory, it would be interesting to find one bio-sourced to increase the rate of extraction. Moreover, bio-surfactants must be tested to get a totally environmentally friendly process.

Acknowledgment This work was financially supported by the "Conseil Général de la Marne" in France.

References

1. Anastas PT, Warner JC (1998) Green chemistry theory and practice. Oxford University Press, Oxford
2. Salic A, Tusek A, Fabek D, Rukavina I, Zelic B (2011) Aqueous two-phase extraction of polyphenols using a microchannel system – process optimization and intensification. Food Technol Biotechnol 49:495–501
3. Filipa A, Claudio M, Ferreira AM, Freire CSR, Silvestre AJD, Freire MG, Coutinho JAP (2012) Optimization of the gallic acid extraction using ionic-liquid-based aqueous two-phase systems. Sep Purif Technol 97:142–149

4. Saraji M, Mousavi F (2010) Use of hollow fiber-based liquid–liquid–liquid microextraction and high-performance liquid chromatography–diode array detection for the determination of phenolic acids in fruit juices. Food Chem 123:1310–1317
5. Rydberg J, Cox M, Musikas C, Choppin GR (2004) Solvent extraction: principles and practice. Marcel Dekker, New York
6. Li NN (1968) Separating hydrocarbons with liquid membranes. US Patent 3,410,794
7. Yankov D, Molinier J, Albet J, Malmary G, Kyuchoukov G (2004) Lactic acid from aqueous solutions with tri-n-octylamine dissolved in decanol and dodecane. Biochem Eng J 21:63–71
8. Yunhai S, Houyoung S, Deming L, Qinghua L, Dexing C, Yongchuan Z (2006) Separation of glycolic acid from glycolonitrile hydrolysate by reactive extraction with tri-n-octylamine. Sep Purif Technol 49:20–26
9. Jun YS, Lee EZ, Huh YS, Hong YK, Hong WH, Lee SH (2007) Kinetic study for the extraction of succinic acid with TOA in fermentation broth; effect of pH, salt and contaminated acid. Biochem Eng J 36:8–13
10. Keshav A, Wasewar KL, Chand S (2008) Extraction of propionic acid with tri-n-octyl amine in different diluents. Sep Purif Technol 63:179–183
11. Uslu H, Kirbaslar SI (2010) Extraction of malic acid by trioctylamine extractant in various diluents. Fluid Phase Equilib 287:134–140
12. Alkaya E, Kaptan S, Ozkan L, Uludag-Demirer S, Demirer GN (2009) Recovery of acids from anaerobic acidification broth by liquid-liquid extraction. Chemosphere 77:1137–1142
13. Keshav A, Wasevar KL, Chand S (2009) Recovery of propionic acid from an aqueous stream by reactive extraction: effect of diluents. Desalination 244:12–23
14. Patnaik PR (1995) Liquid emulsion membranes: principles, problems and applications in fermentation processes. Biotechnol Adv 13:175–208
15. Nakano K, Kato S, Noritomi H, Nagahama K (1996) Extraction of polyunsaturated fatty acid ethyl esters from sardine oil using Ag+-containing o/w/o emulsion liquid membranes. J Membr Sci 110:219–227
16. Nakano K, Kato S, Noritomi H, Nagahama K (1997) Extraction of eicosapentaenoic acid ethyl ester from a model media using Ag(I)-containing o/w/o-type emulsion liquid membranes. J Membr Sci 136:127–139
17. Reis MTA, De Freitas OMF, Ferreira LM, Carvalho JMR (2006) Extraction of 2-(4-hydroxyphenyl)ethanol from aqueous solution by emulsion liquid membranes. J Membr Sci 269:161–170
18. Lee SC (2008) Development of a more efficient emulsion liquid membrane system with a dilute polymer solution for extraction of penicillin G. J Ind Eng Chem 14:207–212
19. Lee SC (2011) Extraction of succinic acid from simulated media by emulsion liquid membranes. J Membr Sci 381:237–243
20. Chanuya BS, Rastogi NK (2013) Extraction of alcohol from wine and color extracts using liquid emulsion membrane. Sep Purif Technol 105:41–47
21. Lee SC (2013) Development of an emulsion liquid membrane system for removal of acetic acid from xylose and sulfuric acid on a simulated hemicellulosic hydrolysate. Sep Purif Technol 118:540–546
22. Berrios J, Pyle DL, Aroca G (2010) Gibberellic acid extraction from aqueous solutions and fermentation froths by using emulsion liquid membranes. J Membr Sci 348:91–98
23. Vicente G, Paiva A, Fornari T, Najdanovic-Visak V (2011) Liquid-liquid equilibria for separation of tocopherol from olive oil using ethyl lactate. Chem Eng J 172:879–884
24. Wang W, Yang H, Liu Y, Cui H, Chen J (2011) The application of biodiesel and sec-octylphenoxyacetic acid (CA-12) for the yttrium separation. Hydrometallurgy 109:47–53
25. Datta S, Bals BD, Lin Y, Negri MC, Datta R, Pasieta L, Ahmad SF, Moradia A, Dale BE, Snyder SW (2010) An attempt towards simultaneous biobased solvent based extraction of proteins and enzymatic saccharification of cellulosic materials from distiller's grains and soluble. Bioresour Technol 101:5444–5448
26. Graille J (2003) Lipides et corps gras alimentaires. Tec et Doc, Paris

Chapter 11
Gluconic Acid as a New Green Solvent for Recovery of Polysaccharides by Clean Technologies

Juan Carlos Contreras-Esquivel, Maria-Josse Vasquez-Mejia,
Adriana Sañudo-Barajas, Oscar F. Vazquez-Vuelvas,
Humberto Galindo-Musico, Rosabel Velez-de-la-Rocha, Cecilia Perez-Cruz,
and Nagamani Balagurusamy

Abstract The gluconic acid is an inexpensive and bio-based organic compound with new insights to drive growth in the eco-friendly industries. In organic chemistry, the gluconic acid is considered as a sustainable medium for organic reactions; meanwhile, natural product technologies suggest their potential as green solvents for extraction. In this chapter, advances of use of gluconic acid as a green solvent are presented in combination with green technologies for production of polysaccharides from biomasses from animal (chitin), microbial (chitosan-glucan), or vegetal (pectin) origins. Furthermore, this weak organic acid is capable of depolymerizing chitosan under microwave radiation for the production of water-soluble chitosan. The use of gluconic acid in combination with biomasses and clean technologies offers new green processes for the production of specialty polysaccharides and its derivatives under environmentally friendly process.

J.C. Contreras-Esquivel (✉) • M.-J. Vasquez-Mejia
Laboratory of Applied Glycobiotechnology, Food Research Department, School of Chemistry, Universidad Autonoma de Coahuila, Saltillo 25280, Coahuila, Mexico

Research and Development Center, Coyotefoods Biopolymer and Biotechnology Co., Simon Bolivar 851-A, Saltillo 25000, Coahuila, Mexico
e-mail: carlos.contreras@uadec.edu.mx; coyotefoods@hotmail.com

A. Sañudo-Barajas • R. Velez-de-la-Rocha
Laboratory of Food Biochemistry, Centro de Investigación en Alimentacion y Desarrollo (CIAD)-AC, Culiacan 80129, Sinaloa, Mexico

O.F. Vazquez-Vuelvas
School of Chemistry, Universidad de Colima, Coquimatlan 28400, Colima, Mexico

H. Galindo-Musico • C. Perez-Cruz
Laboratory of Applied Glycobiotechnology, Food Research Department, School of Chemistry, Universidad Autonoma de Coahuila, Saltillo 25280, Coahuila, Mexico

N. Balagurusamy
School of Biological Sciences, Universidad Autonoma de Coahuila, Torreon 27000, Coahuila, Mexico

11.1 Introduction

Bio-based technologies for the recovery of structural polysaccharides from biomasses represent a generation of high-value multipurpose bio-refining and sustainable companies. The market of polysaccharides has grown rapidly in recent years as a result of new developments for biomaterials. Naturally occurring polysaccharides can be extracted from animal, marine, microbial, or terrestrial biomasses through extractive environmentally friendly technologies.

Chitin and chitosan represent a group of structural polysaccharides obtained from crustacean [1] or fungal biomasses [2]. Terrestrial polysaccharides comprise pectic substances, hemicelluloses (arabinan, arabinoxylan, galactan, glucomannan, xylan, xyloglucan, etc.), and celluloses, which are obtained from wood or food wastes [3]. Seaweed polysaccharides comprise old and emergent biopolymers (alginate, carrageenan, fucoidan, porphyran, etc.), which are recovered from promising and poorly explored biomasses [4].

A key issue in recovery of polysaccharide technology is the process of extraction from the raw materials [5]. For the extraction of polysaccharides from vegetal biomasses, the use of chemical, enzymatic, or physical hydrolyzing agents is a necessary. At present, at industrial scale, the use of mineral acids are employed for the polysaccharide recovery, however during the process is generated hazardous emissions, polluted wastewater and damage to extraction reactors [4, 6]. Several organic acids are intended to be used as a substitute of mineral acids for production of polysaccharides from animal, marine, microbial, or terrestrial biomasses. Pectin polysaccharides have been extracted with organic acids such as citric [7–9], malic, lactic [10, 11], or oxalic acid [12]. Crustacean biomass has been demineralized with lactic acid in combination with emergent technologies for chitin production [13, 14]. The use of organic acids in the production of polysaccharides has shown an increase due to the environmental benefits offered during the extraction step.

Carbohydrate platform provides a supply chain of aldonic acids (i.e., gluconic and xylonic acids) which are obtained by oxidation of the aldehyde functional group of an aldose by chemical [15, 16] or biotechnological [17] methods. Gluconic acid and its derivatives (salts and glucono delta-lactone) are used in the animal feed, biotechnology, cosmetics, ceramic, dentistry, foods, pharmaceutical, and other industries [18], and it has been identified in the list of building compounds that can be produced from vegetal biomass in the future biorefineries [19]. Gluconic acid has been recognized as new green solvent for use in organic synthesis [20, 21].

Extractive capacity of gluconic acid in the industrial sector has been identified for their polysaccharide extraction capacity as a new emergent green solvent [10, 22]. In this chapter, we describe some recent advances of our research group about extraction of selected polysaccharides with aid of gluconic acid for development of clean technologies for sustainable production of biopolymers.

11.2 Gluconic Acid Production

Gluconic acid (pentahydroxycaproic acid) and its salts are produced through chemical [16] or biotechnological [23] methods subsequent to oxidation of glucose obtained generally from starch hydrolyzates. The procedures involve an oxidation process of the aldehyde group of the D-glucose unit to produce a carboxylic salt, and both methods generally employ glucose itself or a resource of this saccharide [24]. Nonetheless, the production of gluconic acid by chemical methods presents disadvantages for industrial goals, such as the low selectivity of the non-favored reactions and the high cost of noble metals used as oxidizing reaction catalyst [25]. As a consequence, the most efficient, non-expensive, and safest techniques to produce gluconic acid are by biotechnological methods [26, 27].

The biotechnological methods of gluconate production make use of fermentation or enzyme technologies. The fermentation processes, in solid and liquid media, have been extensively used employing different microbial species. Fungi and bacteria correspond to the widely utilized microorganisms to produce efficiently gluconic acid [24]. Detailed information about biotechnological production of gluconate salts can be found in recent reviews [24, 26]. After fermentation, gluconic acid is obtained from gluconates by using electrodialysis as well electrodialysis with bipolar membranes [28].

11.3 Use of Gluconic Acid for Pectin Production

Pectin is a polymer having properties of interest for the manufacture of food, cosmetics, and medicine applications and, in the last years, to enhance quality and/or functionality of those products. Chemically, complex mixtures of anionic polysaccharides naturally are present in plant cell walls, especially in the middle lamella and primary cell wall [29].

Pectins contain a high proportion of acidic and neutral sugar moieties shaping their conformation [30], but generally, they do not possess either exact structures or specific composition. The molecule does not behave as a polymer of straight conformation in solution and generally adopts a wormlike conformation. Most common substituent groups are acetyl, feruloyl, and/or methyl esters that cover a variable proportion of carbonyl-free groups depending on the development stage, tissue, or type of cell [31]. The central region of the pectin contains a long homopolymeric chain of $(1 \rightarrow 4)$ α-D-galacturonic acid, which may contain up to 200 units of length usually called homogalacturonan region [32]. Homogalacturonan linked to two pectic structures are recognized: the rhamnogalacturonan-type I, which constitutes, together with homogalacturonan, the fundamental components of pectic substances, and homogalacturonan modified, which may be of type xylogalacturonan or rhamnogalacturonan-type II. The rhamnogalacturonan-I has branches to other types of polymers of varying length, comprising the neutral sugars arabinose and galactose, so that depending on the predominant sugar, the biopolymers may be called arabinans, galactans, and arabinogalactans. The rhamnogalacturonan-II is also branched to neutral sugars and other acidic components that confer a high complexity [33, 34].

Structure and chemical composition determines the feasibility of pectin extraction as well as their properties [35]. Pectins have been extracted typically using physical [36], chemical [37], microbial, or enzymatic methods [6, 38–40]. Physical methods generally accompany all others since heating is considered, in one or other way, for pectin solubilization either biological transformation. More commonly, a combination of those extraction methods has shown success [41]. The raw material, in addition to the pectin fine structure and chemical composition, influences the major pectin quality or yield. Depending on the favorable action of the extraction method for disrupting the plant tissue (leading to a partial chemical disintegration of the polysaccharide matrix), as well as the protopectin solubilization (controlled beta-elimination and/or pectin enzymatic hydrolysis), the optimal method for pectin extraction might be improved and optimized varying parameters such as temperature, pressure, pH, ionic strength, or use of pretreatment, among others. Unfortunately, some extreme conditions lead to protein or carbohydrate degradation and should be controlled to prevent it at maximum.

Conventionally, extraction of pectin at industrial scale is performed using hot acidified water (pH 1.0–2.5; temperature 60–90 °C; time of at least 1 h), although this condition affects the degree of polymerization of pectin [29]. According to some electron micrographs of orange peel (rich in pectin), it presents a microporosity that can be increased by heating, hydration, and pressure; the higher this microporosity, the higher the water diffusion and positive linear effect in extraction of pectin [42]. In consequence, the use of acidic solutions can also lead to improve solvent accessibility to soluble pectins and/or induce a hydrolysis of insoluble protopectins.

Manufacturers have developed thermal processes using diluted mineral acids (nitric, sulfuric, or hydrochloric) [29], although organic acids (citric, gluconic, malic, and lactic) have also been explored as an alternative [10, 11]. The raw material or preprocessed source of pectin might provide the pectin into the hot solution to produce slurry containing the highest yield of the hydrocolloid with the best quality in terms of molecular weight, composition, purity, and gelling properties. The extracted liquid pectin should be fractionated by filtering or spinning the slurry and must be concentrated using ethanol or isopropanol and finally washed, dried, and milled to obtain the solid powder.

Heating technologies may vary at industrial and laboratory scale extractions. Conventional heating, autoclave, conventional soxhlet, electric pulses, and microwave technologies have been explored to determine optimal extraction conditions in different pectin sources [37, 43]. Microwave-assisted extraction is an alternative to reduce times of extraction and energy consumption, in addition to improve yield because this methodology can achieve higher cell disruption. As reported by Huang et al. [44], microwave-assisted extraction combined with the use of ionic liquids, neoteric solvents composed of organic cations and inorganic or organic anions, increases the ability of pectin extraction and may also be of interest for environmentally friendly methods.

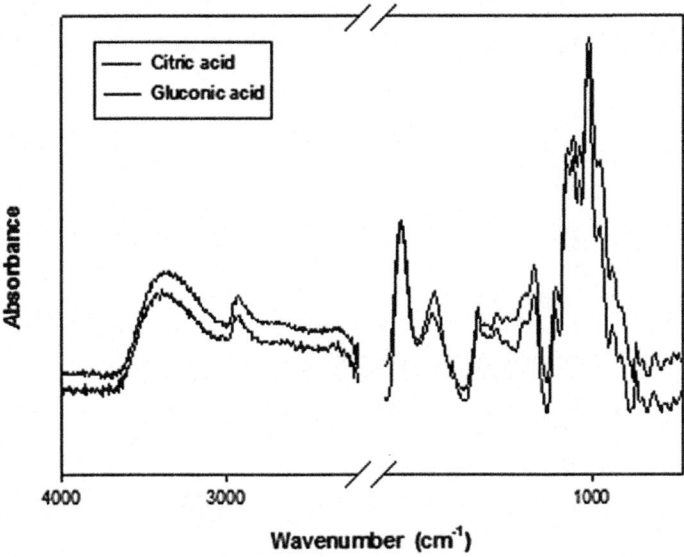

Fig. 11.1 Infrared spectra of the pectic polysaccharides extracted from citrus peel pomace by autoclave treatment with organic acids

Acidification for hydrolysis is an important exploring field since mineral acid is corrosive, non-environmentally friendly, and not suitable for human consumption. Inorganic acid method is more widely used commercially; however, organic acids or enzymatic processes are more suitable and must gain application in the next years. In particular, enzymatic extraction exhibits improved mild conditions, low energy consumption, and no pollution, and combined with a microwave heating method, it produces higher yields as compared with conventional methods based in its ability to disrupt cells [41]. Limited studies have been conducted on the feasibility of the use of organic acids for pectic extraction [7, 8, 10, 11].

Vazquez-Mejia [10] evaluated the effect of using 0.5 % citric or gluconic acid on the extraction process by autoclave treatment (121 °C/10 min) from lime citrus pomace. Similar pectin yields (dry basis) were obtained by using 0.5 % citric (22.16 ± 3.40 %) or gluconic acid (22.47 ± 0.94 %). In Fig. 11.1, infrared spectra for both extracted pectins with organic acids from industrial lime pomace are shown. The vibrational characteristics of polysaccharides generally dominate the region between 1,200 and 900 cm^{-1} because of the C–O stretch bonds related to sugars [7]. Both extracted pectins obtained with gluconic acid, as well as with citric acid, showed a significant absorption in the wavenumber in 1,750 cm^{-1} corresponding to high methoxyl-pectic polysaccharides. The results indicate that the extraction process in the presence of organic acid is not affected by the moieties of methoxyl groups in pectin.

The viscosity of the citric acid-extracted pectin was 7.87 ± 0.04 mPa × seg, while that pectin extracted by gluconic acid was 7.60 ± 1.44 mPa × seg. Both types of

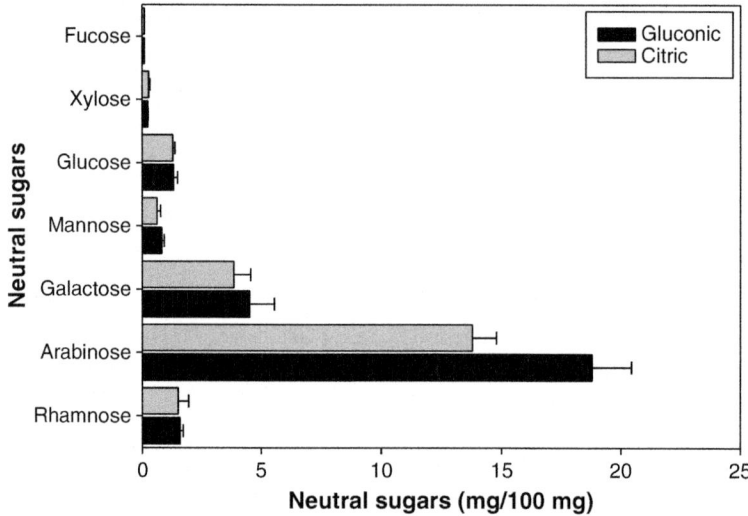

Fig. 11.2 Neutral sugar compositions (% w/w, dry weight) of lime pectin using organic acid under autoclave

lime pectin extracts showed an uronic acid content about 50–60 %. The type of organic acid used in the extraction of pectin showed differences in the quantity of neutral sugars present in each sample (Fig. 11.2). The pectin extracted with gluconic acid was characterized by a high arabinose content compared with the sample extracted with citric acid. These results suggest that the use of gluconic acid can maintain a large amount of neutral sugars of side chains of pectin after extraction. Differential gelling properties of the pectin was in function of the physical, chemical and compositional characteristics of the polymer and the solvent [33].

11.4 Use of Gluconic Acid for Crustacean Chitin Production

After cellulose, chitin is the second most abundant natural biopolymer on earth [45]. The crustacean shells are known to be constituted mainly for chitin, protein, and calcium carbonate [2]. Chitin is highly hydrophobic and insoluble in water, and it is constituted mainly by β-1,4-linked *N*-acetyl-D-glucosamine and minor proportion of D-glucosamine [46].

Currently, chitin is commercially produced by a thermochemical process based on demineralization and deproteinization of crustacean wastes. Chitin is obtained after removal of protein, calcium carbonate, and other minor components by treatment with sodium hydroxide and hydrochloric acid [47].

The use of harsh chemicals has motivated the development of biotechnological (microbial or enzymatic) methods to decrease large amounts of energy and

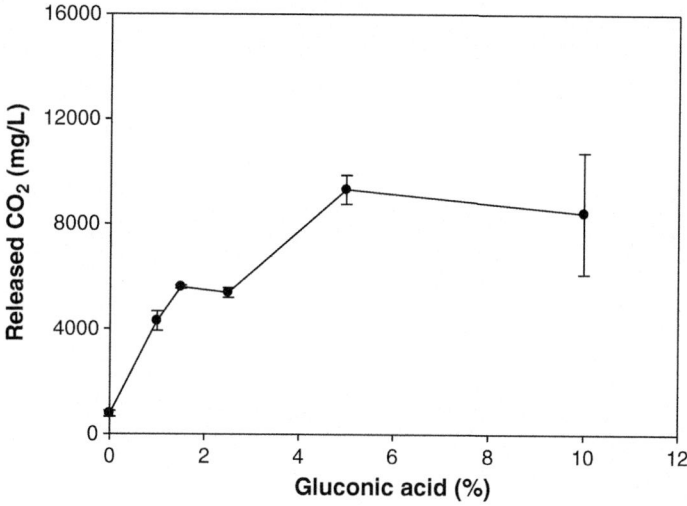

Fig. 11.3 Effect of contact time and gluconic acid concentration on release of carbon dioxide from shrimp shells. Dried shrimp shells (300 mg) were mixed with 15 mL of distiller water or gluconic acid at 24 °C for 60 min in closed container and capped with CO_2 gas analyzer. All experiments were made by triplicate

pollutants during chitin production [48]. The microbiological method for chitin recovery involves lactic acid fermentation, which is produced by lactic acid bacteria supplemented with exogenous carbon sources. In this process, there is simultaneous, but incomplete demineralization and deproteinization of crustacean wastes [49–51]. Another biotechnological approach is the use of enzymes, which are specific for the release of chitin-associated protein [52]. A variety of enzymatic procedures for deproteinization has been developed over the years [50]. This method does not allow the removal of minerals during the process [13]. Nevertheless, biotechnological methods for chitin bioproduction from crustacean wastes are still limited to industrial scale due to long processing times.

Crustacean wastes are generally demineralized with HCl under different reaction conditions for production of marine chitin [53]. An alternative for mineral acids is the use of organic acids from agricultural origin, which are generally safe, produce low hazardous emissions, are easy to degrade in the environment, and allow conservation of natural resources [13].

In our laboratory, we have explored the use of gluconic acid as green solvent for shrimp and crab shells demineralization under closed reactors at room (24 °C) or high (121 °C) temperature. Figure 11.3 shows the course of the release of carbon dioxide from shrimp shell by several concentrations of gluconic acid and water as a control, in a laboratory closed reactor. The release of carbon dioxide from dried shrimp shells by using 5 or 10 % of gluconic acid after 60 min reaction was more than 800 mg/L. Such results show that more than 5 % of gluconic acid is enough for carrying out the process of demineralization of shrimp wastes. No release of

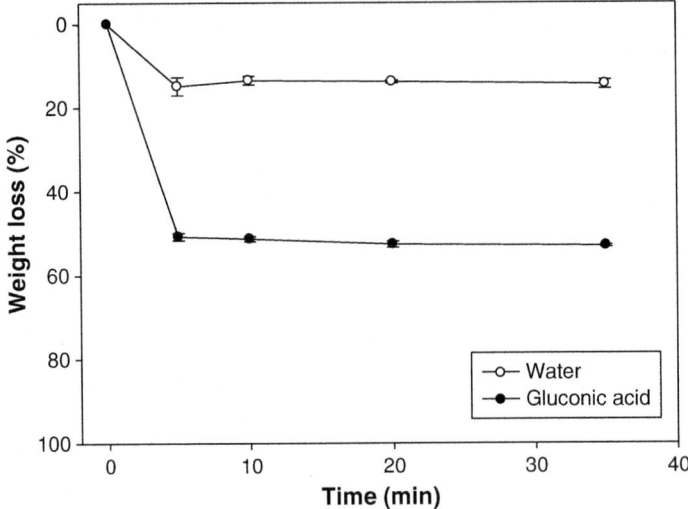

Fig. 11.4 Effect of contact time and type of demineralizing agent on weight loss of shrimp shells. Shrimp shells (3 g) were suspended in extractant solution (150 mL) and then autoclaved for correspondent time. After processing, slurry was filtered, and insoluble biomass was washed with water and dried (60 °C) until constant weight. Insoluble biomass corresponds to demineralized shrimp shells or chitin

carbon dioxide was observed when distilled water was used. These results suggest that gluconic acid is an efficient demineralizing agent at low temperature conditions for the recovery of chitin from wastes and conversion ratio of chitin depending on the concentration of acid.

Recently, environmental-friendly methods have been described for reducing the process time in manufacturing of chitin. Great efforts have been focused on reducing the processing for crustacean wastes for chitin production with thermochemical processes assisted by microwave, ultrasound, or autoclaving [13, 14, 46, 54].

Microwave-assisted technology, which has been recently successful for processing biopolymers, has been studied for chitin and chitosan recovery. Microwave-assisted heating for chitin production involves the use of microwave energy to heat the solvents (demineralizing agents) that are in contact with crustacean materials. A promising method involves the use of microwave radiation to crustacean wastes in an acidic environment. Although there are several papers and patents on the use of microwave heating for synthesis of chitosan, microwave heating has only recently been applied to deproteinization/demineralization for chitin recovery [13, 14]. Microwave heating technology helped in saving time and energy during deproteinization and demineralization steps of crustacean wastes. Recently, we have developed a patented process for chitin production using a microwave-assisted technology in combination with organic acids [14].

Figure 11.4 shows the effect of type or demineralizing agent (5 % gluconic acid or water) and contact time on weight loss of shrimp shells by autoclave-assisted

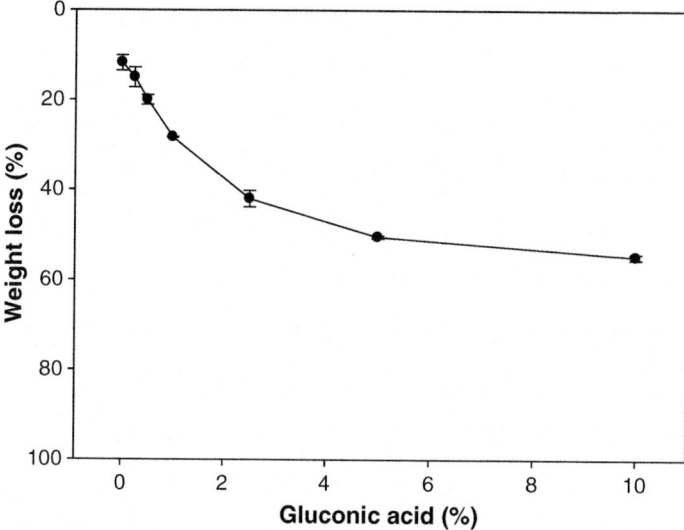

Fig. 11.5 Effect of gluconic acid concentration on demineralization process of shrimp shells by autoclave processing for 5 min at 121 °C

processing (121 °C). The weight loss of the shrimp shells treated for 10 min in aqueous media under autoclave was 10 %. At the same time, the samples treated with 5 % gluconic acid lost 50 % of weight.

It was also studied the effect of gluconic acid concentration on weight loss of shrimp shells by autoclaving processing as an index of the demineralization process (Fig. 11.5). As the concentration of gluconic acid in the medium increases, the weight loss of the shrimp shell decreased due to demineralization process. A plateau state was produced after the use of 5 % gluconic acid. Based on the results described above, the use of gluconic acid as demineralizing agent in the environmentally friendly chitin industry is suitable.

11.5 Use of Gluconic Acid for Chitosan Processing

Chitosan is a cationic biopolymer composed of glucosamine and N-acetyl-glucosamine units associated by β-1,4-glycosidic linkages [55]. This macromolecule is not soluble in water, which limits its wide application in the medicine and food industry [56]. In previous research on chitosan, the most popular solvent for dissolution has been acetic acid solution; however, this acid has a strong unpleasant smell [55]. Dissolved chitosan has antimicrobial and metal-binding properties and form beads, gels, fibers, films, and scaffolds [57]. Under mild

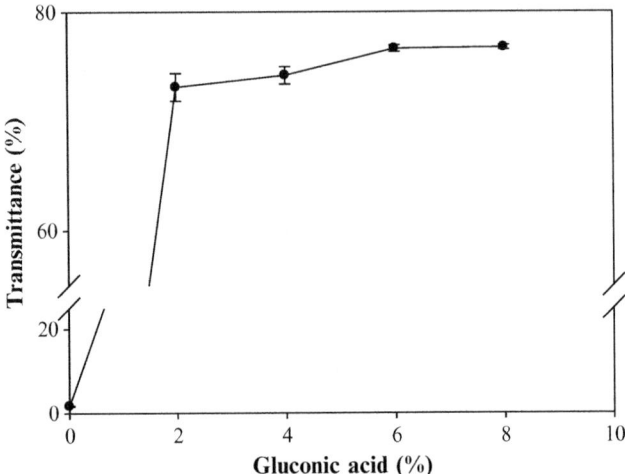

Fig. 11.6 Effect of gluconic acid concentration on the transmittance of chitosan solution

acidic conditions, chitosan can be depolymerized to yield water-soluble derivatives (i.e., glucosamine, chitosan oligomers, or water-soluble chitosan) by chemical, enzymatic, or physical methods.

In our laboratory, we tested the possibility of using gluconic acid as emergent greener solvent for chitosan dissolution [22]. The effect of gluconic acid concentration on the transmittance at 610 nm of chitosan (2 %, w/v) solution is shown in Fig. 11.6. As the gluconic acid concentrations increases, the intensity of the transmitted light of chitosan solution increased. Transmittance above 70 % in the region of 610 nm of chitosan solution was achieved with gluconic acid ranging between 6 and 8 % (v/v). Based on the results, gluconic acid is also suitable as a green solvent for the dissolution of chitosan.

Dissolved chitosan in 5 % (v/v) gluconic acid was depolymerized in a safe closed microwave pressure vessel (Nordic Ware Co., Minneapolis, MN, USA). The study was conducted to evaluate different microwave heating times (0, 1, 5, and 10 min) under pressure on the loss of viscosity of chitosan. The results showed that microwave irradiation causes decreasing of viscosity of chitosan solutions as the contact time increased (Fig. 11.7). The viscosity of irradiated mild-acidified chitosan decreased 35 % after 10 min compared with control solution. In the presence of 3 % (v/v) H_2O_2 together with chitosan dissolved in 5 % (v/v) gluconic acid, the extent of viscosity decreased about 85 % after 5 min of microwave irradiation versus a control solution (data not shown). Tian et al. [58] evaluated chitosan depolymerization by addition of hydrogen peroxide along with HCl in a wide range of temperature from 25 to 70 °C for incubation periods of 1, 2 or 3 h. In our depolymerization process, the chitosan dissolved with gluconic acid, hydrogen peroxide, and pressure is rapidly degraded by the promotion of these environmentally friendly conditions.

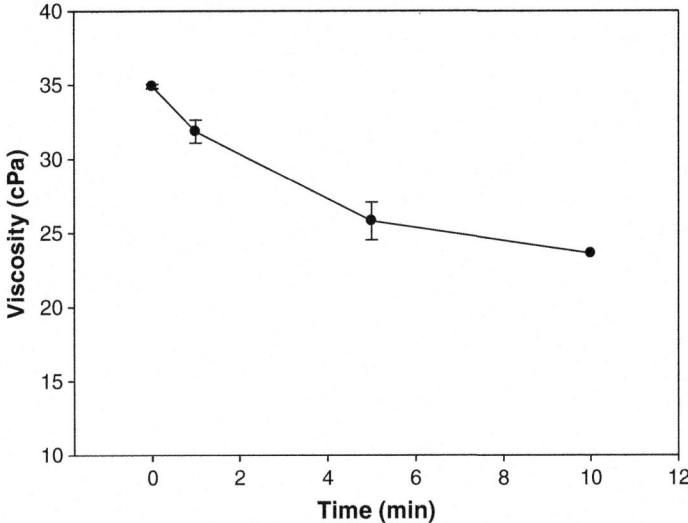

Fig. 11.7 Course of change of viscosity of chitosan dissolved in gluconic acid by microwave heating reaction. Viscosity was evaluated with portable vibrational viscometer (Viscolite 700, Hydramotion Ltd, UK). All experiments were made on triplicate

11.6 Use of Gluconic Acid for Fungal Chitosan-Glucan Production

Aspergillus niger is one of the most important microorganisms used in biotechnology for many decades to produce extracellular enzymes and organic acids [59, 60]. This industrial mold is considered "as generally recognized as safe" (GRAS) by the US Food and Drug Administration [61] for bioprocessing. Large amounts of *A. niger* wastes are generated in dry basis every year which are industrially considered for the production of animal feed, glucosamine, and biopolymers [62]. The structural biopolymers present in rigid cross-linked network from *Aspergillus* spp. cell walls are chitin, chitosan, glucan, and proteins [63–65]. In the *A. niger*, glucans are covalently associated with chitosan [66]. The reported glucan content of fungal cell wall has ranged from 30 to 60 %, depending on cultural conditions. The glucan component is responsible for the tensile strength, rigidity, and shape of the cell [67]. Most glucans are water-insoluble linear polymers made of glucose units joined through β-1,3 bonds, which are present in fungal cell wall [62].

Gluconic acid (6 %, v/v) has been used as green solvent to release and dissolve chitosan-glucan biopolymer from *A. niger* biomass by pressurized microwave-assisted extraction [22]. The effect of heating temperature (110, 120, and 130 °C) and contact time (0.42, 30, 60, and 90 min) was evaluated for maximizing the extraction of chitosan-glucan biopolymer. The solubilization rate of acidic-soluble biopolymer increased as temperature and contact time raise. The best conditions for chitosan-glucan extraction were heating temperature of 130 °C and heating time of 60 min within a maximal yield of 10.5 % (dry basis) and dynamic viscosity of 0.69 mPa × s. The green process based on the use of gluconic acid as green solvent enhanced the release of chitosan-glucan complex from a rigid cross-linked network present in *A. niger* biomass under environmentally friendly process. Extraction assisted by microwave and the use of an extracting agent acid (gluconic acid) are suitable for the recovery of a fungal polysaccharide with characteristics of chitosan-glucan.

11.7 Conclusions

The gluconic acid has recently emerged as an important alternative to mineral acids in the new development of extraction methodologies of structural polysaccharides. The combination of employing gluconic acid with green-energy sources has resulted in a variety of technologies capable of substituting the highly pollutant chemical production processes of specialty polysaccharides. Since the gluconic acid is advantageously produced by biotechnological methods, its wastes industrially generated might be easily biodegraded. The synergy offered by the gluconic acid and extraction technologies as microwave, hydrothermal pressurizing, or sonication is promising when considering eco-friendly technological developments of industrial polysaccharide manufacturing. Criteria such as productivity, cost, and bio-based issues are the value-added advantages of the use of gluconic acid.

Acknowledgments The authors express their gratefulness to the Mexican National Council for Science and Technology (CONACyT) for postdoctoral fellowship program to Oscar Fernando Vazquez Vuelvas and master in science scholarship to Cecilia Perez Cruz.

References

1. Barber PS, Shamshina JL, Rogers RD (2013) A "green" industrial revolution: using chitin towards transformative technologies. Pure Appl Chem 85:1693–1701
2. Hirano S (1996) Chitin biotechnology applications. Biotechnol Annu Rev 2:237–258
3. Dey PM, Brinson K (1984) Plant cell-walls. Adv Carbohydr Chem Biochem 42:265–382
4. Hernandez-Carmona G, Freile-Pelegrin Y, Hernandez-Garibay E (2013) Conventional and alternative technologies for the extraction of algal polysaccharides. In: Functional ingredients from algae for foods and nutraceuticals. Woodhead Published Limited, Cambridge, pp 475–516
5. Panchev IN, Kirtchev NA, Kratchanov C (1989) Kinetic model of pectin extraction. Carbohydr Polym 11:193–204
6. Sakai T, Sakamoto T, Hallaert J, Vandamme E (1993) Pectin, pectinase and protopectinase: production, properties and applications. Adv Appl Microbiol 39:213–294

7. Contreras-Esquivel JC, Espinoza-Pérez JD, Aguilar CN, Montañez JC, Charles-Rodríguez A, Renovato J, Aguilar CN, Rodríguez-Herrera R, Wicker L (2006) Extraction and characterization of pectin from novel sources. In: Advances in biopolymers, molecules, clusters, networks and interactions, ACS symposium series. American Chemical Society, Washington, DC, pp 215–227
8. Kurita O, Fujiwara T, Yamazaki E (2008) Characterization of the pectin extracted from citrus peel in the presence of citric acid. Carbohydr Polym 74:725–730
9. Contreras-Esquivel JC, Aguilar CN, Montanez JC, Brandelli A, Espinoza-Perez JD, Renard CMGC (2010) Pectin from passion fruit fiber and its modification by pectinmethylesterase. J Food Sci Nutr 15:57–66
10. Vasquez-Mejia MJ (2013) Recovery and characterization of pectic polysaccharides by employ of emergent technologies from tejocote pulp and citrus, pomegranate and mango peels. BSc thesis, School of Chemistry, Universidad Autonoma de Coahuila, Saltillo
11. Ma S, Yu SJ, Zheng XI, Wang XX, Bao QD, Guo XM (2013) Extraction, characterization and spontaneous emulsifying properties of pectin from sugar beet pulp. Carbohydr Polym 98:750–753
12. Jensen SV, Sorensen SO, Rolin C (2012) Process for extraction of pectin. US Patent 20120309946 A1
13. Valdez-Peña AU, Espinoza-Perez JD, Sandoval-Fabian GC, Balagurusamy N, Hernandez-Rivera A, De-la-Garza-Rodriguez IM, Contreras-Esquivel JC (2010) Screening of industrial enzymes for deproteinization of shrimp head for chitin recovery. Food Sci Biotechnol 19:553–557
14. Contreras-Esquivel JC (2012) Obtainment of chitin from shrimp waste by means of microwaves and/or autoclaving in combination with organic acids in a single stage. Mexican Patent 298224, Mar 2009
15. Onda A, Ochi T (2008) A new chemical process for catalytic conversion of D-glucose into lactic acid and gluconic acid. Appl Catal A 343:49–54
16. Le ZP, Wang LL, Huang XG, Huang YQ (2011) Study on the oxidation of glucose to gluconic acid by ozone under microwave. J Nanchang Univ (Eng & Technol) 33:217–221
17. Fan Z, Wu W, Hildebrand A, Kasuga T, Zhang R, Xiong X (2012) Novel biochemical route for fuels and chemicals production from cellulosic biomass. PLoS One 7:e31693
18. Hustede H, Haberstroh HJ, Schinzig E (2005) Gluconic acid. In: Ullmann's encyclopedia of industrial chemistry. Wiley-VCH Verlag GmbH & Co, Weinheim, pp 1–9
19. Werpy T, Petersen G (2010) Top value added chemical from biomass: volume 1 – results of screening potential candidates from sugar and synthesis gas. In: Pacific Northwest National Laboratory and the National renewable Energy Laboratory, U.S. Department of Energy. http://www.nrel.gov/docs/fy04osti/35523.pdf. Accessed 20 Feb 2014
20. Zhou B, Yang J, Li M, Gu Y (2011) Gluconic acid aqueous solution as a sustainable and recyclable promoting medium for organic reactions. Green Chem 13:2204–2211
21. Yang J, Zhou BH, Li MH, Gu YL (2013) Gluconic acid aqueous solution: a task-specific bio-based solvent for ring-opening reactions of dihydropyrans. Tetrahedron 69:1057–1064
22. Perez-Cruz C (2013) Use of biomass of *Aspergillus niger* using enzymatic hydrolysis and heating assisted by microwaves. MSc thesis, School of Chemistry, Universidad Autonoma de Coahuila, Saltillo
23. Richter G, Heinecker H (1979) Conversion of glucose into gluconic acid by means of immobilized glucose oxidase. Starch-Starke 31:418–422
24. Ramachandran S, Fontanille P, Pandey A, Larroche C (2006) Gluconic acid: properties, applications and microbial production. Food Technol Biotechnol 44:185–195
25. Lemeune S, Barbe JM, Trichet A, Guilard R (2000) Degradation of cellulose models during an ozone treatment. Ozonation of glucose and cellobiose with oxygen or nitrogen as carrier gas at different pH. Ozone-Sci Eng 22:447–460
26. Singh O, Kumar R (2007) Biotechnological production of gluconic acid: future implications. Appl Microbiol Biotechnol 75:713–722

27. Onda A, Ochi T, Yanagisawa K (2011) New direct production of gluconic acid from polysaccharides using a bifunctional catalyst in hot water. Catal Commun 12:421–425
28. Novalic S, Kongbangkerd T, Kulbe KD (1997) Separation of gluconate with conventional and bipolar electrodialysis. Desalination 114:45–50
29. May CD (1990) Industrial pectins: sources, production and applications. Carbohydr Polym 12:79–99
30. Hwang JK, Kokini JL (1995) Changes in solution properties of pectins by enzymatic hydrolysis of sidechains. J Korean Soc Food Sci Nutr 24:389–395
31. Marry M, McCann MC, Kolpak F, White AR, Stacey NJ, Roberts K (2000) Extraction of pectic polysaccharides from sugar-beet cell walls. J Sci Food Agric 80:17–28
32. Hoefler AC (1999) Pectin: chemistry, functionality and applications. Hercules Inc., Wilmington
33. Schols HA, Coenen GJ, Voragen AGJ (2009) Revealing pectin's structure. In: Visser J, Voragen AGJ (eds) Pectins and pectinases. Wageningen Academic Publishers, Wageningen, pp 17–33
34. Ralet MC, Thibault JF (2009) Hydrodynamic properties of isolated pectin domains: a way to figure out pectin macromolecular structure. In: Visser J, Voragen AGJ (eds) Pectins and pectinases. Wageningen Academic Publishers, Wageningen, pp 35–48
35. Van Buren JP (1991) Function of pectin in plant tissue structure and firmness. In: Walter RH (ed) The chemistry and technology of pectin. Academic, San Diego, pp 1–23
36. Hwang JK, Kim CJ, Kim CT (1998) Extrusion of apple pomace facilitates pectin extraction. J Food Sci 63:841–844
37. Fishman ML, Chau HK, Hoagland PD, Hotchkiss AT (2006) Microwave-assisted extraction of lime pectin. Food Hydrocolloids 20:1170–1177
38. Contreras-Esquivel JC, Hours RA, Aguilar CN, Reyes-Vega ML, Romero J (1997) Microbial and enzymatic extraction of pectin. A review. Arch Latinoam Nutr 47:208–216
39. Shkodina OG, Zeltser OA, Selivanov NY, Ignatov VV (1998) Enzymatic extraction of pectin preparations from pumpkin. Food Hydrocolloids 12:313–316
40. Min B, Lim JB, Ko SH, Lee KG, Lee SH, Lee SY (2011) Environmentally friendly preparation of pectins from agricultural byproducts and their structural/rheological characterization. Bioresour Technol 102:3855–3860
41. Yu X, Sun D (2013) Microwave and enzymatic extraction of orange peel pectin. Asian J Chem 25:5333–5336
42. Rezzoug SA, Maache-Rezzoug Z, Sannier F, Karim A (2008) A thermomechanical preprocessing for pectin extraction from orange peel. Optimisation by response surface methodology. Int J Food Eng. doi:10.2202/1556-3758.1183
43. Liu Y, Shi J, Langrish TAG (2006) Water-based extraction of pectin from flavedo and albedo of orange peels. Chem Eng J 120:203–220
44. Huang G, Shi J, Zhang K, Huang X (2012) Application of ionic liquids in the microwave-assisted extraction of pectin from lemon peels. J Anal Meth Chem. doi:10.1155/2012/302059
45. Yang HC, Hon MH (2010) The effect of degree of deacetylation of chitosan nanoparticles and its characterization and encapsulation efficiency on drug delivery. Polym-Plast Technol 49:1292–1296
46. Flores R, Barrera-Rodriguez S, Shirai K, Duran-de-Bazua C (2007) Chitin sponge, extraction procedure from shrimp wastes using green chemistry. J Appl Polym Sci 104:3909–3916
47. Aye KN, Stevens WF (2004) Improved chitin production by pretreatment of shrimp shells. J Chem Tech Biotechnol 79:421–425
48. Cira LA, Huerta S, Hall GM, Shirai K (2002) Pilot scale lactic acid fermentation of shrimp wastes for chitin recovery. Process Biochem 37:1359–1366
49. Jo GH, Jung WJ, Kuk JH, Oh KT, Kim YJ, Park RD (2008) Screening of protease-producing *Serratia marcescens* FS-3 and its application to deproteinization of crab shell waste for chitin extraction. Carbohydr Polym 74:504–508
50. Arbia W, Arbia L, Adour L, Amrane A (2013) Chitin extraction from crustacean shells using biological methods – a review. Food Technol Biotechnol 51:12–25

51. Kaur S, Dhillon S (2013) Recent trends in biological extraction of chitin from marine shell wastes: a review. Crit Rev Biotechnol. doi:10.3109/07388551.2013.798256
52. Gildberg A, Stenberg E (2001) A new process for advanced utilization of shrimp waste. Process Biochem 36:809–812
53. Jung WJ, Jo GH, Kuk JH, Ki KY, Park RD (2005) Demineralization of crab shells by chemical and biological treatments. Biotechnol Bioprocess Eng 10:67–72
54. Kwon KN, Choi HS, Cha BS (2009) Effect of microwave and ultrasonication on chitin extraction time. Korea J Food & Nutr 22:8–13
55. Chen PH, Hwang YH, Kuo TY, Liu FH, Lai JY, Hsieh HJ (2007) Improvement in the properties of chitosan membranes using natural organic acid solutions as solvents. J Med Biol Eng 27:23–28
56. Xia Z, Wu S, Chen J (2013) Preparation of water soluble chitosan by hydrolysis using hydrogen peroxide. J Biol Macromol 59:242–245
57. Zivanovic S, Li JJ, Davidson M, Kit K (2007) Physical, mechanical, and antibacterial properties of chitosan/PEO blend films. Biomacromolecules 8:1505–1510
58. Tian F, Liu Y, Hu K, Zhao B (2004) Study of the depolymerization behavior of chitosan by hydrogen peroxide. Carbohydr Polym 57:31–37
59. Gouka R, Punt P, Van-den-Hondel C (1997) Efficient production of secreted proteins by *Aspergillus*: progress, limitations and prospects. Appl Microbiol Biotechnol 47:1–11
60. Roukas T (2000) Citric and gluconic acid production from fig by *Aspergillus niger* using solid state fermentation. J Ind Microbiol Biotechnol 25:298–304
61. Schuster E, Dunn-Coleman N, Frisvad J, Van-Dijck M (2002) On the safety of *Aspergillus niger*. Appl Microbiol Biotechnol 59:426–435
62. Cai J, Yang J, Du Y, Fan L, Qiu Y, Li J, Kennedy JF (2006) Enzymatic preparation of chitosan from the waste *Aspergillus niger* mycelium of citric acid production plant. Carbohydr Polym 64:151–157
63. Muzzarelli R, Boudrant J, Meyer D, Manno N, Demarcáis M, Paoletti M (2012) Current views on fungal chitin/chitosan, human chitinases, food preservation, glucans, pectins and inulin: a tribute to Henri Braconnot, precursor of the carbohydrate polymers science, on the chitin bicentennial. Carbohydr Polym 87:995–1012
64. Ruiz-Herrera J (2012) Cell wall composition. In: Group TF (ed) Fungal cell wall structure, synthesis, and assembly, 2nd edn. CRC Press, Boca Raton, pp 7–21
65. Vries RP, Visser J (2001) *Aspergillus* enzymes involved in degradation of plant cell wall polysaccharides. Microbiol Mol Biol Rev 65:497–522
66. Kogan G, Rauko P, Machova E (2003) Fungal chitin–glucan derivatives exert protective or damaging activity on plasmid DNA. Carbohydr Res 338:931–935
67. Fleet GH, Phaff HJ (1981) Fungal glucans-structure and metabolism. In: Tanner W, Loewus F (eds) Encyclopedia of plant physiology new series. Springer, Berlin, pp 416–440

Chapter 12
2-Methyltetrahydrofuran: Main Properties, Production Processes, and Application in Extraction of Natural Products

Anne-Gaëlle Sicaire, Maryline Abert Vian, Aurore Filly, Ying Li, Antoine Bily, and Farid Chemat

Abstract 2-Methyltetrahydrofuran (MeTHF) is a solvent produced from renewable raw materials by the hydrogenation of products obtained from carbohydrate fractions of hemicellulose from various feedstocks. MeTHF has the advantages to be biodegradable and has a promising environmental footprint, good preliminary toxicology assessments, and an easy recycling. An experimental study was conducted with MeTHF, in comparison to n-hexane, for the extraction of carotenoids and aromas. In parallel to this experimental study, a HSP (Hansen solubility parameters) theoretical study has been realized for the evaluation and the understanding of the interactions between the solvent and different compounds such as triglycerides contained in canola oil, carotenoids, and aromas. The results of these studies show that MeTHF appears to be a potential alternative solvent to n-hexane for the extraction of various products.

12.1 Introduction

Extraction processes appear to take a very large part in industrial processes and produce not only by-products but also waste solvent or wastewater to recycle or eliminate. Extraction solvents generally are organic volatile compounds produced from nonrenewable resources, such as petroleum, and may be harmful for human health and environment. For example, one of the extraction solvents most commonly

---

A.-G. Sicaire • M. Abert Vian (✉) • A. Filly • Y. Li • F. Chemat
Green Extraction Team, Université d'Avignon et des Pays de Vaucluse, INRA, UMR 408, F-84000 Avignon, France
e-mail: maryline.vian@univ-avignon.fr; aurore.filly@alumni.univ-avignon.fr; Ying.li@univ-avignon.fr; Farid.chemat@univ-avignon.fr

A. Bily
R&D Director, Nutrition & Health, Naturex, F-84000 Avignon, France
e-mail: a.bily@naturex.com

Fig. 12.1 MeTHF structure

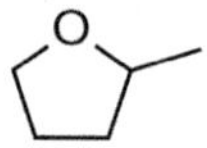

Table 12.1 MeTHF major benefits

Biodegradable	Renewable – biomass derived
Noncarcinogenic	Noncorrosive
High solvency power for resins, polymers, and dyes	Stable to acids and bases
Quite low boiling point	Easy to recycle
Low vapor pressure	Not a hazardous air pollutant
Low VOC	Not an ozone-depleting chemical

VOC volatile organic compound

used industrially is n-hexane, a fraction of petroleum. It has the advantage to be quite easy to obtain and to have chemical properties that provide ideal functionalities in terms of solubility for various products such as vegetable oils. Moreover, it is very easy to recycle considering its very low miscibility with water. Nevertheless n-hexane is produced from fossil energies and has recently been classified as CMR 3 which means that it is suspected to be reprotoxic [1].

Biomass-derived chemicals appear to be in accordance with several of the 12 principles of green chemistry described by Anastas and Warner [2] such as the third principle concerning the reduction of hazardous chemical syntheses, the fourth concerning the use of safer chemicals, the tenth concerning the degradation, or the seventh suggesting the use of renewable feedstock. These principles give suggestions for the design of greener products and processes.

In fact, 2-methyltetrahydrofuran (MeTHF) represented in Fig. 12.1 is a solvent produced from renewable raw materials as its reactants can be obtained from biomass by the hydrogenation of products obtained from carbohydrate fractions of hemicellulose from various feedstocks [3–5]. It is biodegradable, has a promising environmental footprint and good preliminary toxicology [6] assessments, and is easy to recycle. Considering all the advantages of this solvent, summarized in Table 12.1, several applications of MeTHF can be found in the literature especially as green solvent in organic chemistry [3, 7], but it can also be considered as an interesting solvent for the extraction of bioactive components from natural sources [8, 9].

12.2 MeTHF Properties

MeTHF represented in Fig. 12.1 (CAS No. 96-47-9, IUPAC name 2-methyltetrahydrofuran), with molecular formula $C_5H_{10}O$, is a clear liquid that is derived from renewable resources as corncobs or sugarcane bagasse. MeTHF is biodegradable,

Table 12.2 Basic properties of MeTHF

Properties	MeTHF
Molecular weight – M (g/mol)	86.1
Melting temperature – T_f (°C)	−136
Normal boiling temperature – T_b (°C)	80.2
Vapor pressure at 20 °C (mm)	102
Density at 20 °C (g/mL)	0.854
Viscosity at 25 °C (cp)	0.46
Flash point (°C)	−11
Enthalpy of vaporization (kJ·kg^{-1})	375
Dielectric constant at 20 °C	6.97
Solubility in water at 20 °C (g/100 g)	14

nontoxic, and non-ozone depleting; indeed, it is still not approved yet by the Food and Drug Administration (FDA) to be used for food contact. General properties of MeTHF are presented in Table 12.2. All the presented properties [3, 7, 10] make MeTHF as a suitable compound for several applications especially in the extraction field.

12.3 Production Processes

12.3.1 Raw Materials

Building blocks for the synthesis of MeTHF are issued of carbohydrates derived from lignocelluloses' biomass, which represent the largest terrestrial biomass resources. Although MeTHF can be produced thanks to catalytic processes from furfural or levulinic acid. Both compounds are obtained from the implementation of the concept of biorefinery with the retreatment of by-products such as corncobs or sugarcane bagass generated by agricultural industry. Furfural, a heterocyclic aldehyde with molecular formula $C_5H_4O_2$, is a versatile compound in the fragrance industry. It is a colorless oily liquid with an almond smell. Lignocellulose material can lead to furan molecules by dehydration reactions of carbohydrates from biomass origin. Furfural can be isolated from polysaccharide hemicellulose, polymers of C5 sugars, contained in many plant materials. The pentosan contained in hemicellulose is hydrolyzed in pentose carbohydrates which are dehydrated to furfural in acid conditions using sulfuric or phosphoric acid as catalysts [11]. Levulinic acid is a keto acid with molecular formula $C_5H_8O_3$. It appears as a white crystalline solid, and it is a highly versatile compound with several applications like in resins or plasticizers industry but also as precursor for pharmaceuticals [12]. Levulinic acid can be produced by acid hydrolysis at high temperature of carbohydrates [13], which are C6 sugars such as glucose, galactose, or sucrose, isolated from wood-based feedstock.

Fig. 12.2 Production of MeTHF

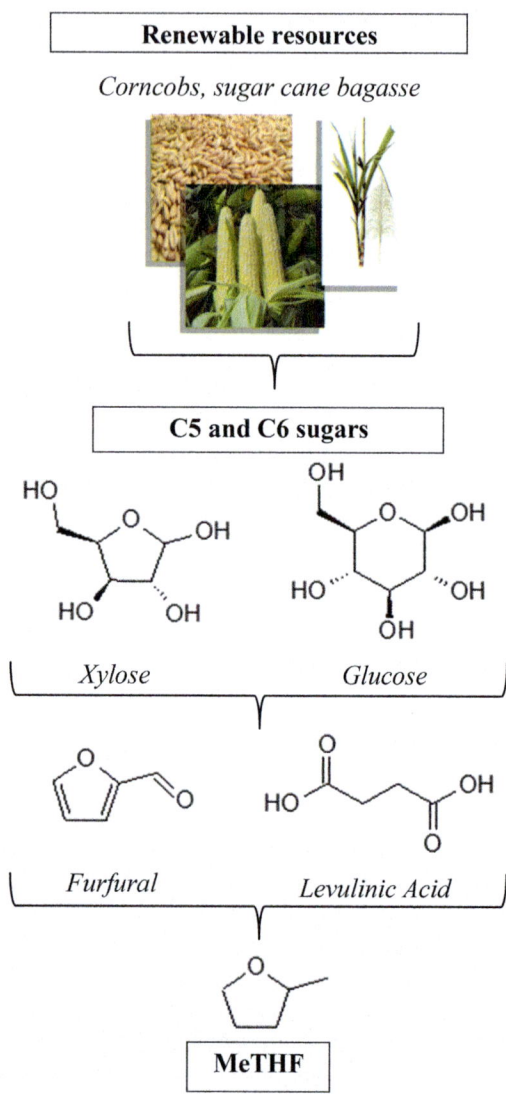

12.3.2 Synthesis

MeTHF can be produced from furfural or levulinic acid as shown in Fig. 12.2.

12.3.3 Synthesis from Furfural

MeTHF results from successive hydrogenations, as shown in Fig. 12.3, of furfural (and reactions intermediates) over Ni-Cu, Fe-Cu, Cu-Zn, or Cu-Cr alloy catalysts as

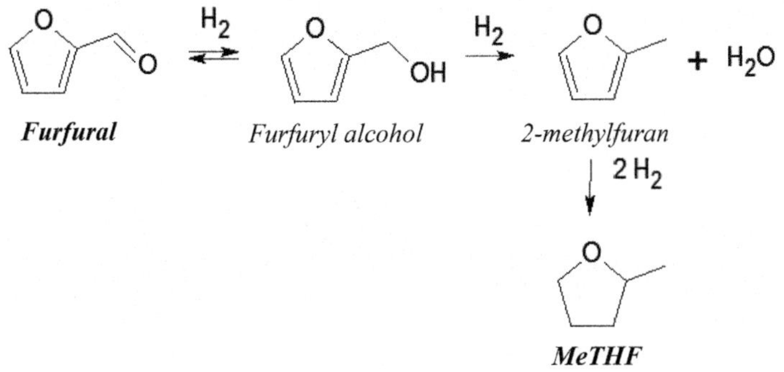

Fig. 12.3 Production of MeTHF from furfural

reported in the literature [14–18]. First, two successive hydrogenations of furfural on the Cu-Zn catalyst allow nearly a complete conversion of the compound to furfuryl alcohol directly converted to 2-methylfuran in a range of temperature of 200–300 °C with a yield higher than 95 % [5].

MeTHF results in the conversion of 2-methylfuran at lower temperature (100 °C) over Ni-based catalyst with a yield around 86 %. Increasing the temperature would decrease the quantity by conversion in 2-pentanone. The choice of the catalyst also has a great importance in the conversion yield. In fact, MeTHF is the main product of the hydrogenation of furfural depending on the catalyst and on reaction conditions [5].

12.3.4 Synthesis from Levulinic Acid

The synthesis of MeTHF from levulinic acid consists in consecutive catalyzed hydrogenations and dehydrogenations as shown in Fig. 12.4. The catalyzed hydrogenation of the keto group of levulinic acid leads to a hydroxyl acid that results in γ-valerolactone. Further hydrogenation of the keto bond of γ-valerolactone allows the formation of the cyclic hemiacetal in equilibrium with the aliphatic hydroxyl aldehyde. The hydrogenation of the last carbonyl group leads to 1,4-pentanediol that is etherified in MeTHF by dehydration in acid conditions.

The reactions are conducted with a ruthenium catalyst complex with tridentate phosphine ligands and acidic ionic additives in conditions described by Geilen et al. [4].

12.3.5 Recovery of MeTHF

MeTHF can be recovered by conventional distillation, thanks to the solubility behavior of MeTHF/water mixtures and the formation of a favorable azeotrope between the two compounds.

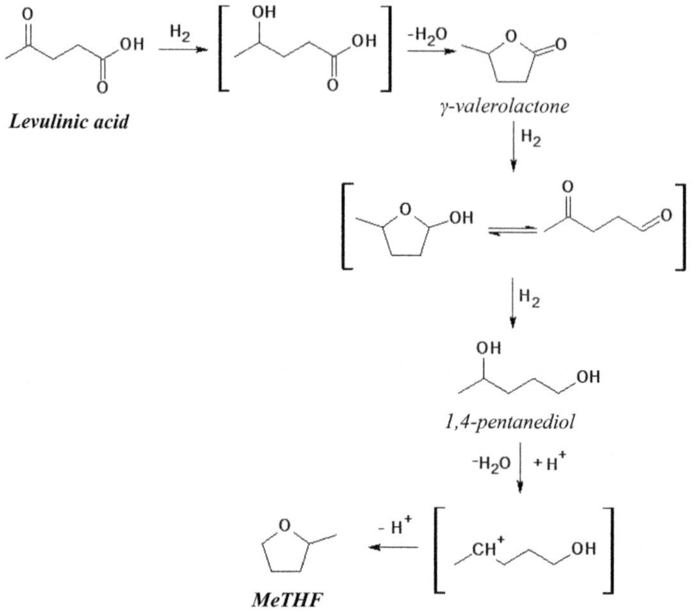

Fig. 12.4 Production of MeTHF from levulinic acid

Table 12.3 Solubility of water in MeTHF and MeTHF in water

Solubility of water in MeTHF		Solubility of MeTHF in water	
Temperature (°C)	wt% water	Temperature (°C)	wt% water
0.0	4.0	0.0	21.0
9.5	4.1	9.5	17.8
19.3	4.1	19.3	14.4
29.5	4.2	29.5	11.4
39.6	4.3	39.6	9.2
50.1	4.4	50.1	7.8
60.7	4.6	60.7	6.6
70.6	5.0	70.6	6.0

As can be seen in Table 12.3, the solubility of water in MeTHF varies very slightly between 0 to 70 °C, whereas the solubility of MeTHF in water decreases a lot as the temperature is increased. Considering these properties, MeTHF/water mixtures need to be separated at least at 60 °C in order to minimize the amount of MeTHF in water [10]. The azeotrope formed between MeTHF and water contains 10.6 % water and so 89.4 % MeTHF. Though MeTHF can be recovered at atmospheric pressure in batch or continuous distillation processes at 60 °C, as shown in Fig. 12.5, considering the recycling of a mixture containing 100 parts MeTHF and 100 parts water [3, 7, 10].

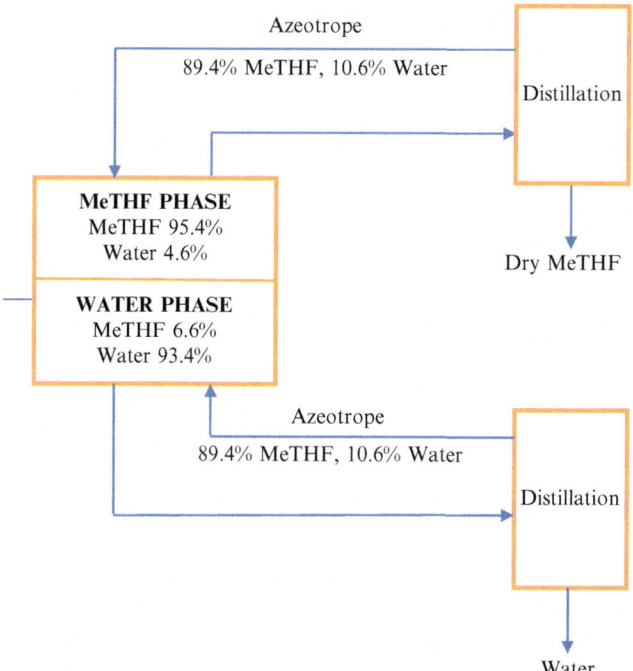

Fig. 12.5 Recovery of MeTHF at 60 °C

12.4 Applications

MeTHF has many applications as solvent for organic synthesis [3, 7, 10], but considering its various properties it looks very promising as alternative to commonly used petroleum solvent for extraction of natural products [8, 9]. A literature search did not yield any reference to earlier reports on using MeTHF as solvent for the extraction of natural products, although using MeTHF for the extraction of carotenoids or aromas that usually imply solvents as n-hexane can be considered.

12.4.1 Extraction of Carotenoids

Solvents issued from petroleum are the one currently used for the extraction of carotenoids. A study was conducted to evaluate the potential of MeTHF compared to n-hexane, the solvent currently used by industrials.

A kinetic study was conducted during 3 h with n-hexane and MeTHF. Figure 12.6 gives the yield of total carotenoids extracted for 1 g of dry carrots.

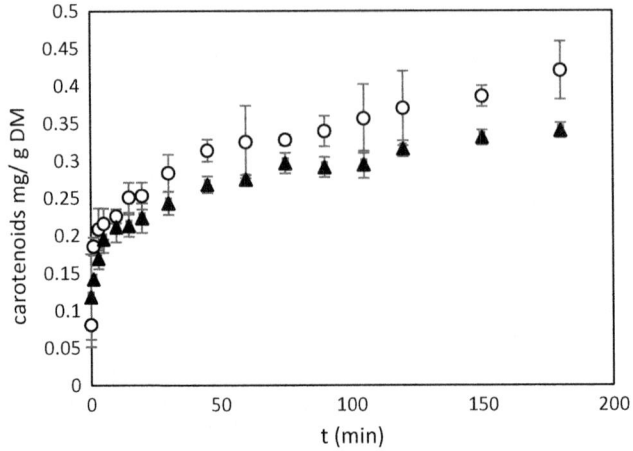

Fig. 12.6 Kinetic of total carotenoids extracted from carrots (○ MeTHF, ▲ n-hexane)

As can be seen from Fig. 12.6, the extraction is realized in two steps. A first solvent-exchange surface interaction takes place for a short time frame. Thus, starting accessibility δX_s (in mg carotenoids/g dry material) corresponding to a "washing" step reveals the amount of carotenoids obtained in a very short time frame (t near 0) through the convection of solvent interacting with the exchange surface. Afterward, the main part of the extraction is controlled through various penetration processes of the solvent within the carrot particles (capillarity, molecular diffusivity, etc.). The driving force of the global operation is the gradient of concentration and the model can be similar to Fick's law with an effective diffusivity D_{eff} (m$^2 \cdot$s^{-1}) as the process coefficient [19].

According to Fick's first law [19]

$$\frac{\rho_s}{\rho_d}(\vec{v}_s - \vec{v}_d) = -D_{eff}\vec{\nabla}\left(\frac{\rho_s}{\rho_d}\right)$$

It can be assumed the absence of expansion or shrinkage of the solid particles which are not moving, i.e., =0 and =constant.

$$\rho_s \vec{v}_s = -D_{eff}\vec{\nabla}\rho_s$$

Crank's solution for a sphere

$$\frac{X_\infty - X}{X_\infty - X_{t_0}} = \sum_{i=1}^{\infty} \frac{6}{i^2 \pi^2} \exp\left(-\frac{i^2 \pi^2 D_{eff}}{r_d^2}(t - t_0)\right)$$

$$\frac{X_\infty - X}{X_\infty - X_{t_0}} = A \, \exp(-k(t - t_0))$$

12 2-Methyltetrahydrofuran: Main Properties, Production Processes...

Table 12.4 Starting accessibility and effective diffusivity for the extraction of carotenoids from carrots with n-hexane and MeTHF

Solvent	δXs (mg/g DM)	D_{eff} (.10^{10} m²/s)
n-Hexane	0.220	0.009
MeTHF	0.269	0.013

$$\mathrm{Ln}\left(\frac{X_\infty - X}{X_\infty - X_{t_0}}\right) = -k\,(t - t_0)$$

$$D_{eff} = k\frac{r_d^2}{\pi^2}$$

Starting accessibility: value obtained by extrapolating diffusion model to $t = 0$: $X_0 \neq (X_i = 0)$

$$X_0 - X_i = X_0 = \delta X_s$$

with

δX_s : starting accessibility (g of extract/g of dry material)
D_{eff} : effective diffusivity (m² s^{-1})
ρ_s : apparent density of the solute within the solid matrix (kg m^{-3})
ρ_d : apparent density of the solid dry material (kg m^{-3})
v_s : velocity of the solute (m s^{-1})
v_d : velocity of the solid dry material (m s^{-1})
X_∞ : amount of solute within the matrix (mg · g^{-1} dry material)
d_p : radius
X : amount of solute extracted at time t (mg · g^{-1} dry material)
k : transfer coefficient (m · s^{-1})

Starting accessibility and diffusivity were calculated using previous equations. Starting accessibility is determined by extrapolating the value for $t = 30$ min at $t = 0$. Calculated values are listed in Table 12.4. Starting accessibility appears to be higher for MeTHF than for n-hexane, and D_{eff} is 1.5 times higher with MeTHF than with n-hexane. This means that the washing step with MeTHF permits to solvate a higher amount of carotenoids at the surface of the matter than n-hexane does which suggests that there is a part of the solute extracted almost instantly that comes from the layers of cells the most exposed to the solvent. The same trend is observed for the effective diffusivity D_{eff} where the value for n-hexane is $0.009 \times 10-10$ m²/s and $0.013 \times 10-10$ m²/s for MeTHF. This means that the extraction is faster with MeTHF than with n-hexane. This is probably due to the difference of boiling point (69 °C vs 80 °C); a higher temperature increases the extraction, even if a higher temperature can increase the risk of a degradation of the extract. The carotenoid content in dry carrots was determined by HPLC analysis. After 6 h extraction

MeTHF permits to extract 419 μg/g dry matter, whereas n-hexane permits to extract 338 μg/g dry matter which represents 23 % more carotenes extracted with MeTHF compared to n-hexane.

12.4.2 Extraction of Aromas

n-Hexane has been used as an extraction solvent for aromas of buds black currant. Recent regulation have banned numerous common organic solvents, as n-hexane, that have been recognized as hazardous to human health and environment. It is urgently to replace n-hexane by alternative solvents that minimize health and environmental risk. MeTHF can be considered as a good solvent from a (HSE) Health Safety Environment point of view which makes it a good candidate for the replacement of organic solvents. Performance trials using MeTHF and n-hexane were conducted in bud black currant.

The quality of oils extracted by these solvents was evaluated. Compounds were identified using GC-MS, while the content of separated components was measured by GC-FID (Table 12.5). A total of 30 major compounds (in agreement with the literature [20, 21]) were identified in n-hexane extract. The results reported in Table 12.5 show that the relative proportions of these compounds are similar for both solvents.

Extracts obtained with n-hexane and MeTHF, respectively, contain 47.73 and 42.88 % of non-oxygenated compounds, while the amounts of oxygenated compounds, respectively, were 25.26 and 17.06 %. The principal volatile compounds were δ-3-carene (21.45 % with n-hexane and 17.55 % with MeTHF) and terpinolene (11.34 % with n-hexane and 8.16 % with MeTHF), followed by other important compounds as sabinene, β-caryophyllene, caryophyllene oxide, p-cymen-8-ol, trans β-ocimene, β-phellandrene, β-myrcene, α-humulene, cis β-ocimene, spathulenol, humulene epoxide, limonene, terpinolene epoxide, and 3-caren-5-one. Some of these main compounds are represented in Fig. 12.7. Besides, the higher extraction yield (7.10 %) obtained with MeTHF compared to n-hexane (3.87 %) is probably due to the fact that MeTHF allows the extraction of other compounds as amino acids, flavonoids, and phospholipids.

12.4.3 Comprehension of Solubility of Primary and Secondary Metabolites of Various Natural Products by Using Hansen Theoretical Prediction

Hansen solubility parameters (HSP) of solvents and solutes have been studied using HSP theoretical prediction [22]. HSP provides a convenient and efficient way to characterize solute/solvent interactions according to the general "like

Table 12.5 Major compounds extracted with n-hexane and MeTHF

Compounds	RI	n-hexane (%)	MeTHF (%)
α-Thujene[a]	929	0.25	0.28
α-Pinene[a]	938	0.93	0.79
Camphene[a]	945	–	–
Sabinene[a]	**968**	**4.93**	**3.88**
β-Pinene[a]	973	0.85	0.75
β-Myrcene[a]	**983**	**1.86**	**1.46**
2-Carene[a]	1,001	0.31	0.27
Alpha phellandrene[a]	998	0.17	0.15
δ-3-Carene[a]	**1,002**	**21.45**	**17.55**
α-Terpinene[a]	1,009	0.81	0.69
p-Cymene[a]	1,013	0.82	0.56
β-Phellandrene[a]	**1,022**	**2.06**	**1.57**
Limonene[a]	**1,025**	**1.41**	**1.07**
*trans*β-ocimene[a]	**1,026**	**2.79**	**2.24**
*cis*β-ocimene[a]	**1,037**	**1.69**	**1.35**
γ-Terpinene[a]	1,040	0.25	0.29
4-Terpinyl acetate	1,030	–	–
cis sabinene hydrate[a]	1,068	0.35	0.22
α-p-Dimethylstyrene[a]	1,071	–	–
Terpinolene[a]	**1,080**	**11.34**	**8.16**
trans sabinene hydrate[a]	1,097	0.33	0.23
Terpinolene epoxide		**1.22**	**0.46**
cis verbenol	1,110	–	–
Dehydrolinalool[a]	1,124	–	–
trans sabinol[a]	1,130	–	–
Pinocarvone	1,135	–	–
p-Cymen-8-ol[a]	**1,163**	**2.85**	**1.73**
Terpin-4-ol[a]	1,173	–	–
α-Terpineol[a] (p-menth-1-en-8-ol)	1,180	–	–
Eucarvone	1,240	–	–
Bornyl acetate[a]	1,270	0.36	0.35
3-Caren-5-one		**1.13**	**0.69**
α-Terpinyl acetate[a]	1,340	0.32	0.35
Sesquiterpenes			
β-Elemene[a]	1,380	0.22	–
α-Humulene[a]	**1,452**	**1.82**	**2.32**
α-Muurolene[a]	1,477	–	–
Germacrene D[a]	1,480	0.82	1.29
γ-Cadinene[a]	1,513	–	–
δ-Cadinene[a]	1,524	–	–
Spathulenol[a]	**1,576**	**1.61**	**1.14**
Caryophyllene oxide[a]	**1,562**	**4.16**	**2.69**
Aromadendrene oxide (1)	1,595	–	–
Humulene epoxide II[a]	**1,600**	**1.59**	**1.04**
Hardwickic acid		5.1	11.06

(continued)

Table 12.5 (continued)

Compounds	RI	n-hexane (%)	MeTHF (%)
Yields		3.87 %	7.10 %
Total non-oxygenated		47.95 %	42.88 %
Total oxygenated		25.26 %	17.06 %
Global identification		73.21 %	59.94 %

RI retention indices
The percentage correspond to percent of total peak area (%)
[a]Aromas already known in this matrix

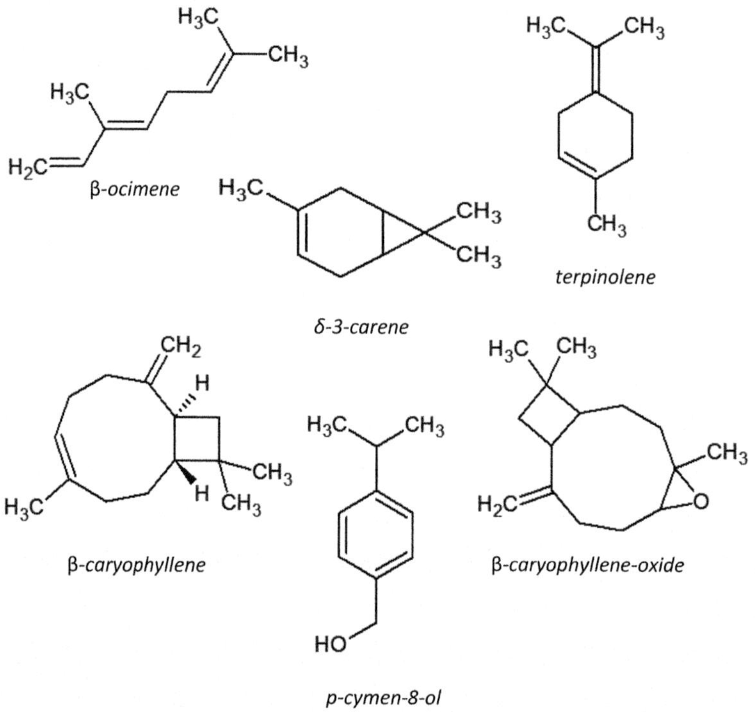

Fig. 12.7 Principal volatile compounds in bud black currant

dissolves like" rule. HSP has been found to be superior for more applications to acknowledged Hildebrand parameter, which the fundamental total cohesive energy density is partitioned by atomic dispersion forces (δ_d^2), molecular polar forces arising from dipole moments (δ_p^2), and hydrogen bonds (exchange of electrons, proton donor/acceptor) between moleculars (δ_h^2), as given in Eq. (12.1):

$$\delta_{total}^2 = \delta_d^2 + \delta_p^2 + \delta_h^2 \tag{12.1}$$

Table 12.6 Hansen solubility parameters of various solvents and extracts

	Compounds	δ_d (MPa$^{1/2}$)	δ_p (MPa$^{1/2}$)	δ_h (MPa$^{1/2}$)	δ_{total} (MPa$^{1/2}$)
Solvent	n-Hexane	15	0	0	15
	MeTHF	16.8	4.8	4.6	17.7
Carotenoids in carrots	β-Carotene	17.4	0.8	1.7	17.5
	Lutein	17.8	1.3	4.5	18.4
	Zeaxanthin	17.8	1.4	4.8	18.5
Main aroma compounds in black currant	Terpinolene	17.2	1.9	4.3	17.8
	δ-3-Carene	16.4	1.1	2.2	16.6
	β-Caryophyllene	16.8	0.7	2.2	16.6
	Sabinene	16.5	1.6	2.1	16.7
	trans β-ocimene	16.4	1.5	2.6	16.7
Possible triglycerides in rapeseed oil	Rapeseed oil[1]	16.4	4.7	4.2	17.6
	Rapeseed oil[2]	16.6	4	4.1	17.5
	Rapeseed oil[3]	16.6	4.1	3.6	17.5
	Rapeseed oil[4]	16.5	4.2	4.6	17.6
	Rapeseed oil[5]	16.4	4	4.5	17.5

[1]Triglyceride (R_1 C18:1n9, R_2 C18:2n6, R_3 C18:3n3)
[2]Triglyceride (R_1 C18:1n9, R_2 C18:1n9, R_3 C18:2n6)
[3]Triglyceride (R_1 C18:1n9, R_2 C18:2n6, R_3 C18:2n6)
[4]Triglyceride (R_1 C18:1n9, R_2 C18:1n9, R_3 C18:3n3)
[5]Triglyceride (R_1 C18:2n6, R_2 C18:2n6, R_3 C18:3n3)

where δ total is the Hansen total solubility parameter, which now consists of its three partitioned HSP in terms of dispersion (δ_d), polar (δ_p), and hydrogen-bonding (δ_h) force, respectively.

In general, the more similar the two δ_{total} are, the greater the affinity between solutes and solvents. The chemical structures of the solvents and solutes discussed in this article could be mutually transformed by JChemPaint ver. 3.3 to their Simplified Molecular Input Line Entry Syntax (SMILES) notations, which were subsequently used to calculate the solubility parameters of extracts and extractants. These solubility parameters were further modeled to a frequently used two-dimensional HSP graph for better visualization of the solute/solvent interaction due to insignificant differences between δ_ds (HSPiP Version 4.1.03, Denmark).

Since MeTHF was firstly used as extractant for natural products, it is interesting to introduce methods for predicting its physiochemical properties and the solubility parameters of its extracts, most of which do not exist in HSP database. The useful prediction method proposed by Yamamoto was applied to calculate HSP of solutes and solvents only through their chemical structures due to its stability and high accuracy comparing to other HSP estimation methods. Yamamoto-Molecular Break (Y-MB) method breaks SMILES into corresponding functional groups and thus estimates various physicochemical properties. The theoretical physicochemical properties of extractants and the HSP of their main extracts were obtained by Y-MB calculation through their chemical structures, which were represented in Table 12.6.

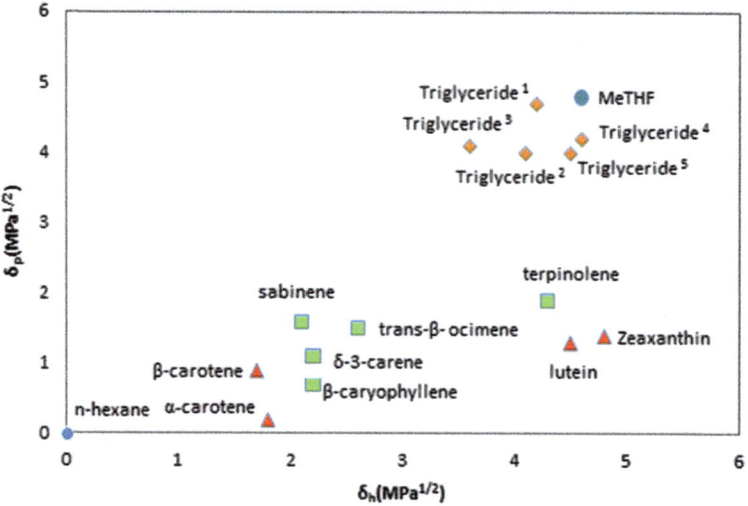

Fig. 12.8 General two-dimensional diagram of Hansen solubility parameters for all solutes (♦ possible triglycerides in rapeseed oils, ■ aroma compounds in black currant, and ▲ carotenoids in carrot) in solvents (• n-hexane and MeTHF)

MeTHF has nearly the same molecular weights (≈86.1 g/mol) as that of n-hexane. Moreover, it has higher boiling (82.4 °C) and flash point (−1.9 °C) in comparison to n-hexane (69 °C and −23.3 °C, respectively), which signifies MeTHF is a less flammable and less hazardous azeotrope with water. MeTHF with higher HSP values is considered higher polarity than n-hexane. The major drawback of using MeTHF is its higher viscosity and density than n-hexane, which can induce lower global identification in the component analysis, as well as its relatively high boiling point which may lead to higher energy consumption in its recovery. In addition, MeTHF with higher dielectric constant (6.97) has also found to be stable in acids and base, which allows the extraction more efficient and stable. Besides, this agro-solvent from renewable resources also has low volatility, which improves safety and reduces solvent consumption and CO_2 emissions. Considering all these aspects, MeTHF could be theoretically considered a better alternative solvent to n-hexane in various extractions.

The two-dimensional (2-D) graph of δ_p versus δ_h has usually been used as references for easy understanding of solubility in solvent extraction as the result of the insignificant δ_d. Figure 12.8 illustrated that all combination possibilities of triglycerides in rapeseed oils were distinguished by small variations depending on their constituent fatty acids. According to the "like extracts like" principle, the triglycerides of rapeseed oil may be the most possible solutes in MeTHF, while the main aroma compounds in black currant seemed more likely to dissolve in n-hexane. Regarding carotenoids in carrots, lutein and zeaxanthin were more soluble in MeTHF, whereas α- and β-carotene were closed to n-hexane. These predicted results were generally in accordance with experimental results even though MeTHF gave higher total extraction yield than n-hexane in all independent extraction of rapeseed oil, aroma compounds, and carotenoids, which have further proved MeTHF as the alternative solvent to n-hexane for the extraction of natural compounds from plants.

References

1. Liu J, Huang L, Sun Y, Li YC, Zhu JL, Wang WX, Zhang WC (2013) N-hexane alters the maturation of oocytes and induces apoptosis in mice. Biomed Environ Sci 26(9):735–741
2. Anastas PT, Warner JC (1998) Green chemistry: theory and practice. Oxford University Press, Oxford/New York
3. Pace V, Hoyos P, Castoldi L, Domínguez deMaría P, Alcántara AR (2012) 2-methyltetrahydrofuran (2-MeTHF): a biomass-derived solvent with broad application in organic chemistry. ChemSusChem 5(8):1369–1379
4. Geilen FMA, Engendahl B, Harwardt A, Marquardt W, Klankermayer J, Leitner W (2010) Selective and flexible transformation of biomass-derived platform chemicals by a multifunctional catalytic system. Angew Chem Int Ed 49(32):5510–5514
5. Zheng H-Y, Zhu Y-L, Teng B-T, Bai Z-Q, Zhang C-H, Xiang H-W, Li Y-W (2006) Towards understanding the reaction pathway in vapour phase hydrogenation of furfural to 2-methylfuran. J Mol Catal Chem 246(1–2):18–23
6. Antonucci V, Coleman J, Ferry JB, Johnson N, Mathe M, Scott JP, Xu J (2011) Toxicological assessment of 2-methyltetrahydrofuran and cyclopentyl methyl ether in support of their use in pharmaceutical chemical process development. Org Process Res Dev 15(4):939–941
7. Aycock DF (2007) Solvent applications of 2-methyltetrahydrofuran in organometallic and biphasic reactions. Org Process Res Dev 11(1):156–159
8. vom Stein T, Grande PM, Kayser H, Sibilla F, Leitner W, Domínguez de María P (2011) From biomass to feedstock: one-step fractionation of lignocellulose components by the selective organic acid-catalyzed depolymerization of hemicellulose in a biphasic system. Green Chem 13(7):1772
9. Saunois A, Legrand J, Mercier E (2011) Extraction solide/liquide. WO 2011092334 A2
10. Penn Specialty Chemicals Inc. (2005) Methyltetrahydrofuran. Penn Specialty Chemicals Inc., Memphis
11. Adams R, Voorhees V (2003) Furfural. In: Organic syntheses. Wiley, New York: Hoboken, NJ
12. Ghorpade V, Hanna M (1997) Industrial applications for levulinic acid. In: Campbell GM, Webb C, McKee SL (eds) Cereals. Springer, New York, pp 49–55
13. Upare PP, Yoon J-W, Kim MY, Kang H-Y, Hwang DW, Hwang YK, Kung HH, Chang J-S (2013) Chemical conversion of biomass-derived hexose sugars to levulinic acid over sulfonic acid-functionalized graphene oxide catalysts. Green Chem 15(10):2935–2943
14. Zhu Y-L, Xiang H-W, Li Y-W, Jiao H, Wu G-S, Zhong B, Guo G-Q (2003) A new strategy for the efficient synthesis of 2-methylfuran and γ-butyrolactone. New J Chem 27(2):208–210
15. Yang J, Zheng H-Y, Zhu Y-L, Zhao G-W, Zhang C-H, Teng B-T, Xiang H-W, Li Y (2004) Effects of calcination temperature on performance of Cu–Zn–Al catalyst for synthesizing γ-butyrolactone and 2-methylfuran through the coupling of dehydrogenation and hydrogenation. Catal Commun 5(9):505–510
16. Lukes RM, Wilson CL (1951) Reactions of furan compounds. XI. Side chain reactions of furfural and furfuryl alcohol over nickel-copper and iron-copper catalysts. J Am Chem Soc 73(10):4790–4794
17. Thomas HP, Wilson CL (1951) Reactions of furan compounds. XV. Behavior of tetrahydrofurfuryl alcohol over iron-copper catalysts. J Am Chem Soc 73(10):4803–4805

18. Wilson CL (1948) Reactions of furan compounds. X. Catalytic reduction of methylfuran to 2-pentanone. J Am Chem Soc 70(4):1313–1315
19. Allaf T, Mounir S, Tomao V, Chemat F (2012) Instant controlled pressure drop combined to ultrasounds as innovative extraction process combination: fundamental aspects. Procedia Eng 42:1061–1078
20. Dvaranauskaitė A, Venskutonis PR, Raynaud C, Talou T, Viškelis P, Sasnauskas A (2009) Variations in the essential oil composition in buds of six blackcurrant (Ribes nigrum L.) cultivars at various development phases. Food Chem 114(2):671–679
21. Píry J, Príbela A, Ďurčanská J, Farkaš P (1995) Fractionation of volatiles from blackcurrant (Ribes nigrum L.) by different extractive methods. Food Chem 54(1):73–77
22. Hansen CM (2012) Hansen solubility parameters: a user's handbook, 2nd edn. CRC Press, Boca Raton

Chapter 13
Innovative Technologies Used at Pilot Plant and Industrial Scales in Water-Extraction Processes

Linghua Meng and Yves Lozano

Abstract Water remains the cheapest and the safest solvent to eco-friendly extract number of biogenic substances from the worldwide biodiversity to produce natural water-soluble extracts containing several biomolecule families such as polysaccharides, proteins, polyphenols, glycosides, etc. Among these water-soluble compounds, some showed potential free-radical scavenging capacity and antioxidant activity. As extraction processes were often time consuming, mechanical operations can be added to the extraction process to speed up water diffusion of valuable compounds from raw material. Apart from using conventional operating techniques such as mechanical stirring coupled with extraction medium heating, newly developed ones may increase efficiency of water-extraction processing. These innovative techniques include ultrasound-assisted extraction, pressurised hot water extraction, negative pressure cavitation-assisted extraction and pulsed electric field-assisted extraction. Some of these techniques are still under development at various scales, from the laboratory to the pilot plant, but others are already operational and used in industrial processes. After water-extraction step, purification and concentration of extracted products is often needed. Additional process steps are added, including membrane separation technology and gel column chromatography. They are already used at industrial scales and are preferred to heat-based separation techniques. They are claimed to better preserve biological activity of most of the heat-sensitive water-extracted compounds as they efficiently operate and avoid compound liquid–gas phase transition. They remain among the most energy-saving technological

L. Meng (✉)
Department of Pharmacy, School of Medicine, Shanghai Jiao Tong University, Shanghai, China
e-mail: linghua_meng@163.com

Y. Lozano
CIRAD, UMR CIRAD-110 INTREPID, Montpellier, France
e-mail: Yves.lozano@cirad.fr

systems and are frequently called 'green technologies'. In this chapter, some of these innovative technologies involved in water-extraction processes are presented, and when applicable, pilot plant- or industrial-scale applications are described.

Green technologies aim to preserve the quality of the environment and the rational utilisation of natural resources when they are used in transformation processes of vegetal raw materials. To implement these objectives in practice, they are used to prevent or to minimise the negative impact that some extraction processes may have on the surrounding environment. They encourage the development of extraction procedures that avoid, or reduce as far as possible, using extraction medium made of hazardous substances or that produce new pollutants and waste.

To adapt the general principles of the green chemistry to plant extraction processing, a similar approach has to be extended to the development of technologies associated in the extraction processes, leading to the production of new added-value natural water extracts obtained from various sources of vegetal raw material. These principles are as follows:

1. To substitute the use of dangerous organic solvents with number of less hazardous alternatives
2. To encourage the use of emerging extraction technologies
3. To make maximum use of high-efficiency separation techniques involving less toxic organic solvents or using only products that are least harmful for the environment

Solid–liquid extraction is a separation process involving transfer of solutes from a solid matrix into a liquid, named solvent. Water is recognised as the most 'green' solvent because it is non-flammable, non-toxic, readily available and eco-friendly compatible with the environment. Because of the polarity of water, many hydrophilic compounds such as polyphenols, protein, glycosides, polysaccharides, etc., can be easily dissolved in water.

The traditional water-extraction techniques commonly used at laboratory level or industrial scale include maceration with or without stirring, mild heating or heating under reflux. These techniques require generally long extraction times and large amounts of sample and water. Especially, the heating process may destroy the thermal-sensitive compounds which are normally bioactive ones.

To improve efficiency of water extraction, several innovative technologies have been developed. Among them, the most popular and technologically advanced are ultrasound-assisted extraction, pressurised hot water extraction, negative pressure cavitation-assisted extraction and pulsed electric field-assisted extraction. Some of them are still under development at various scales, but the others are already involved in industrial processes.

Depending on the technological water-extraction process applied, additional steps may be needed such as purification and concentration of the compounds extracted in the aqueous solvent. Membrane separation technology and gel column chromatography are already used at these process steps. They are preferred to the classical heat-based separation/concentration techniques because they better preserve most part of the biological activity of the heat-sensitive extracted compounds, as they can also operate efficiently at room temperature. They remain among the most energy-saving technologies; this is why they are frequently classified as 'green technologies'.

Some of these innovative technologies involved in water-extraction processes will be described in this chapter, and some applications at industrial and pilot plant scales will be also presented.

13.1 Ultrasound-Assisted Water Extraction (UAWE)

Extraction of bioactive compounds using ultrasound (US) is one of the upcoming extraction techniques that can offer high operation reproducibility with short extraction times, simplified manipulation, reduced solvent consumption, low-temperature uses and reduced energy usage.

Ultrasounds are mechanic waves that necessitate an elastic medium to spread over. The difference between sounds and ultrasounds is the frequency of the wave of the signal: ultrasounds can be defined as vibrations of the same kind as 'normal' sound but of such a high frequency (higher than 20 kHz) that they cannot be heard by the human ear. The field of the ultrasounds is limited in the upper frequencies by those of the microwaves, starting at a frequency around 10 MHz.

As a sound wave passes through an elastic medium such as water, it induces a longitudinal displacement of particles, as if the source of the sound wave acted as a piston on the surface of this medium [1]. This action generates successive compression and decompression cycles within the medium. If cycles follow each other rapidly in a high frequency, small gas or vapour bubbles are formed, developed and collapsed almost immediately and violently within the medium. This phenomenon, called cavitation, created locally very intense physical or chemical effects, as a result of extreme conditions of temperature and pressure produced when the cavitation bubbles implodes. The temperature and the pressure have been estimated to be up to 5,000 K and 2,000 atm in an ultrasonic bath at room temperature, which can lead to the disruption of biological membranes placed in this water bath, making easier mass transfers by increasing diffusion of water-soluble compounds of the material into the water and by enhancing water penetration into the cellular material. Carla Da Porto et al. [2] have compared the efficiency of the ultrasound-assisted extraction (UAE) with conventional extraction applied to plant material. They showed that 15 min treatment grape samples using UAE lead to the same amounts of total polyphenols and total tannins than water maceration of the sample during 12 h. The published work of Corrales [3] showed also that total phenolic content obtained after 1 h time water UAE of a grape sample was 50 % higher than the one obtained using classical water maceration. Antioxidant activity of the ultrasound-assisted extract was about twofold higher than those of the extract obtained by maceration.

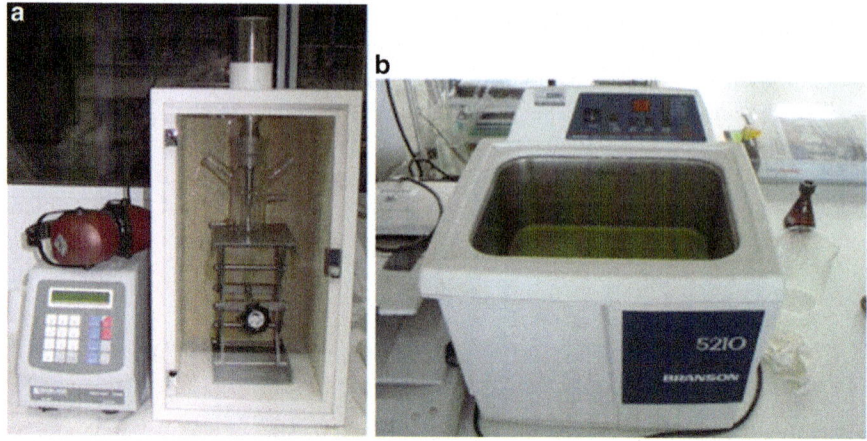

Fig. 13.1 Independent US probe (**a**) and built-in US water bath (**b**) for laboratory-scale uses (Reprint from [1] with permission from Elsevier)

13.1.1 Characteristic Parameters of UAWE

The specific parameters used to describe an UAWE procedure are (1) the frequency, (2) the power and (3) the ultrasonic intensity of the ultrasounds used.

The frequency of the ultrasounds is currently included in the range of 15 kHz to 60 kHz. Usually, only the 20–25 kHz frequency range is used for ordinary extractions, cleaning or degassing operations. Ultrasound power applied is generally less than 500 W for a laboratory-scale extraction apparatus and can be within 1–3 kW for pilot plant- or industrial-scale systems [4].

The ultrasonic intensity (UI) generated by an ultrasound probe in a cylindrical reactor tank is calculated from the following equation:

$$\mathrm{UI} = \frac{4P}{\pi D^2} \quad (13.1)$$

where UI (W·cm^{-2}) is the ultrasonic intensity, P (W) is the ultrasound power and D (cm) is the internal diameter of the reactor tank.

13.1.2 Laboratory- and Industrial-Scale UAWE Apparatus

Ultrasound-assisted extraction is also called sonication or ultrasonication. At the laboratory level, it is usually applied using an ultrasonic probe (Fig. 13.1a) or an ultrasonic bath (Fig. 13.1b). For the ultrasonic probe system, sonication is generally made using a single ultrasonic probe equipped with its electronic regulation. The

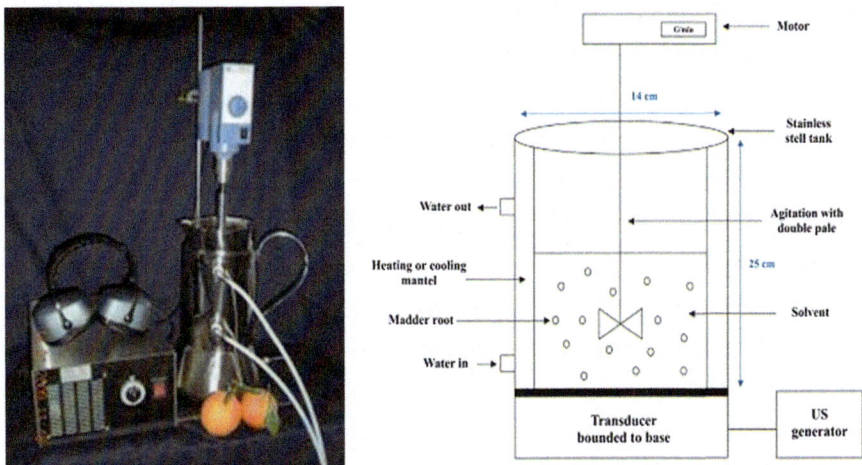

Fig. 13.2 Laboratory-scale UAWE reactor with controlled temperature and stirring-speed (Reprint from [1] with permission from Elsevier)

probe is immersed in a beaker where the material is soaked in solvent. The ultrasonic frequency can be regulated of some models and the ultrasonic intensity is delivered on a small surface (only the tip of the probe) which is more powerful than the ultrasonic bath. This system of probe is widely used for sonication of small volumes of sample, but it causes a temperature rise significantly. The US bath is more frequently found in laboratories and used as an all-purpose extraction or cleaning apparatus. The US frequency cannot be adjusted manually and the delivered intensity is low and often highly attenuated by the water volume used or by some adapted operating conditions to the sample quantity available. When the sample quantity of the material to be extracted is small, it is a common practice to dip the small sample into the water contained by a small beaker, which in turn placed into the liquid of the water bath.

A more sophisticated laboratory-scale UAE reactor was developed by REUS Co (www.etsreus.com, FRANCE), as shown in Fig. 13.2. This reactor can be used to study UAE of different types of samples, under various controlled operating conditions. This apparatus can be used to perform preliminary extraction experiments for new applications of UAE of material samples. It helps to determine the best extraction conditions in view to scale up the operation and to develop new extraction process to operate at a higher scale using pilot plant- or industrial-scale UAE extractors. This laboratory-scale apparatus consists of a double-walled bowl (0.5–3 L) made of stainless steel that allows the thermoregulation of the extraction medium. The bowl stands on base which contained the ultrasonic probes which is controlled by an external electronic regulation. The ultrasonic intensity delivered is about 1 W/cm^2 with a constant frequency of 25 kHz. A mobile stirring device is added to the system and is composed of a propeller moved by a variable speed

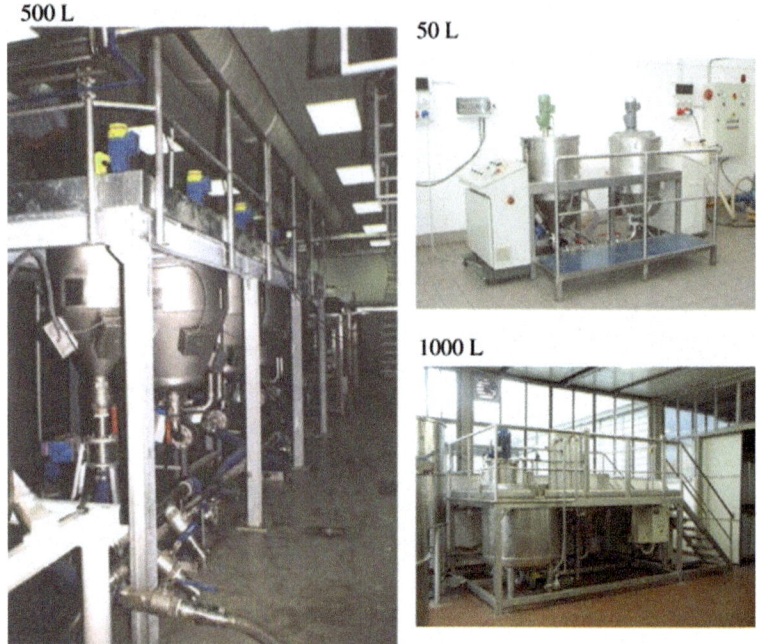

Fig. 13.3 Industrial-sale UAWE equipments with 50, 500, and 1,000 L reactor tank capacities (Reprint from [1] with permission from Elsevier)

electric motor. The stirring system can be adjusted in direction of rotation and speed, and the propeller can be set at the required depth in the extraction medium.

To run out pilot plant- and industrial-scale trials or to scale up UAE from laboratory experiments, REUS offers UAE reactors with a tank capacity ranging from 30 to 1,000 L (Fig. 13.3). A pump system is adapted to the ultrasonic tank in order to fill it at the start of the operation. The pumping system can also be used, to stir the extraction medium by continuous feeding back the tank with the water-sample mixture collected at the bottom of the tank. It is also used at the end of the UAE operation to empty the tank.

Industrial-scale UAWE equipments are generally preferred to conduct extraction operations in a continuous countercurrent flow mode which is a more suitable mode at this level. For this reason, an industrial extractor, using a counterflow extractor assisted by a sound transduction system, was invented by Pacheco et al. [5]. According to the description given in the patent, the equipment (Fig. 13.4) included an inclined casing (12) containing a helical screw conveyor (13) equipped with lots of blades (17), a hopper (21) used to feed the extractor with the raw material at the lower end of the casing (15). A second helical screw conveyor (22) forming a specific angle θ with the axis of the main screw of the casing allowed to continuously load the raw material in the extractor. An outlet hopper was located at the upper end

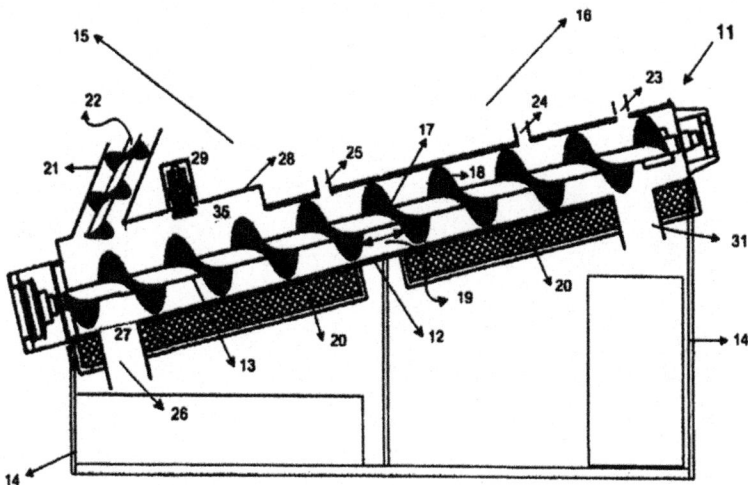

Fig. 13.4 Schematic diagram of the ultrasound-assisted countercurrent flow extractor (According to [5])

(31) of the casing to release the extracted material. The equipment was completed with a water load line (16) with different water upper inlets (23, 24, 25) to fill the casing with water for material extraction which was recovered with the extracted water-soluble compounds from the bottom outlets (19, 20). Two discharge lines (27, 31) were placed at the front and the end of the casing to continuously recover the extracted material during the process and to empty the casing at the end of the operation. The extracted material was collected into two boxes (14, 26). During the extraction, liquid and solid media were separated through a built-in filter (20). The US transducer (29) is located at the lower end (15) of the casing, close to the material feeder to provide ultrasounds as soon as the material entered the casing and put into contact with extraction water.

Based on the principle of this invention, many small Chinese factories produced similar equipments for industrial continuous countercurrent flow extraction (Fig. 13.5). According to the production levels targeted by the factories, various sizes of such equipments were built to process from 33 to 1,350 kg/h of raw material (Sinobest, www.sinobest.com.cn; Jining Tianyu www.tychaosheng.com). Some of these extractors can operate using at the same time two different levels of ultrasonic frequency: 20 kHz are first applied at the casing entrance (material feeding section) and 45 kHz are then applied around the upper casing outlet (material discharge section). The relative low ultrasonic frequency favours initiation of rapidly formation and growing cavitation bubbles, and the high frequency allows a better compound diffusion between the raw plant material and the extraction water [6, 7].

Fig. 13.5 A general view of a dynamic countercurrent UAWE Chinese plant

13.1.3 Application of UAWE at Pilot Plant and Industrial Scales

Ultrasound-assisted water extraction (UAWE) is still considered as an emerging technology that begins to have several new applications in the sector of the food and pharmaceutical industries [7]. After laboratory-scale feasibility and reproducibility studies have been undertaken, pilot plant experiments in batch mode have been tested and validated, using pre-industrial equipments. Optimisation of operation parameters of the process, conducted in a semi-continuous mode, was completed. Up-scaling was then made to validate the process development at an industrial scale and in a continuous production mode.

Several teams of French researchers [1, 8–10] proceeded to such UAWE developments, in accordance with the principles stated by the green chemistry. They applied this green technology to apple pomace processing, a waste from the apple juice and cider manufacturing processes. This waste contained polyphenol molecules showing antioxidant properties [9]. The studies were first performed using a laboratory-scale ultrasound-assisted extractor (PEX1, R.E.U.S., Contes, France), equipped with a 1 L volume beaker (i.d. = 14 cm, 10 cm height). The ultrasonic probe delivered a 25 kHz ultrasound frequency and a 150 W power. The UAWE parameters (ultrasonic intensity, temperature and sonication duration) were optimised using a central composite design within the following parameter ranges: 0.335–$0.764 \, W \cdot cm^{-2}$ for ultrasonic intensity and 10–40 °C for temperature and 5–55 min for sonication time. The extraction medium used was a 50 mM malate buffer solution at pH = 3.8. The best extraction conditions were found: $0.764 \, W \cdot cm^{-2}$ for

Fig. 13.6 Pilot plant-scale model of 30 L volume UAWE apparatus: general view, inside of the reactor tank, recirculating pump and outside bottom of the reactor, top of the reactor with opened protection door (Reprint from [11] with permission from Elsevier)

ultrasound intensity, 40 °C for temperature and 40 min for extraction time. The solid–liquid ratio was optimised as 150 g·L^{-1} dry material/water in function of total polyphenols obtained by a conventional maceration method.

The optimised UAWE lead to a total amount of extracted polyphenol compounds 30 % higher (555 mg of catechin equivalent per 100 g of dry pomace weight) than the one obtained by the conventional extraction, simple water maceration.

Then, the first scaling-up study is undertaken at pilot plant scale, using an ultrasound-assisted extractor equipped with a 30 L volume extraction tank and four probes delivering a ultrasonic frequency of 25 kHz with a total power of 4 × 200 W (Fig. 13.6). The polyphenol extraction yield was about the same level as those obtained by the extraction trial run at the lab scale using the optimised experimental conditions. The extraction was completed within 40 min time, and the total polyphenol of the water extract obtained was 560 mg catechin equivalent per 100 g of dry pomace weight.

13.2 Microwave-Assisted Extraction (MAE)

Microwaves are non-ionising electromagnetic waves that combine the use of an electric field and of a magnetic field, each one oscillating in a perpendicular plane to the other at a frequency range of 0.3–300 GHz. Microwaves are characterised by the three properties: transmittance, reflectance and absorption. Microwaves almost go through materials such as glass, plastic or porcelain or even some organic liquid without being absorbed (transmittance is total). However, metallic material can totally reflect microwaves (reflectance is total), but liquid media, such as water or polar molecule solutions, can absorb microwaves to a certain extent, resulting in heat production within the media.

Considering that water molecules are known as dipoles, a sort of bar magnet, with a positive and a negative pole. The electromagnetic field produced by the microwaves oscillates as it passes through the water molecules, changing the polarity of the field and causing the dipole/water molecules to flip themselves in order to be aligned with the direction-reversing polarity millions of times a second. At the microwave frequency used in commercial systems (2,450 MHz), the dipoles align themselves and randomise at a speed of 4.9×10^9 times per second [12]. This molecular agitation and the friction of the water molecules reversing direction generate heat within the water medium. In the case of solution with ionic solutes, the ions create heat by colliding in the rapidly oscillating electromagnetic field, leaving less microwave energy available for dipole/water molecule to generate heat.

Water molecules are the most polar ones and absorb most of the emitted microwaves. Therefore, water appears to be the best solvent for MAE. The extraction occurs when the water absorbs energy coming from the microwaves and increases the pressure inside the plant material causing the cell structure to break allowing water to penetrate into the matrix and subsequently increases the extraction yield. This additional physical force to heat produced in the extraction medium contributes to increase the extraction yield [13].

Working on MAE Lucchesi et al. [14] observed by scanning electronic microscopy that the husk of cardamom seeds was clearly damaged after 1 h extraction time. At the end of the extraction, the authors noticed a great number of perforations on the external surface of the seeds and that some starch was also dispersed in the water-extraction medium. In the same way, Zhang et al. [15] showed that MAE created interstices on leaves of *Epimedium koreanum* and numbers of chloroplasts have been released from the plant cells, leading the water extract to turn from colourless to green.

MAE is now widely applied to extract from plant material many chemical compounds such as phenolic compounds [16–19], polysaccharides [20], terpenoids or essential oils [21, 22], alkaloids [15], saponins [23] and pectins [24]. MAE was also proposed as a new alternative to the conventional hydrodistillation technique for essential oil extraction, as this technology preserves heat-sensitive compounds because water heating is quicker than hydrodistillation. Extraction yields are generally found to be similar with those obtained with both techniques, but MAE may be completed in shorter operation times, several folds lower than those required for hydrodistillation, for example. Essential oil composition obtained using these two techniques may be slightly different because of water solubility of some essential oil components that play a significant role in hydrodistillation but have no real effect when using the solvent-free MAE directly with fresh plant material.

Therefore, MAE has attracted significant attention in research and development of innovative applications for the extraction of natural products due to its special heating mechanism, moderate capital cost, high-throughput capability and good performance under atmospheric conditions.

13.2.1 Parameters of MAE

Besides the conventional parameters that characterise the operational conditions of a MAE trial (volume of extraction medium, solid–liquid ratio, trial duration, extraction temperature, etc.), microwave power is a specific one which has to be taken under consideration [15].

Usually, microwave power is a parameter, manually controlled, that affects directly the temperature of the extraction medium during MAE. Increasing the microwave power leads generally to increase the extraction liquid medium temperature, which in turn modifies the liquid physical properties, such as viscosity and surface tension. Compound extraction yield is modified and if heat-sensitive compounds are present in the sample, they may also be affected by an extraction temperature increase. Power level has to be adjusted to the mass amount of sample to be extracted at the same time because the total microwave energy provided is shared among the different pieces of samples put at the same time in the extraction reactor, and each piece has to receive enough microwave energy so that extraction can occur. Power level applied has to be adapted to the sample weight involved and to the sensitivity of the compounds to be extracted [25].

Extraction time setting depends on the level of the microwave power applied. Combination of the two settings has in turn an effect on the temperature reached during the extraction trial and on the total energy delivered per sample weight unit. This combination has to be fine-tuned to speed up sample extraction by increasing as far as possible cell wall rupture and avoiding degradation of heat-sensitive extracted compounds. Therefore, MAE apparatus shows generally two manual settings:

- One setting to adjust the microwave power according to the total volume of extraction liquid and to the total weight of the sample to be extracted (weight/volume ratio or number of sample pieces)
- One setting to fix the extraction time to take into account the acceptable temperature rise during extraction [13]

Compared to conventional extraction techniques, MAE generally requires shorter extraction time to reach the same extraction results. When using a traditional MAE apparatus operating at 2,450 MHz frequency [12], it is widely recognised that a 10 min time is enough to complete the extraction [18, 26].

Fig. 13.7 Domestic microwave oven used for MAE laboratory-scale studies

13.2.2 Laboratory- and Pilot Plant-Scale MAE Apparatus

Most of the researches on MAE undertaken at laboratory level used a simple domestic microwave oven (Fig. 13.7). The apparatus is composed of a 20–25 L volume cavity/oven and uses a probe system delivering a microwave frequency of 2,450 MHz, and the variable power is generally limited within the range 500–1,000 W only. Power and time settings are provided to set the desired values before operation, according to the nature of the sample to be extracted.

Today, professional equipments are available from manufacturers and can be used for laboratory-scale researches. These apparatus provide regulation functions and various types of captors to control accurately and safely parameters and operation conditions during the extraction process. Milestone Inc. in Italy (www.milestonesci.com), CEM Co. (www.cem.com) in the USA and Sineo Microwave Chemistry Technology Co. in China (www.sineomicrowave.com) are examples of private sector developer manufacturers. These instruments can be classified into two types of systems: closed vessel and open vessel.

- The closed-vessel systems used for MAE consist of a magnetron tube placed on a turntable in an oven and equipped with only the necessary parameter controls. Generally, one manual setting is provided to adjust the maximum power to be delivered during the extraction trial and one setting to adjust the extraction duration, also displaying the running extraction time, and to end the process in case the maximum acceptable temperature in the extraction medium is reached, and a pressure control is coupled with the on/off control of the magnetron tube (Fig. 13.8). MAE closed-vessel systems, allowing sample extraction in pressurised vessels, were developed and marketed by Milestone Inc. These apparatus allow operating securely at higher temperature. They are equipped with a specific extraction vessel with a vent-and-reseal technology security (US Patent 5,270,010) that operates according to the three steps as shown (Fig. 13.9). The vessel cap is held in place by a dome-shaped spring (sketch 1). When an uncontrolled overpressure appends inside the vessel, the spring placed over the

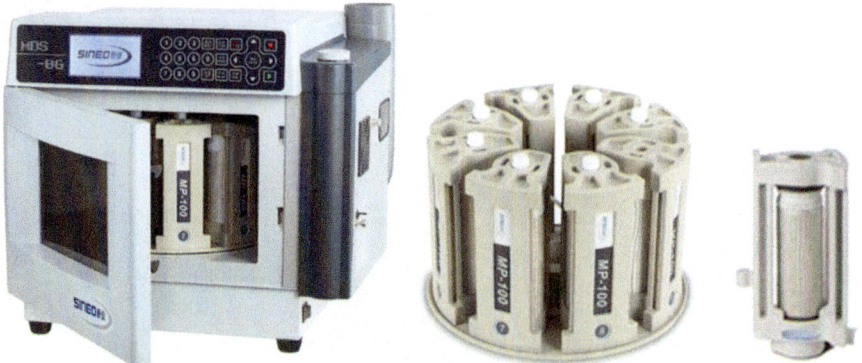

Fig. 13.8 Closed-vessel microwave chemistry workstation (MDS-8G model, Sineo Co., China)

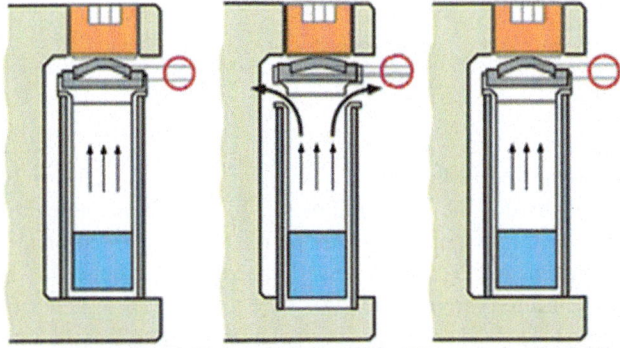

1. Vessel sealed. 2. Pressure released. 3. Vessel resealed.

Fig. 13.9 Vent-and-reseal technology to secure the MAE vessel (According to [27])

vessel cap is flattened by the overpressure and the cap lifts up slightly, releasing the overpressure (sketch 2). Immediately after the overpressure released, the spring pushes down the cap that reseals the vessel (sketch 3), while the extraction process carries on and the vent-and-reseal system is ready to operate when needed. This technical improvement eliminates potent risks of vessel failure or explosion in case of a possible happening of an out-of-control exothermic reaction during MAE.

- MAE open-vessel systems are solvent-free technology. Such apparatus are mainly proposed by Milestone Inc. (NEOS and NEOS-GR models) (Fig. 13.10). The two models are equipped with a chamber with door, built in a material that does not allow microwave leakage to outside. This chamber is equipped with a magnetron probe and its built-in electronic control device. MAE takes place in this chamber. The vessel containing the sample to be extracted is placed in this MAE chamber. The NEOS model is equipped with a glass distillation system,

Fig. 13.10 NEOS (**a**) and NEOS-GR (**b**) models for MAE (Milestone Inc.)

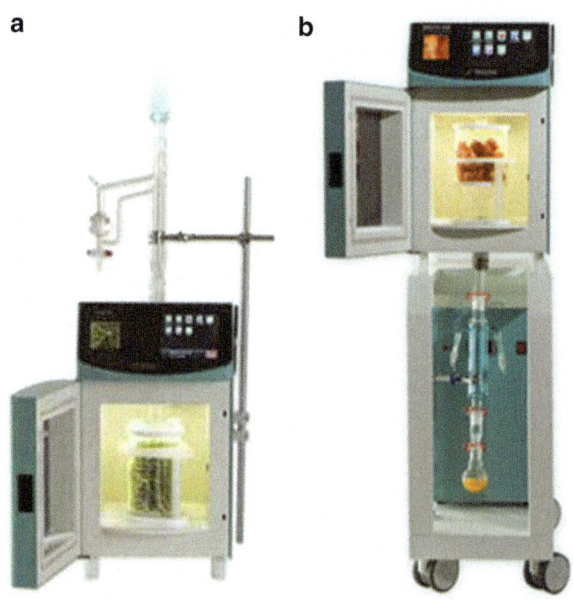

watertight connected to the vessel placed in the MAE chamber (sketch a). In the NEOS-GR model, the MAE chamber is placed above another chamber. The facing bases of the two chambers are pierced to allow a communication glass tube to be placed. An extraction vessel placed into the upper MAE chamber is watertight connected through the glass tube to the glass system placed into the lower chamber. This system consists of glassware to collect the liquid extracted from the raw material in the upper MAE chamber. The NEOS-GR system is somewhat the upside down design of the NEOS system. In both systems, glassware is put in direct connection with the open air, avoiding any possible overpressure to occur during MAE.

The NEOS system was scaled up at pilot plant level by Milestone Inc. A semi-industrial-scale apparatus, MAC-75 model as shown in Fig. 13.11, was manufactured and is now marketed. The MAC-75 apparatus is a multimode microwave reactor. It is equipped with 4 magnetrons ($4 \times 1,500$ W total power, 2,450 MHz frequency) that can be set to various power levels set by 500 W increment levels. The stainless steel extraction chamber has a capacity of 150 L and contains a removable, rotating PTFE drum that allows up to 75 L of sample to be loaded in.

The industrial microwave apparatus are designed more often for dying and sterilisation. At a larger scale, industrial MAE line is mainly composed of microwave generators (Fig. 13.12a), such as those provided by Synotherm Co. (www.synotherm.net). These generators are attached with a specific design to open- or closed-type microwave chambers. A model of such an apparatus is developed by Yueneng Microwave Co. (www.pin-ba.com), as shown by Fig. 13.12b.

Fig. 13.11 Semi-industrial-scale MAE apparatus MAC-75 model (Milestone Inc., reprint from [28] with permission from Elsevier)

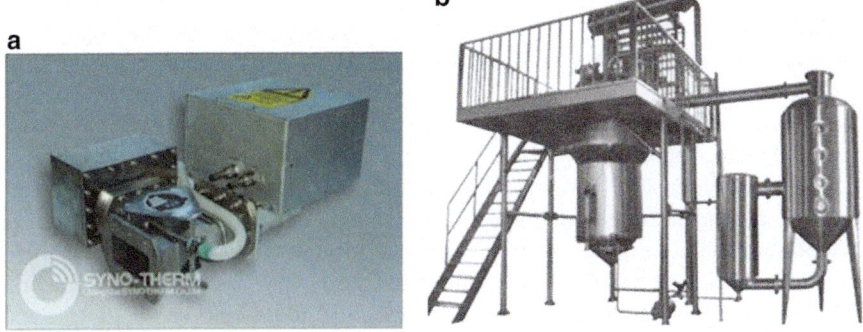

Fig. 13.12 Industrial-scale microwave generator (**a**) and MAE apparatus (**b**)

13.3 Pulsed Electric Fields Extraction (PEFE)

Pulsed electric fields (PEF) are a non-thermal emerging technology based on the application of external electric fields that induce damage to cell membranes (electroporation) with preservation of the intrinsic quality of the product processed, including purity, colour, texture, aroma, flavour and other nutritional components.

This technology has made its way from the laboratory to market. It opens new perspectives for the food industry. It can be used to preserve liquid bulk products such as fruit juice, milk, yoghurt and soup, by inactivating the microbial organisms they may contain [29]. The electric impulses can be applied homogeneously through the product, and the technology is readily applicable for the pasteurisation of liquid foods at low temperature. Moreover, PEFE can speed up extraction of natural compounds, compared to traditional solid/liquid extraction [30]. Many researches have already demonstrated the progress brought to the efficiency of plant extraction using PEF technology [31–33]. Yin Yongguang et al. [34] reported that polysaccharide extraction using PEFE method gave higher yields compared to the others conventional extraction methods, such as alkali-based or enzyme-assisted extraction techniques. Eduardo Puértolas et al. [35] showed that grape treated by PEF before alcoholic fermentation gave a wine with higher colour intensity, better total polyphenol index (TPI) and higher total anthocyanin content (TAC) than wine obtained without PEF pretreatment. Moreover, coupling PEF with the fermentation process shortened about 48 h the grape maceration step, compared to the control classical process.

PEFE proceeds through mass transfer between the raw material to be extracted and the solvent. As the material to be extracted is generally from plant or animal origin, the valuable compounds to be extracted are generally enclosed in the cell tissues. The cell membrane is as a semipermeable barrier, playing an important role in compound exchanges towards the membrane. The membrane structure affects the selectivity and the speed of these exchanges. PEF treatment of such raw material reduces selectivity and increases permeability of the cell membranes of the material. PEF can also partially or totally destroy the cell membrane integrity, leading to higher and quicker mass transfer between the cell content and the surrounding solvent. This phenomenon, actively developed for applications in molecular biology and in medicine, is called electropermeabilisation or more commonly electroporation.

When a biological cell is exposed to external electric field strength, a time- and position-dependent transmembrane potential is induced across the non-conductive cytoplasmic membrane. This is the result of the accumulation of oppositely charged ions on both sides of the membrane. Under the effect of the electric fields, attraction between these ions occurs and causes reduction of the membrane thickness and even the formation of pores. A critical value of the external electric field is required to induce a transmembrane potential (0.2–1.0 V) that leads to the formation of reversible pores in the membrane. When a more intense PEF is applied, irreversible electroporation takes place, resulting in cell membrane disintegration as well as loss of cell viability [30].

In the food industry, products are generally preserved by heat treatments that killed the bacterial flora they may contain. PFE can produce irreversible electroporation if the level of PEF is high enough. Therefore, PEF can be advantageously used to kill all kinds of these undesirable endogenous biological cells. This is the reason why PEF has been extensively studied for non-thermal food processing providing microbiologically safe and minimally processed foods. As transmembrane potential is proportional to the radius of cell, the larger the cell, the greater the transmembrane potential. Thus, the electric field strength level required for electroplasmolysis has to be adapted to the cell size. It was also successfully applied to disintegrate biological

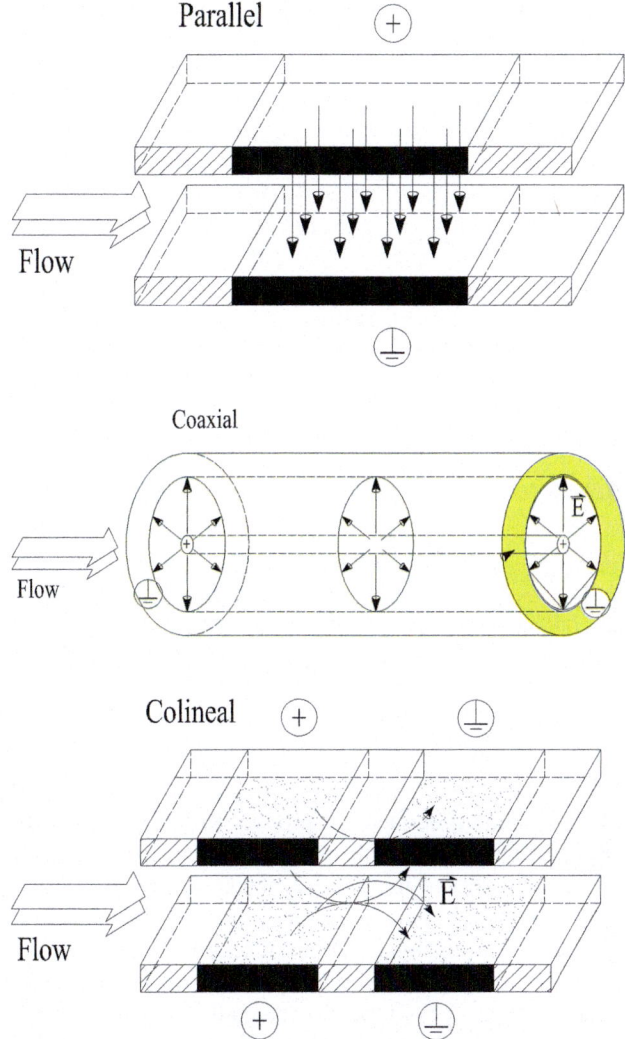

Fig. 13.13 Designs of treatment chambers built-in commercial PEFE apparatus

tissue to improve the release of intracellular compounds during extraction of vegetal sample. Generally, for plant cell, the electric field strength required is about 0.5–5 kV/cm [36].

The two components of a PEF-based apparatus are the pulse generator and the treatment chamber. Different treatment chamber designs have been developed in the past few years. Today, the three most important chamber designs kept in the development of commercial PEF apparatus are configurations showing parallel electrodes, coaxial electrodes and colinear electrodes (Fig. 13.13) [36].

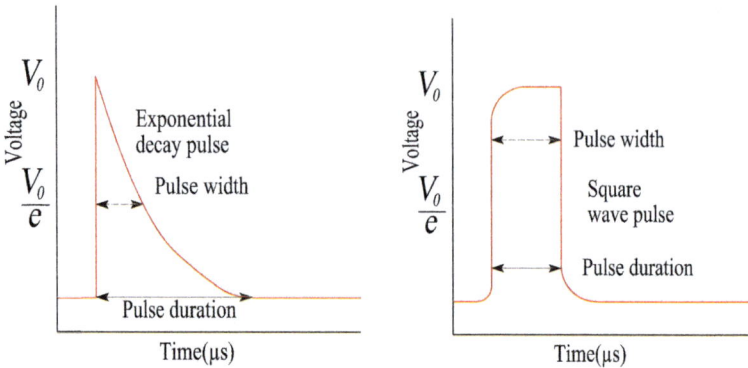

Fig. 13.14 Most usual pulse shape uses in common PEFE applications

13.3.1 Parameters of PEFE

The most typical process parameters that characterise PEF technology are electric field strength, pulse shape, pulse width, number of pulses, frequency and pulse-specific energy. The distance between the electrodes of the treatment chamber and the voltage delivered define the strength of the electric field (E, in kV/cm units). The most usually used pulse shapes are those with exponential decay or with square waveform, as shown in Fig. 13.14. Square waveform geometry has been determined to be the ideal pulse shape for PEF processing because in this configuration, the electric field intensity remains constant within the pulse duration. The treatment time for a PEF application is defined as a product of the pulse width and number of pulses applied. The frequency (f) is the number of pulse per second (Hz).

The specific energy (W) of the pulse depends on the voltage applied, on the treatment duration and on the ohmic resistance of the volume of the product filling the treatment chamber, limited by the electrode length. This resistance is a function of the geometry and the conductivity of this product volume. The specific energy is calculated according to Eq. (13.2), and its value allows to evaluate the energy cost of the PEFE process:

$$W = \frac{1}{m} \int_0^\infty k \cdot E(t)^2 dt$$

m = material mass (kg); $\quad t$ = treatment time (s) \hfill (13.2)
E = strength of the electric field $(\text{kV} \cdot \text{cm}^{-1})$
k = eletrical conductivity of the material treated $(\text{ms} \cdot \text{cm}^{-1})$

PEFE using water as solvent was studied to extract various bio-compounds from different natural substrates, such as colourant compounds [33, 37, 38], sucrose [32], polysaccharides [34], phenols [3, 39], podophyllotoxin [40], water-soluble compounds from microalgal biomasses [41], fennel [42] and chicory [31].

13 Innovative Technologies Used at Pilot Plant and Industrial Scales... 287

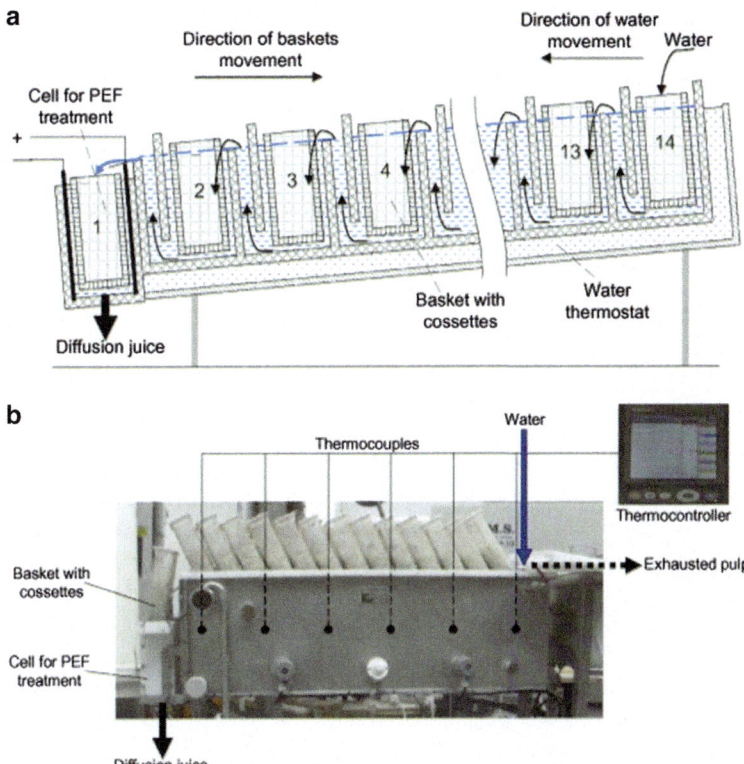

Fig. 13.15 Countercurrent PFE water extractor: (**a**) schematic representation and functioning principle, (**b**) picture of a pilot-scale apparatus (Reprint from [43] with permission from Elsevier)

13.3.2 Applications of PEFWE at Pilot Plant Scale

Pulsed electric field-assisted water extraction (PEFWE) is not yet applied at the industrial scale because there is a lack in the technology dealing with high-voltage pulse generation at an industrial scale. Nevertheless, some PEFWE units were yet developed at pilot plant scale.

The team of Eugène Vorobiev [43] developed an application of PEFWE to extract sugar from sugar beets. A pilot plant scale countercurrent cold and mild heat extractor was built. It consists 14 extraction sections set in series, as shown in Fig. 13.15. The PEF treatment chamber (section 1 of the extractor) was specially designed and isolated from the other sections. Section 1 was equipped with two stainless steel electrodes to generate PEF. The sugar beet roots were cut into cossettes and filled in the baskets and then placed in the 14 extraction sections. The 14 sections of the apparatus were filled with running water from the last and upper section (i.e. section 14) down to the lower PEF section (i.e. section 1). The first basket with sugar beet cossettes was submitted to PEF treatment in section 1 for

5 min, and then baskets were moved manually between the neighbouring sections. The process continue in such a way that the baskets with sugar beet cossettes moved from one section to the other and water was added to the last section 14 to contact with most exhausted cossettes every 5 min also in the opposite direction (countercurrent flow). Each basket was filled with 0.5 kg of cossettes, and in each section, the solid/liquid ratio was set as 1:1.2. The output water flow running out from the PEF section 1 was enriched in extracted sugar by water diffusion during its percolation through the 14 sections of the apparatus. This diffusion juice was collected for further treatments.

The total time of diffusion can be calculated as $t_d = 14 \times 5$ min $= 70$ min. Water temperature was varied from 30 to 70 °C, which was controlled by the thermocontroller inside the extractor. The temperature of cossettes at the input of extractor was $T = 10$–13 °C.

This apparatus equipped a pilot plant-scale PEF generator (Hazemeyer, 5,000 V, 1,000 A, France). It provided monopolar pulses with a near-rectangular shape signal and the electric field intensity used was $E = 600$ V $\cdot$ cm^{-1} when operating at 30 °C and was $E = 260$ V $\cdot$ cm^{-1} at 60 °C. A train of pulses consisted of 500 successive pulses of 100 μs duration each and repetition pulse of 5 ms intervals. Only one train of pulses was used for a 50 ms PEF treatment of every set of sugar beet basket, which corresponds to 5.4 kW $\cdot$ h $\cdot$ t^{-1} energy input. Temperature elevation during each PEF treatment cycle did not exceed 3 °C, making this PEFWE to be considered as a low or moderate thermal extraction process. The juice purity (sucrose/total soluble solid content) was not lower than those obtained by conventional hot water diffusion (70 °C) of sugar beet cossettes.

The team of Javier Raso studied the influence of PEF treatment on wine making at laboratory and pilot plant levels [35, 44–47]. They showed that phenol extraction from grapes, specifically anthocyanin compounds, can be improved by using PEF pretreatment. They confirmed that the same results were obtained when scaling up the process at pilot plant level; they thought that the process could be used at an industrial-scale level as an innovative use of PFE-assisted extraction for wine making. Therefore, a PEFE equipment was built (Modulator PG, ScandiNova, Uppsala, Sweden) which generates square waveform pulses of a width of 3 μs and a frequency up to 300 Hz [46]. The maximum output voltage was 30 kV and current intensity was 200 A. Cabernet Sauvignon grapes were processed using a continuous flow PEF treatment.

The extractor was equipped with a colinear type treatment chamber showing two successive treatment zones of 2 cm long and 2 cm inner diameter each, positioned between ground and high-voltage electrodes, as shown in Fig. 13.16a. The applied electric field strength is not uniform along the 2 treatment zones but shown symmetrically placed force lines, as shown in Fig. 13.16b. The grape pomace was pumped in the colinear treatment chambers at a mass flow rate of 118 kg $\cdot$ h^{-1}, using a progressive cavity pump (Rotor-MT, Bominox, Gerona, Spain). PEF treatment consisted in an average of 50 pulses of an electric field strength of 5 kV $\cdot$ cm^{-1} (total specific energy: 3.67 kJ $\cdot$ kg^{-1}) at a frequency of 122 Hz.

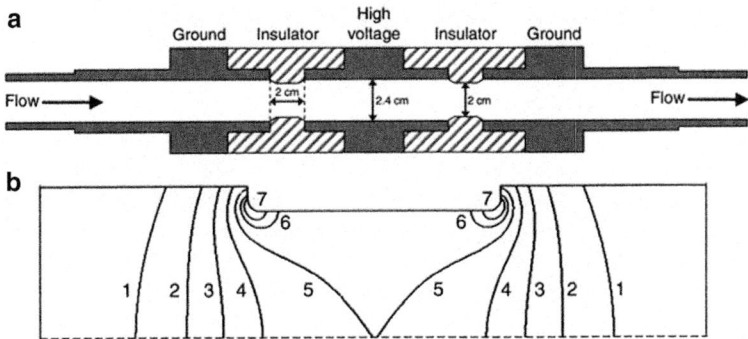

Fig. 13.16 (a) Scheme of the colinear PEF treatment chambers, (b) finite elements method simulation of the electric field line distribution from the weakest (1 kV·cm^{-1}) to the strongest electric field strength (7 kV·cm^{-1}), input voltage = 14.2 kV of the upper half part treatment zone of one chamber (Reprint from [46] with permission from Elsevier)

In this work, it was shown that an increase of the electric field from 2 to 7 kV·cm^{-1} leads to an extraction rate increase of both anthocyanins and total phenols.

13.4 Negative Pressure Cavitation Extraction (NPCE)

The negative pressure cavitation (NPC) technology was invented by the team of Yu-jie Fu, at Northeast Forestry University, Haibin, China [48].

Cavitation is produced by pressure forces acting upon the liquid leading to the formation of vapour cavities (small liquid-free volumes or bubbles) within this liquid medium. This phenomenon occurs when a liquid is subjected to rapid changes of pressure leading to the formation of cavities where the pressure inside is low. When subjected to higher pressure, the cavities collapse rapidly and generate an intense shockwave. Cavitation serves as a means to concentrate in very short time in a region that diffused fluid energy to create a zone of intense energy dissipation. Ultrasounds applied to a liquid medium can generate such a phenomenon. Cavitation effects can also be produced by acoustic and hydrodynamic means.

NPC is another type of technique to generate cavitation. It uses negative pressures, and its intensity is not weaker than that produced by ultrasounds. Zhang et al. [49, 50] compared the morphological change of pigeon pea roots treated by these different techniques. They showed that the root cell walls were more destroyed using NPC cavitation than using ultrasounds. Liu et al. [51] compared the extraction efficiency obtained using four techniques: NPC, ultrasounds, microwaves and reflux. They concluded that NPCE showed an equivalent extraction efficiency as UAE which was more effective than MAE and reflux extraction. Moreover, the UAE

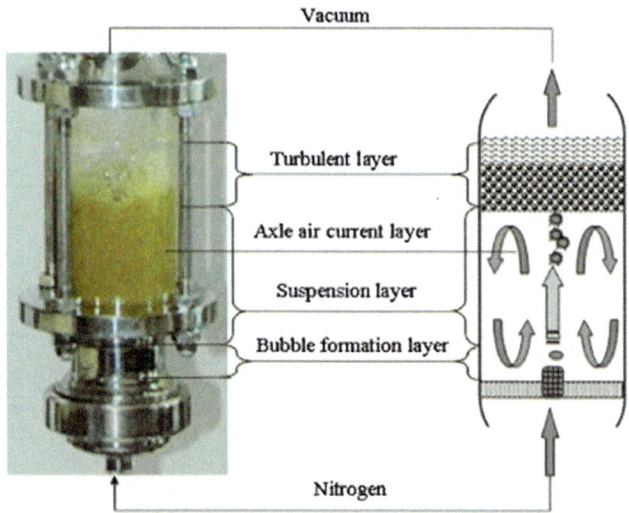

Fig. 13.17 Layer distribution within an extraction medium submitted to NPC (Reprint from [52] with permission from Elsevier)

produced a great amount of heat, as NPCE kept the extraction medium temperature at its initial level, which was in favour of the extraction of heat-sensitive products [50]. NPCE appeared also to be a cheap and energy efficient process to extract natural products.

13.4.1 Mechanism and Parameters of NPCE

The mechanism that occurs during NPCE consists of successive formation and collapse of tiny bubbles under the action of vacuum within the liquid extraction medium placed in an extraction vessel along with the solid sample to be extracted (pieces or powder of raw material). Four layers can be distinguished from bottom to top of the extraction vessel: bubble formation layer, suspension layer, axle air current layer and turbulent layer. When nitrogen gas is continuously introduced into the extraction vessel, small nitrogen bubbles appear under the action of negative pressure caused by light vacuum applied into the vessel (Fig. 13.17) and ascend through the liquid–solid medium. This results in the formation of a highly instable gas–liquid–solid system. The suspension layer is formed and is located a little higher than the bubble formation layer. The tiny bubbles enter this suspension layer and grow rapidly because of the negative pressure created locally in this area, until they suddenly collapse, producing a cavitation phenomenon with intense collisions so that the surface of the surrounding raw material particles is corroded. The extraction liquid can diffuse more easily into the solid particles, enhancing diffusion of extractible compounds. Higher in the vessel, intense vertical motion created in the axle air current layer helps extracting compounds from the sample material which is completed in the turbulent upper layer situated near the source of vacuum. Thereby, NPCE creates intensive cavitation-collision, turbulence, suspension and interface effects that combine to form a dynamic mass transfer enhancing extraction and accelerate mass transfer of targeted compounds from the sample solid matrix to the solvent [49, 50, 52].

The parameters that affect the efficiency of NPCE are the negative pressure, the nitrogen-gas flow, the particle size of the solid sample to be extracted, the extraction time and the liquid–solid ratio.

According to the NPCE mechanism, successive formations and bursts of bubbles in the liquid medium depend on the vacuum level created (negative pressure). The cavitation phenomenon generally occurred when the vacuum is set between −0.01 and −0.09 MPa. Zhang et al. [52] reported that a decrease of the negative pressure from −0.02 to −0.05 MPa enhanced the extraction yield of flavonoids from *Dalbergia odorifera*, and once the pressure was set lower than −0.05 Mpa, the extraction yield decreased slightly.

Nitrogen-gas flow is another key parameter in cavitation technique that affects extraction efficiency. Liu et al. [51] investigated the effect on the extraction yield of five flavonoids extracted from pigeon pea leaves submitted to NPCE. They observed a yield increase when the nitrogen-gas flow was set within the range of 10–40 mL · min^{-1}, with an optimised gas flow 30 mL · min^{-1}, to get the highest extraction yield.

13.4.2 Laboratory- and Pilot Plant-Scale NPCE Apparatus

The NPCE laboratory-scale device was designed and patented (Patent CN2597047) by the team of Fu [51]. It consisted of an extraction pot (1), a collection pot (2), a vacuum pump (4) and a nitrogen stock vessel (9). The extraction pot and the condenser (5) were glass made, as the other parts were steel made. The 400 mL volume extraction pot was a 17 cm high and 5.5 cm inner diameter cylinder (Fig. 13.18). Solid samples and solvent were added into the extraction pot through the inlet (3). The negative pressure was generated by a vacuum pump, and the nitrogen gas was introduced through the bottom valve (4). After NPCE, the solvent was collected in the collection pot (2) and filtered, and the residue was discarded.

A NPCE pilot plant-scale apparatus was developed by the same team [48]. The apparatus showed the same types of elements as those described for the laboratory-scale device, and a heating system was added to allow working at high temperature (Fig. 13.19). The working extraction volume was brought to 10 L.

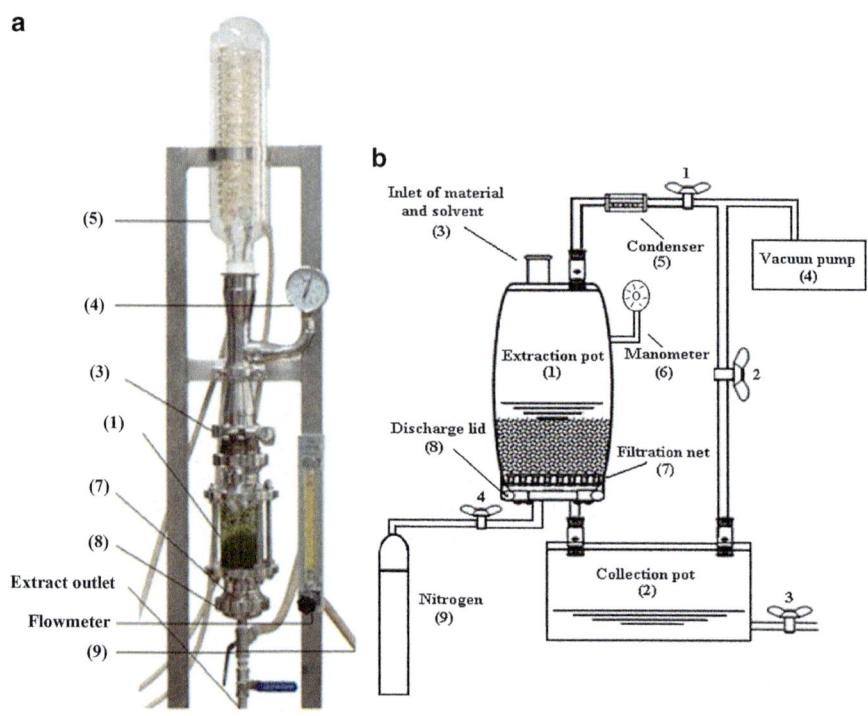
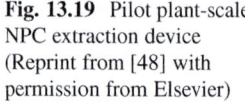

Fig. 13.18 NPCE apparatus: schematic representation (**a**) and laboratory-scale unit (**b**) (Reprint from [50] with permission from Elsevier)

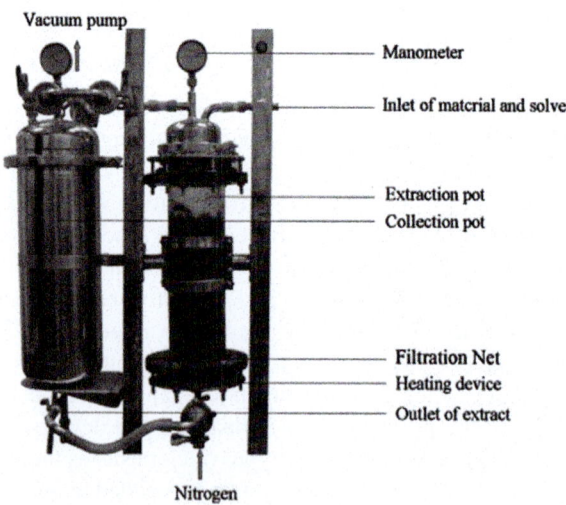

Fig. 13.19 Pilot plant-scale NPC extraction device (Reprint from [48] with permission from Elsevier)

13.4.3 Application of NPCE at Pilot Plant Scale

Since the NPCE technology has appeared for only 4 years, NPC applications were not largely developed and NPCE is not widely used today. Only laboratory-scale extractions of flavonoids and of other phenol compounds with antioxidant capacities were studied and recently reported in the literature [49–52].

The only application of NPCE conducted at pilot plant level was the extraction of the three main flavonoids (genistin, genistein and apigenin) from the pigeon pea roots [48], using the device descried in the previous chapter (13.4.2). The extraction conditions were firstly optimised at the laboratory scale using an optimisation experimental design (Box–Behnken design). An ionic solution was used as the extraction medium. Combinations of five kinds of anions (Cl^-, Br^-, $H_2PO_3^-$, HSO_4^- and BF_4^-) associated with 1-R-3-methylimidazolium cations (radicals R being alkyl groups which chain length increased from ethyl to octyl) were compared [48]. [C_8mim]Br ionic aqueous solution with a concentration of 0.53 $mol \cdot L^{-1}$ was chosen as solvent. The optimised conditions for NPCE were: temperature = 74 °C, negative pressure = −0.07 MPa, liquid–solid ratio = 20:1 $mL \cdot g^{-1}$ and the extraction time was 15 min. Five hundred grams of pigeon pea roots were extracted in the pilot plant NPCE apparatus. Extraction yields for genistin, genistein and apigenin were, respectively, 0.477 ± 0.013, 0.480 ± 0.014 and 0.271 ± 0.021 $mg \cdot g^{-1}$. As these yields were similar to those obtained using the smaller laboratory-scale device, the authors concluded that the NPCE method could be scaled up for applications at industrial-scale level.

13.5 Pressurised Hot Water (PHW) Extraction (PHWE)

Pressurised liquid extraction, using heated water as the extraction medium (solvent), is frequently named pressurised hot water extraction (PHWE). It is considered as another emerging green extraction technology for different classes of compounds present in numerous kinds of matrices such as environmental, food and botanical samples. This technique is also known under different names, such as pressurised solvent extraction (PSE), accelerated solvent extraction (ASE) or enhanced solvent extraction (ESE).

Water is a highly polar solvent (high relative permittivity $\varepsilon_r = 80$), and when it is heated at high temperature with enough pressure to maintain water in liquid form, which is named pressurised hot water (PHW), its physical properties are changed. Its relative permittivity falls ($\varepsilon_r = 35$ at 200 °C), near those of simple alcoholic solvents, such as ethanol ($\varepsilon_r = 24$) or methanol ($\varepsilon_r = 33$ at 25 °C), making water as good solvent as some organic ones. Thus, water becomes to be able to dissolve a wider range of compounds including low-polarity organic ones. PHWE stands as a good alternative to reduce utilisation of some organic solvents for liquid extractions that traditionally used them.

The term 'pressurised hot water' (PHW) is used to denote the region of condensed phase of water between the temperature ranges from 100 °C (boiling point of water) to 374 °C (critical point of water). It has been reported as subcritical water, superheated water, near-critical water and pressurised low polarity water.

In the PHWE technique, the raw material sample, placed in a metallic pressure-resistant cylindrical cell, is put into contact with the PHW, playing the role of a chromatography eluant. As pressure has to be increased to maintain water in its liquid state at the high working temperature, the targeted compounds present in the sample may partition themselves between the sample matrix and the percolating liquid (PHW). They are chromatographically eluted out from the cell into the pressurised collection vial.

Lots of published papers showed that PHWE appeared as powerful technique for water extraction of essential oils [53], proteins [54], polysaccharides [55, 56], anthraquinones [57], lignans [58], terpenes [59], low-polarity flavonoids [60], phenolics [61], microbial lipids [62] and organic pollutants such as PAHs, PCBs, pesticides, herbicides, etc. [63]. Efficiency of PHWE was also compared with other extraction technologies, such as microsound-assisted extraction, sonication-assisted extraction, Soxhlet extraction or other traditional reflux mode extractions. It was reported that PHWE was as powerful as these traditional technologies and, in some cases, was even more efficient [64].

13.5.1 Parameters of PHWE

The parameters that affect extraction efficiency of PHW technique include temperature of the liquid-state pressurised water, extraction time, water flow rate and use of extraction technical helps such as a small percentage of organic solvents or surfactants [65].

Water temperature used for extraction is the most important parameter which could affect extraction efficiency and selectivity of PHWE. When increasing water temperature, water physicochemical properties change significantly: decrease of its relative permittivity and reduction of its viscosity and surface tension. Hence, raising the working temperature modified the polarity of the extracting water, which turns it into a specific solvent for low-polarity compounds. As water viscosity is reduced and water surface tension increased, diffusivity of the extracting water is significatively enhanced, allowing water to enter more easily the sample matrix, leading to a better mass transfer between water and the sample. But decomposition of heat-labile extracted solutes can occur due to the applied high temperature and pressure.

Working temperature was found to be a selectivity factor in PHWE as shown by M. J. Ko et al. [60] in studying its application for flavonoids extraction. The optimal extraction temperature for flavonoid aglycone with an OH side chain, such as quercetin, was found to be 170 °C. For aglycone compounds with an $O-CH_3$, as in isorhamnetin, or with a H group, as in kaempferol, or for apigenin, with double bonds, the optimised extraction temperature was found to be 190 °C. Flavonoid glycoside forms were better extracted at lower temperatures: 110 °C for quercitrin, a glycoside form of quercetin and 150 °C for spiraeoside and isoquercitrin.

Alicia Gil-Ramírez et al. [66] compared the extraction yields of isoxanthohumol at different extraction temperature (50, 100, 150 and 200 °C). They found that the highest yield was obtained at 150 °C.

Benito-Román et al. [55] showed that 155 °C was the optimal temperature in PHWE of β-glucans of high molecular weight. Above 160 °C, the yield of β-glucans dissolved in extracting water decreased. Cacace et al. [58] found that maximum amounts of lignans and other flaxseed bioactive compounds, including proteins, were best extracted at 160 °C and phenolic compounds at 140 °C.

Yu Yang et al. [59] studied the stability of five terpenes during PHWE. They showed that terpene degradation became more serious when water temperature increased, and there was a significant drop of the extraction yield when water temperature was set around 200 °C, but yields were quite similar at both temperatures 100 and 150 °C. Chunhui Deng et al. [53] observed that the best extraction efficiencies for three active terpenoids compounds, camphor, borneol and borneol acetate, present in *F. amomi* samples were obtained at 160 °C.

Effect of pressure on PHWE yields is more limited than that of temperature [55]. In general, liquids are highly incompressible in their subcritical states. At constant temperature, pressure variation does not modify so much water solvation power, making pressure parameter to have a lesser effect than temperature in PHWE processes. Pressure is only used to maintain extraction water in its liquid state, according to the working temperature used. Moderate pressures such as 15 bar at 200 °C or 85 bar at 300 °C are enough to maintain water in its liquid state. Within this pressure range, Chunhui Deng et al. [53] did not found much change in extraction yields of the three terpenoids extracted from *Fructus amomi*, a traditional Chinese medicinal plant. In most published works on PWHE of natural products, the working pressure was kept within the range 10–50 bar [67–69].

PHWE can be performed in two modes: (1) the static mode where the sample is just put into contact with water under the working temperature and pressure chosen and (2) the dynamic mode where the water is percolated at a certain flow rate towards the sample.

In the static mode, extraction duration depends strongly on the extraction temperature and on the nature of the sample matrix and of the compounds to be extracted. Extraction time during PHWE of β-glucan from waxy barley flour was practically limited to 45 min when extraction temperature was set between 155 and 160 °C. The highest yield (53.7 %) was obtained after 18 min extraction time [55]. Rovio et al. [70] investigated at different temperatures the extraction kinetics of eugenol and eugenyl acetate from clove. They found that, at 125 °C, 80 min extraction time was needed to completely extract the two terpenes, but only 15 min was enough to obtain the same result when the working temperature was set at 250 °C or at 300 °C.

In the dynamic mode of PHWE conducted at a laboratory scale, water flow rates between 1 and 1.5 ml/min were used [67, 68]. But they may be out of this range in some experimental trials. In PHWE of lignans from flaxseed (*Linum usitatissimum* L.), using three extraction cells of 7.0, 9.4 and 19.3 mm i.d. and 10 cm long, the optimal water flow rate was found to as low as 0.5 mL/min [58]. On the opposite, the water flow rate has to be set at a higher value of 5 mL/min to obtain the best extraction yield for anthraquinone extraction [57].

In some cases, organic or inorganic additives have been used along with water in PHWE to improve compound recovery. Ju and Howard [61] have compared grape skin PHWE with and without adding sodium metabisulfite in the extraction water to obtain anthocyanins and other antioxidant-active compounds. They showed that 1,400 μg/mL sodium metabisulfite added to the extraction water improved extraction contents of total anthocyanins and total phenolic. Eng et al. [71] evaluated the assistance of surfactant for glycyrrhizin and ephedrine PHWE. They found that adding anionic surfactant such as SDS to the extraction water enhanced the solubility of the targeted compounds into the mobile phase and therefore higher extraction yields were obtained.

13.5.2 Laboratory- and Pilot Plant-Scale PHWE Apparatus

There is not yet marketed PHWE apparatus to our knowledge. At laboratory scale, this type of equipment is normally designed and built by researcher teams themselves from a specific technical adaptation of already existing commercial equipments, such as for accelerated solvent extraction (ASE) or supercritical fluid extraction (SFE) [67].

Briefly, two major types of PHWE apparatus are built on the operating principles related to the static and the dynamic extraction modes.

Apparatus working on the dynamic extraction mode are composed of the following parts: a pressurised solvent tank to supply water to the system, a pump for pushing the solvent through the extraction cell containing the sample to be extracted, a heater device to provide the system with the desired operation temperature, the extraction cell consisting of a high-pressure-resistant cylinder where the solvent and the sample are put together into contact for extraction to occur, a pressure control device coupled with a back-pressure regulator and a collection vessel to recover the extract (Fig. 13.20). The extraction vessel is usually a stainless steel cylinder having 10–15 cm long and an internal diameter of 7–20 mm and 10 mL total volume. The high-pressure pump pressurises the water (extraction solvent) and pushes it through the sample at a constant flow rate. Temperature of the extraction vessel is maintained at the chosen value by various means such as GC ovens, sand baths or resistive heating blocks [69, 72–74].

In the extraction vessel, the sample has to be finely dispersed by mixing it in a powder form with a certain quantity of sand or other inert material to prevent any possible clogging during solvent percolation through this bed of mixed particles. As

Fig. 13.20 Schematic diagram for PHWE (Reprint from [72] with permission from Elsevier)

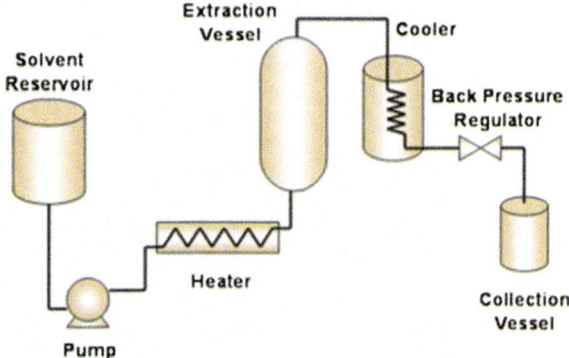

plant particles have a general tendency to absorb some quantity of water leading to some bed compression, the inert material added facilitated water percolation monitored by the pumping system during the course of the extraction.

Apparatus working on the static extraction mode are now marketed by Thermo Scientific (www.thermoscientific.com). An illustration of such apparatus is the commercial laboratory-scale ASE® equipment proposed as one model of the Dionex ASE system (www.dionex.com) (Fig. 13.21). Several models are offered: ASE 100 and ASE 150 system equipped with a single extraction cell, ASE 300 system with 12 cells, ASE 200 and ASE 350 systems with 24 cells. The newest model, Dionex ASE 350 system, is an apparatus that automatically extracts up to 24 samples (of 1–100 g each) and accommodates various cell sizes of 1, 5, 10, 22, 34, 66 and 100 mL volume.

Pilot plant-scale equipments for PHWE were scaled up from the design of the laboratory-scale apparatus [72]. Lagadec et al. [75] scaled up a PHWE system to remove contamination products from soils using a super large extraction vessel of 102 mm i.d. × 1,000 mm long, which size is about 10 times bigger than those of the laboratory-scale extraction vessel. The capacity of such a vessel is about 1,000 times compared to those of a laboratory-scale unit. This allows extraction of more than 8 kg of soil sample per extraction cycle, compared to only 8 g sample that can be extracted in a laboratory-scale apparatus. Water was heated by a propane heater and the extraction cell is maintained in temperature using a thermocouple-controlled heating tapes rather than using an oven. The hot water flow rate was set at 0.5 L/min for the pilot-scale extractor.

13.5.3 Application of PHWE at Pilot Plant Scale

Irene Rodríguez-Meizoso et al. [76] applied PHWE at pilot plant scale to extract antioxidant compounds from rosemary leaves. They developed an original system for PHWE, including an on-line drying system: continuous PHWE of rosemary leaves followed by a continuous production of an aerosol created from the extract

Fig. 13.21 Thermo Scientific™ Dionex™ ASE™ 350 Accelerated Solvent Extractor

by a supercritical CO_2 nebulisation system. This aerosol was injected in the particle formation chamber along with a hot N_2 gas flow which instantaneously dried the aerosol, and particles of extract compounds were immediately formed in the chamber (Fig. 13.22).

Water was pumped as the extraction solvent, into the extraction vessel, using a modified Suprex Modifier pump. Extraction cell and inlet/outlet tubing connected to the vessel were placed inside a GC oven (Carlo Erba Strumentazione, Milano, Italy) to maintain the extraction temperature at the required level (200 °C) that allowed maximal extraction of antioxidant-active compounds. The extraction cell was filled with a mixture of solid particles of grinded rosemary leaves and washed sea sand. The process starts by filling the extraction cell with water at room temperature. Then, the CO_2 injection and the heating systems are started together. When the starting working conditions were reached (80 bar, 2–3 mL/min CO_2 and 200 °C cell-temperature), N_2 injection was started at a pressure of 6–7 bar and water is pumped at a constant flow rate (0.1–0.3 mL/min) through the particle bed placed in the extraction cell. With such an extraction system, 10 g rosemary extract was obtained from 29.4 g of rosemary leaves using only 382 mL water. The extract obtained was enriched in carnosic acid, an antioxidant-active compound of rosemary leaf [75].

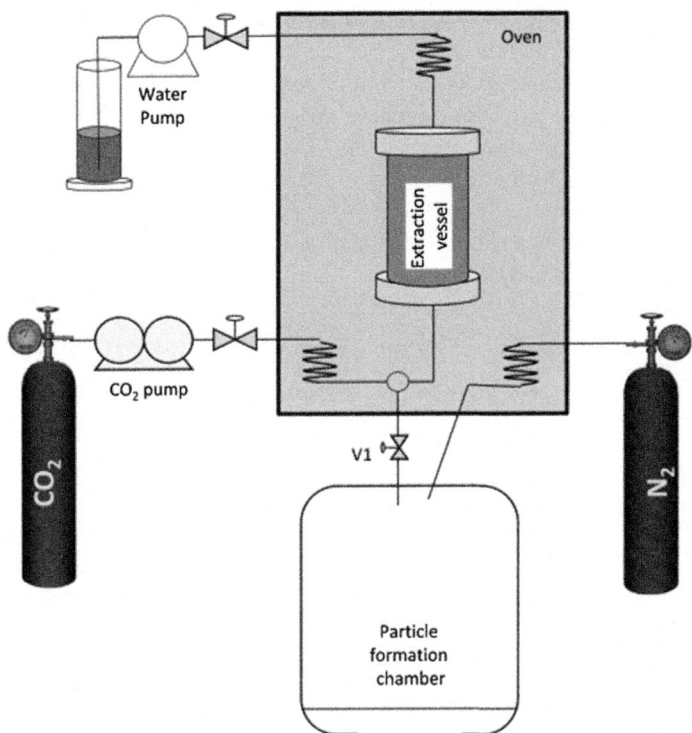

Fig. 13.22 Pilot plant-scale PHWE system developed by Irene Rodríguez-Meizoso et al. (Reprint from [75] with permission from Elsevier)

13.6 Membrane-Based Separation and Extraction

Classical water extraction (maceration, infusion, decoction, etc.) of solid sample material such as plant needs generally large volumes of water, and the bulk extract obtained is generally a mixture of water-soluble compounds, of macromolecules in colloidal state and of water-suspended insoluble particles. If the water-soluble compounds remained not only the interest extracted compounds, they have to be separated from other matters present in the extraction water. In further process steps, they can be purified and concentrated. Several separation techniques are available today to perform successfully all these additional technological steps that are often needed to complete a water-extraction process chain. Among them, membrane separation technology offers a certain number of advantages for the separation, the purification, and the concentration of very valuable and heat-sensitive water-extracted compounds. They can be applied in all these additional process steps. Membrane-based separation is an emerging technology adapted for the posttreatment of a global water extract since this technology can operate at room temperature and avoid any phase change of the extracted products. Moreover, this technology is claimed to be more energy-saving than conventional thermal separation/concentration processes such as evaporation, distillation, sublimation or crystallisation. Major membrane separation techniques (microfiltration, ultrafiltration, nanofiltration, reverse osmosis) have nowadays found lots of specific uses and applications in different industrial sectors. They are based on physical separation technique according to the filtration principle, leading, from a liquid medium or from a raw material extract process, to two output fractions, namely, permeate and retentate. Both of them can be used further to recover the valuable extracted products of interest they may contain.

13.6.1 General Considerations About Membrane Technology

With nearly 50 years of rapid technological development and progress, membrane-based processes enjoy today numerous of industrial applications that have brought great benefits to human life. These applications include water purification, dairy standardisation and stabilisation, sea and brackish water desalination, wastewater reclamation and reuse, food and beverage production, gas and vapour separation, energy conversion and storage, air pollution control and hazardous industrial waste treatment, hemodialysis, protein and microorganism separations, etc. The scope of membrane technology applications is still extending and is stimulated by numerous developments of novel or improved materials and separation membranes with better chemical, thermal and mechanical resistant properties and better permeability and selectivity characteristics, as well as by a significant decrease of capital and operation costs. Development of novel applications using membrane separation technology is however closely dependent of the future development of the heart of the membrane process: the membrane itself with new intrinsic and specific physical characteristics.

Generally speaking, a membrane is a barrier of a few hundred nanometres to several millimetres thick to separate two phases and to be able to allow a selective transfer of various components.

Separation membranes can be classified into two types, according to the internal structure of the material they are made of. The first type is the isotropic membrane group: they are microporous and non-porous membranes characterised by constant structural properties along the entire membrane thickness, i.e. pore sizes are small and relatively constant throughout the membrane thickness. In separation process, these membranes act as depth filters, the solution move by diffusion through the membrane and small particles in suspension in the solution may be retained in their internal structure, resulting in clogging the membrane and reducing filtration fluxes.

The second type is the anisotropic (asymmetric) membranes group: the membrane material shows a composite structure consisting of a number of layers, each with different structures, pore sizes, and permeabilities. The anisotropic membrane has a relatively dense, extremely thin and dense surface layer (i.e. the 'skin', also called the permselective layer) with constant pore sizes, which characterise the average pore size and the selectivity of the overall filtration membrane. The permselective layer is supported on a much thicker porous substructure showing good flux, to withstand the compressive forces used in the separation or filtration process. The thin layer is always on the high-pressure side of the membrane (the feed side). These membranes had the advantage of higher transfer fluxes, and almost all industrial processes use such membranes.

The liquid membranes can be also placed in this group. They consist of a liquid phase (e.g. a thin oil film), either in supported or unsupported forms that serve as a membrane barrier between two phases of aqueous solutions or gas mixtures.

Membrane separations are physical separation, compared to other separation and concentration techniques. Membrane separations are attractive for industrial because (1) membrane processes are suitable for filtration of liquids containing sensitive products. The filter is a physical membrane that operates without addition of any chemicals and is an absolute barrier for many types of compounds. Concentration of biological, nutritional and organoleptic compounds at low temperature by membrane separation is more favourable than thermal evaporation operations; (2) the membrane unit is modular and it is easy to assemble several membrane units to scale up the useable membrane filtration surface from the laboratory-scale equipment (some cm^2) to industrial units with several hundred m^2 of membrane surface. At this operating level, filtration and membrane cleaning can be conducted in a continuous automated process with often efficient energy consumption; (3) membrane separation can operate in different process modes (continuous, batch, multi-stages) that can be also coupled with other technological unit operations.

Transport rate of species through the membrane (permeation) is achieved by applying a driving force across the membrane. The flow across the membrane can be driven by application of mechanical, chemical or electrical forces that can be hydrostatic or vapour pressure, concentration gradient, temperature or electrical potential. The way by which the material and the solution are transported across a membrane gives a broad classification of the separation membranes [77]:

- Pressure-driven processes, such as in microfiltration (MF), nanofiltration (NF), ultrafiltration (UF), reverse osmosis (RO) or in gas separation (GS), or partial-pressure-driven processes, such as in pervaporation (PV)
- Concentration-gradient-driven processes, such as in dialysis
- Temperature-driven processes, such as in membrane distillation (MD)
- Electrical-potential-driven processes, such as in electrodialysis (ED)

Considering the temperature-sensitive biological activities of some water-extracted natural products, pressure-driven membrane processes are preferred for filtrating and concentrating such products. Depending on the membrane performances, often linked to the nominal membrane pore size, pressure-driven membrane separation process can be classified into four categories: microfiltration (MF), ultrafiltration (UF), nanofiltration (NF) and reverse osmosis (RO). Their main characteristics are shown in Table 13.1.

Table 13.1 Classification and general characteristics of filtration membranes

Membrane uses	Microfiltration MF	Ultrafiltration UF	Nanofiltration NF	Reverse osmosis RO
Pore sizes	0.1–10 μm	1–100 nm	≤1 nm	<0.5 nm
	Porous membrane			Dense membrane
Osmotic pressure effect	Negligible	Very weak	Average to weak	Important
Specific transmembrane flux	100–1,500 $l \cdot h^{-1} \cdot m^{-2}$	40–200 $l \cdot h^{-1} \cdot m^{-2}$	50–100 $l \cdot h^{-1} \cdot m^{-2}$	10–60 $l \cdot h^{-1} \cdot m^{-2}$
Transmembrane pressure effect	Weak	Weak	Average	High
Usual operating pressure	0.1–3 bar	0.5–10 bar	4–20 bar	≥20 bar
Retention of →	Large size bacteria, yeast, particles	Bacteria, macromolecules, proteins, large size viruses	Viruses, 2-valent ions and molecules	Salts, small size organic molecules
Energy consumed	<0.5 $kwh \cdot m^{-3}$	<1 $kwh \cdot m^{-3}$	1–2 $kwh \cdot m^{-3}$	2–10 $kwh \cdot m^{-3}$

13.6.1.1 Microfiltration

When membrane filtration is used for the removal of larger particles, microfiltration and ultrafiltration are applied. Because of the open character (pores) of the membranes, the productivity is high, while the pressure difference applied between the membrane sides is low.

MF membranes are used for separation of particles with a size range of 0.1–10 μm (impurities, viruses and bacteria) from a solvent or a water-extract solution. The separation mechanism is based on a sieving effect of the membrane pores, and particles are separated according to their dimensions although some charge or adsorptive separation is possible. In MF process, the pressure applied is quite low ($P < 3$ bar) compared to that used in other filtration processes [78]. MF membranes were mainly used for sterilisation by filtration in the pharmaceutical industry (removal of microorganisms) or for final cleaning of rinse water in the semiconductor industry (removal of undesired particles). MF was also easily and economically used in cold sterilisation of beer and wine, as well as clarification of cider and other cloudy juices. Both organic and inorganic materials can be used for manufacturing microfiltration membranes. Most organic membranes are currently made of organic polymers (cellulose acetate, polysulfone or polyamide) whose qualities confer a great adaptability to different applications. Mineral membranes are totally made of a mineral matter (e.g. ceramic membrane), so they can be used within a large temperature range and a wide domain of mechanical constraints and even aggressive chemical media.

These membranes can be used according to the two main filtration configurations: cross-flow and dead-end filtration modes [79].

In the cross-flow filtration mode, the feed flow is tangential to the surface of membrane, the retained retentate is removed from the same membrane side, whereas the permeate flow, going through the membrane, is recovered on the other membrane side.

When using a dead-end filtration mode, all the fluid passes in a direction substantially perpendicular to the membrane surface, and all particles larger than the pore sizes of the membrane are stopped at its surface. The trapped particles prevent other contaminants from entering and passing through the membrane by building up a 'filter cake' on the surface of the membrane which reduces the efficiency of the filtration process until the filter cake is washed away in back flushing. The main disadvantage of a dead-end filtration is the extensive membrane fouling and concentration polarisation, and the process is a batch-type process which is easy to implement and usually cheaper than the cross-flow membrane filtration [80].

13.6.1.2 Ultrafiltration

UF membranes were firstly manufactured with the initial goal of producing high-flux RO membranes. The first commercial UF membranes were introduced in the mid-1960s by Millipore and Amicon (www.millipore.com) as a spin-off of the development of asymmetric RO membranes. In the UF process, no significant osmotic pressure is generated across the UF membranes because of the membrane porous structure (pore size 1–100 nm) which allows permeation of micro-solutes (molecular weights < 300 Da) through the membranes [81]. UF membranes have an asymmetric porous structure and are often prepared by the phase-inversion process. UF membranes are used to retain macromolecules, colloids and solutes with molecular weight larger than 10,000. These chemical species may produce an osmotic pressure of only a few bar. Thus, the driving force in UF is mainly the hydrostatic pressure applied against one side of the membrane (0.5–10 bar). The selectivity of UF membranes depends on size and surface charge differences among compounds to be separated, on the membrane physical properties and on the hydrodynamic conditions applied.

13.6.1.3 Nanofiltration

The term nanofiltration was introduced by FilmTec (www.dowwaterandprocess.com) in the second half of the 1980s to describe a type of 'RO process' that allows some feed water ionic solutes to permeate selectively through the separation membrane, using a pressure gradient. NF spans the gap in particle size between UF and RO. The size of the solutes excluded in this process is of the order of 1 nm, while water and non-charged compounds with a molecular weight < 200 Da are able to permeate the semipermeable separation layer of the membrane. Different from RO membranes which have a non-porous structure and a transport mechanism of solution-diffusion, NF membranes operate at the interface of porous and non-porous membranes with both sieving and diffusion transport mechanisms. Therefore, it was acknowledged that NF performed an intermediate capability as 'loose' RO (non-porous, diffusion) or 'tight' UF (porous, sieving) [82, 83].

13.6.1.4 Reverse Osmosis

Reverse osmosis (RO) membranes do not work according to the principle of pores governing separations by microfiltration and ultrafiltration. Separation takes place by diffusion through the RO membrane. The pressure that is required to perform RO is much higher than the pressure required for MF and UF, while productivity is much lower.

Reverse osmosis (RO) is based on the diffusion principle and occurs when the water is moved across the membrane against the concentration gradient, from lower concentration to higher concentration. The principle of RO resulted from the application of a pressure against the opposing osmotic pressure generated by a solution containing solutes (P_s) to force the flow of water (P_w) in the opposite direction to the natural direction generated by the difference between osmotic pressures created by the two solutions ($P_s > P_w$). Pure water flows from the more

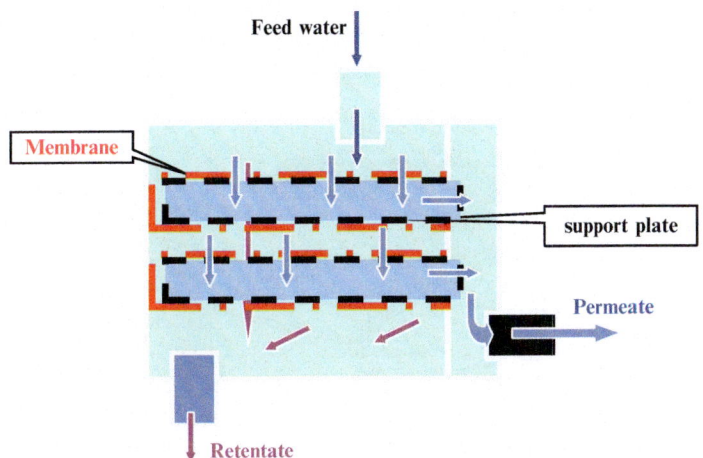

Fig. 13.23 Schema of a plate-and-frame module

concentrated to the less concentrated solution. RO is a membrane technology generally used to concentrate water extracts obtained from previous extraction techniques.

As RO membranes retained most of water-soluble compounds and salts, including the small monovalent ions, it is one of the methods used to desalinate seawater. RO membranes are generally categorised into asymmetric membranes and thin-film or composite membranes. An asymmetric RO membrane shows a multilayer structure made from one polymer material and has a thin, selective skin layer supported by a more porous sub-layer.

13.6.2 Typical Membrane Modules at Pilot Plant and Industrial Scales

Large membrane areas and small volumes are required for industrial applications in membrane processes. Membrane units set together into membrane modules are the practical solution. The module is the base for membrane installation and process design. Four main types of modules, depending on the supported membrane, can be distinguished as follows [84]:

- *Plate-and-frame module* is the oldest and simplest module. Sets of two membranes are placed in a sandwich-like fashion with their feed sides facing each other. In each feed and permeate compartment, a suitable spacer is placed. The number of sets needed for a given membrane area furnished with sealing rings and two end plates is then built up to a plate-and-frame stack. The membrane permeate is collected from each support plate, as shown in Fig. 13.23.

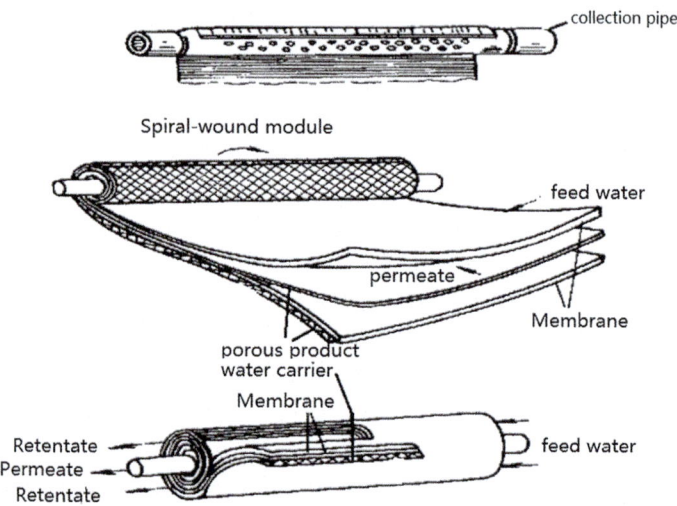

Fig. 13.24 Schema of a spiral-wound module

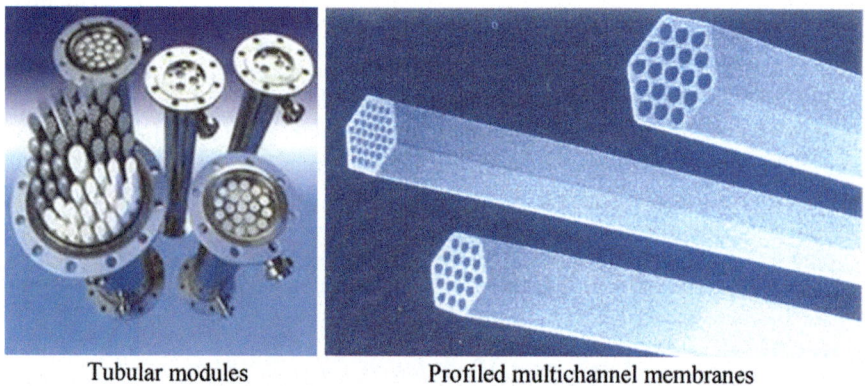

Fig. 13.25 Tubular filtration modules equipped with membranes and ceramic profiled membrane units

- *Spiral-wound module* is a rolled plate-and-frame module around a central collection pipe as shown in Fig. 13.24. Membrane and permeate-side spacer material are then glued along three edges to build a membrane envelope. The feed-side spacer separating the top layer of the two flat membranes acts also as a turbulence promoter. The feed flows axial through the cylindrical module parallel along the central pipe and the permeate flows radially towards the central pipe. The spiral-wound module has a compact structure and large membrane area per unit volume. It is easy to operate. The disadvantage is that the feed water must be clarified to prevent fouling.
- *Tubular module* is shown in Fig. 13.25. The feed solution always flows through the centre of the tubes, while the permeate flows through the porous supporting

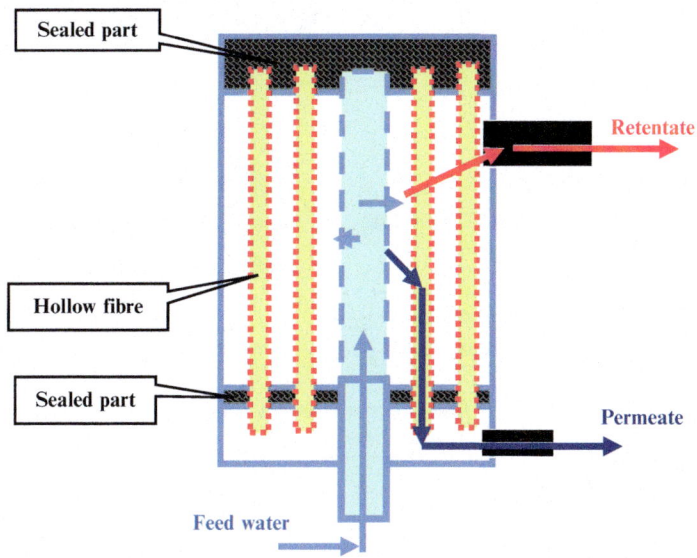

Fig. 13.26 Schema of a hollow-fibre filtration module

tube into the module housing. Profiled and multichannel ceramic membranes are mostly assembled in such tubular module configurations. The main advantages of the tubular module are usefulness and cleanness, but there is a major disadvantage for large energy consumers about its reduced exchange surface per unit volume (reduced compactness).

- *Hollow-fibre module* consists of a set of hollow fibres of diameter less than one micrometre assembled together in a module, as shown in Fig. 13.26. The free ends of the fibres are often potted with agents such as epoxy resins, polyurethanes or silicon rubber. The membranes are self-supporting for this module. This configuration provides the highest flow per module density.

The choice of the module configuration, as well as the arrangement of the modules in a system, depends on economic considerations with correct engineering parameters being employed to achieve this, which include the type of separation problem; the eases of cleaning, maintenance and operation; the compactness and scale of the system; and the possibility of membrane replacement [85].

13.6.3 Application of Membrane Technology at Pilot Scale

Perilla frutescens is an edible plant frequently used as one of the most popular spices and food colourants in some Asian countries such as China and Japan. Water extract of Perilla contains abundant polyphenols including anthocyanins, flavones

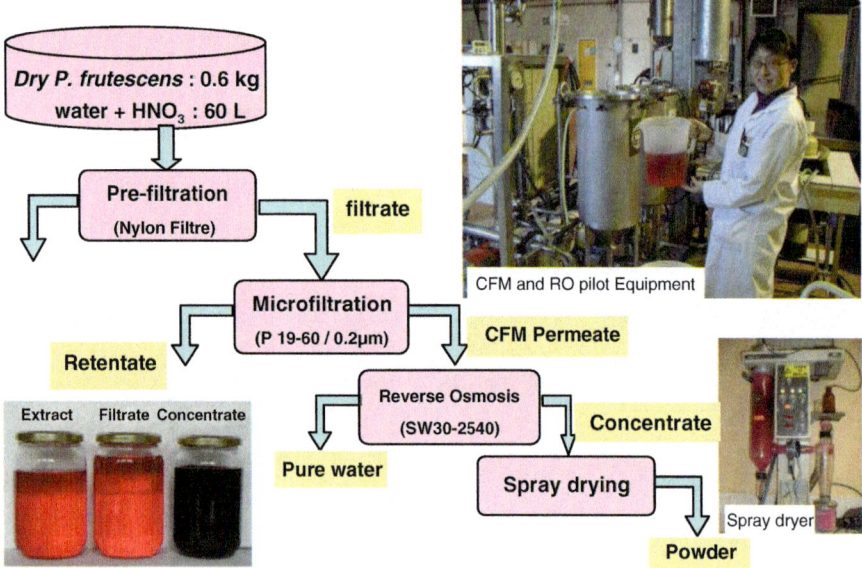

Fig. 13.27 The membrane process of perilla extracts production

and phenylpropanoids [86] that showed antioxidant activity [87]. Considering the heat sensibility of polyphenol compounds contained in this plant, especially anthocyanins, one of the natural colourants which can be used in food and pharmaceutical industry, the sterilisation and concentration of the water perilla extract should avoid the heat treatment. A membrane process including cross-flow microfiltration (CFM) and reverse osmosis (as shown in Fig. 13.27) was developed to clarify, sterilise and concentrate the perilla extract at pilot plant scale by Meng et al. [88].

The dry leaves of perilla were extracted by acidified water using a ratio of 1:100 (g/mL) in room temperature overnight. Then, the extract was pre-filtrated by a nylon cloth filter. After pre-filtration, a system of single-stage continuous feed and bleed loop configuration (TIA, Bollène, France) CFM was applied to clarify and sterilise the perilla extract. The multichannel ceramic membrane used was a P 19–60 (Membralox) industrial-type membrane, 800 mm long, 0.2 μm average pore size, with a total filtration surface of 0.304 m^2 (Pall-Exekia, Tarbes, France). The transmembrane pressure was set at 0.6 bar during the operation and the feed flow was controlled at 4.5 m/s. The CFM permeate flux stabilised rapidly after the start of the CFM to an average value of 150 L/h/m^2. The concentration of extract was realised by RO. The RO membrane used was of an industrial type, SW 30–2540 composite polymeric membrane, packed in a spiral-wound configuration (Filmtec), with 2 m^2 of filtration surface. The process was kept going at a constant transmembrane pressure of 40 bar until the volume of the RO retentate reached the value of the dead volume of the RO unit (3 L). The flux of the RO permeate (pure water) showed an immediate stabilisation at the value of 22 L/h/m^2 and stayed constant at this level for more than 45 min of operation. Finally, the CFM permeate was concentrated 9.4 times by RO. HPLC analysis of the polyphenol compositions in the extracts (before and after concentration) showed CFM, and RO process did not make the degradation of the thermo-sensibility compounds. Finally, using a spray-dryer, the concentrated extract was totally dried and made into powder which was a stable antioxidant-active red product, with a long shelf-life.

The same general process chain was successfully applied to water extraction of different plant materials, including leaves and flowers from tropical trees, plants or herbs traditionally used in the local medicine. Therefore, concentrated water extracts have been prepared from vegetal material used by African traditional healers to prepare some local medicine or healthy beverage. Extraction-concentration process started with plant water diffusion followed by membrane purification–concentration. Membrane technology is used here for two purposes in the process: cleaning the extraction water before using it in the diffusion step and purifying the water extract (bacteria-free) and concentrating it for a better shelf-life in local conditions and for making local marketing easier than using a single-strength water extract.

Hibiscus sabdariffa flowers, *Delonix regia* flowers, *Justicia secunda* leaves and *Tectona grandis* leaves were some of the traditional African plants that have been processed with modern pilot-scale technology, in a way mimicking traditional preparation recipes delivered by local practitioners [89–91].

13.7 Conclusion

Although water is the 'best' and the safest solvent of the world, it was not yet applied usually in industrial extractions because of its low extraction efficiency towards many other non-water-soluble and valuable compounds that can be extracted from the worldwide biodiversity and the difficulty to concentrate water solutions. Many innovative technologies were developed to improve the efficiency of water extractions. Ultrasound-assisted extraction has already been applied in industrial processes such as the extraction of Chinese traditional medicine. Others technologies still remain operational only at the laboratory and pilot plant scales, and should need more research and development efforts to raise them to technically and economically viable applications for the industry. Regarding the problem of water-extract concentration, nanofiltration and reverse osmosis technologies may provide interesting alternative solutions depending on the value added to the water-extracted product and its potential uses in various industrial sectors, as a marketable finished product or as a raw ingredient for manufacturing others final products. With the additional use of membrane technologies in water extraction of various raw materials, the water-extracts can be eco-friendly concentrated at room temperature without extra-use of organic solvents or chemicals.

References

1. Chemat F, Zill-e-Huma, Khan MK (2011) Applications of ultrasound in food technology: processing, preservation and extraction. Ultrason Sonochem 18(4):813–835. doi:http://dx.doi.org/10.1016/j.ultsonch.2010.11.023
2. Da Porto C, Porretto E, Decorti D (2013) Comparison of ultrasound-assisted extraction with conventional extraction methods of oil and polyphenols from grape (Vitis vinifera L.) seeds. Ultrason Sonochem 20(4):1076–1080. doi:http://dx.doi.org/10.1016/j.ultsonch.2012.12.002
3. Corrales M, Toepfl S, Butz P, Knorr D, Tauscher B (2008) Extraction of anthocyanins from grape by-products assisted by ultrasonics, high hydrostatic pressure or pulsed electric fields: a comparison. Innov Food Sci Emerg Technol 9(1):85–91. doi:http://dx.doi.org/10.1016/j.ifset.2007.06.002
4. U&STAR Ultrasonic Technology Co. Ltd (2014) Ultrasonic sonochemistry processor system, lab, bench scale, industrial scale, commercial application machines, equipment, reactors, devices. http://www.ustar-ultrasonic.com/product/Ultrasonic_Sonochemistry_Processor_System_Lab_Bench_Scale_Industrial_Scale_Commercial_Application_Machines_Equipment_Reactors_Devices.html
5. Pacheco MGC, Garreton LFJG, Hernández YDPV, Camacho JGA, Lobo JIS (2011) Device and method for extracting active principles from natural sources, using a counter-flow extractor assisted by a sound transduction system. Spanish Patent WO2011079404 A1
6. Luo D-l, Zeng X-y, Xu B-c, Qiu T-q (2009) Design and analysis of dual-frequency ultrasound dynamic countercurrent extraction apparatus. Tech Acoustics (Chin) 28(4):488–490
7. Knorr D, Froehling A, Jaeger H, Reineke K, Schlueter O, Schoessler K (2011) Emerging technologies in food processing. Annu Rev Food Sci Technol 2:203–235
8. Chemat F, Grondin I, Shum Cheong Sing A, Smadja J (2004) Deterioration of edible oils during food processing by ultrasound. Ultrason Sonochem 11(1):13–15. doi:http://dx.doi.org/10.1016/S1350-4177(03)00127-5
9. Pingret D, Fabiano-Tixier A-S, Bourvellec CL, Renard CMGC, Chemat F (2012) Lab and pilot-scale ultrasound-assisted water extraction of polyphenols from apple pomace. J Food Eng 111(1):73–81. doi:http://dx.doi.org/10.1016/j.jfoodeng.2012.01.026
10. Khan MK, Abert-Vian M, Fabiano-Tixier A-S, Dangles O, Chemat F (2010) Ultrasound-assisted extraction of polyphenols (flavanone glycosides) from orange (Citrus sinensis L.) peel. Food Chem 119(2):851–858. doi:http://dx.doi.org/10.1016/j.foodchem.2009.08.046
11. Virot M, Tomao V, Le Bourvellec C, Renard CMCG, Chemat F (2010) Towards the industrial production of antioxidants from food processing by-products with ultrasound-assisted extraction. Ultrason Sonochem 17(6):1066–1074. doi:http://dx.doi.org/10.1016/j.ultsonch.2009.10.015
12. Sparr Eskilsson C, Björklund E (2000) Analytical-scale microwave-assisted extraction. J Chromatogr A 902(1):227–250. doi:http://dx.doi.org/10.1016/S0021-9673(00)00921-3
13. Costa SS, Gariepy Y, Rocha SCS, Raghavan V (2014) Microwave extraction of mint essential oil – temperature calibration for the oven. J Food Eng 126:1–6. doi:http://dx.doi.org/10.1016/j.jfoodeng.2013.10.033
14. Lucchesi ME, Smadja J, Bradshaw S, Louw W, Chemat F (2007) Solvent free microwave extraction of Elletaria cardamomum L.: a multivariate study of a new technique for the extraction of essential oil. J Food Eng 79(3):1079–1086. doi:http://dx.doi.org/10.1016/j.jfoodeng.2006.03.029
15. Zhang H-F, Yang X-H, Wang Y (2011) Microwave assisted extraction of secondary metabolites from plants: current status and future directions. Trends Food Sci Technol 22(12):672–688. doi:http://dx.doi.org/10.1016/j.tifs.2011.07.003
16. Švarc-Gajić J, Stojanović Z, Segura Carretero A, Arráez Román D, Borrás I, Vasiljević I (2013) Development of a microwave-assisted extraction for the analysis of phenolic compounds from Rosmarinus officinalis. J Food Eng 119(3):525–532. doi:http://dx.doi.org/10.1016/j.jfoodeng.2013.06.030

17. Zhang H-F, Zhang X, Yang X-H, Qiu N-X, Wang Y, Wang Z-Z (2013) Microwave assisted extraction of flavonoids from cultivated Epimedium sagittatum: extraction yield and mechanism, antioxidant activity and chemical composition. Ind Crop Prod 50:857–865. doi:http://dx.doi.org/10.1016/j.indcrop.2013.08.017
18. Dahmoune F, Boulekbache L, Moussi K, Aoun O, Spigno G, Madani K (2013) Valorization of Citrus limon residues for the recovery of antioxidants: evaluation and optimization of microwave and ultrasound application to solvent extraction. Ind Crop Prod 50:77–87. doi:http://dx.doi.org/10.1016/j.indcrop.2013.07.013
19. Ma F-Y, Gu C-B, Li C-Y, Luo M, Wang W, Zu Y-G, Li J, Fu Y-J (2013) Microwave-assisted aqueous two-phase extraction of isoflavonoids from Dalbergia odorifera T. Chen leaves. Sep Purif Technol 115:136–144. doi:http://dx.doi.org/10.1016/j.seppur.2013.05.003
20. Coelho E, Rocha MAM, Saraiva JA, Coimbra MA (2014) Microwave superheated water and dilute alkali extraction of brewers' spent grain arabinoxylans and arabinoxylo-oligosaccharides. Carbohydr Polym 99:415–422. doi:http://dx.doi.org/10.1016/j.carbpol.2013.09.003
21. Bousbia N, Abert Vian M, Ferhat MA, Petitcolas E, Meklati BY, Chemat F (2009) Comparison of two isolation methods for essential oil from rosemary leaves: hydrodistillation and microwave hydrodiffusion and gravity. Food Chem 114(1):355–362. doi:http://dx.doi.org/10.1016/j.foodchem.2008.09.106
22. Périno-Issartier S, Ginies C, Cravotto G, Chemat F (2013) A comparison of essential oils obtained from lavandin via different extraction processes: ultrasound, microwave, turbohydrodistillation, steam and hydrodistillation. J Chromatogr A 1305:41–47. doi:http://dx.doi.org/10.1016/j.chroma.2013.07.024
23. Li J, Zu Y-G, Fu Y-J, Yang Y-C, Li S-M, Li Z-N, Wink M (2010) Optimization of microwave-assisted extraction of triterpene saponins from defatted residue of yellow horn (Xanthoceras sorbifolia Bunge.) kernel and evaluation of its antioxidant activity. Innov Food Sci Emerg Technol 11(4):637–643. doi:http://dx.doi.org/10.1016/j.ifset.2010.06.004
24. Prakash Maran J, Sivakumar V, Thirugnanasambandham K, Sridhar R (2014) Microwave assisted extraction of pectin from waste Citrullus lanatus fruit rinds. Carbohydr Polym 101:786–791. doi:http://dx.doi.org/10.1016/j.carbpol.2013.09.062
25. Teo C, Chong W, Ho Y (2013) Development and application of microwave-assisted extraction technique in biological sample preparation for small molecule analysis. Metabolomics 9(5):1109–1128. doi:10.1007/s11306-013-0528-7
26. Jiao J, Li Z-G, Gai Q-Y, Li X-J, Wei F-Y, Fu Y-J, Ma W (2014) Microwave-assisted aqueous enzymatic extraction of oil from pumpkin seeds and evaluation of its physicochemical properties, fatty acid compositions and antioxidant activities. Food Chem 147:17–24. doi:http://dx.doi.org/10.1016/j.foodchem.2013.09.079
27. Vessel Technology. http://www.milestonesci.com/ethos-ex/ex-vessel-technology.html
28. Filly A, Fernandez X, Minuti M, Visinoni F, Cravotto G, Chemat F (2014) Solvent-free microwave extraction of essential oil from aromatic herbs. From laboratory to pilot and industrial scale. Food Chem 150:193–198. doi:http://dx.doi.org/10.1016/j.foodchem.2013.10.139
29. Huang K, Wang J (2009) Designs of pulsed electric fields treatment chambers for liquid foods pasteurization process: a review. J Food Eng 95(2):227–239. doi:http://dx.doi.org/10.1016/j.jfoodeng.2009.06.013
30. Donsì F, Ferrari G, Pataro G (2010) Applications of pulsed electric field treatments for the enhancement of mass transfer from vegetable tissue. Food Eng Rev 2(2):109–130
31. Loginova KV, Shynkaryk MV, Lebovka NI, Vorobiev E (2010) Acceleration of soluble matter extraction from chicory with pulsed electric fields. J Food Eng 96(3):374–379. doi:http://dx.doi.org/10.1016/j.jfoodeng.2009.08.009
32. López N, Puértolas E, Condón S, Raso J, Ignacio Á (2009) Enhancement of the solid–liquid extraction of sucrose from sugar beet (Beta vulgaris) by pulsed electric fields. LWT Food Sci Technol 42(10):1674–1680. doi:http://dx.doi.org/10.1016/j.lwt.2009.05.015

33. López N, Puértolas E, Condón S, Raso J, Alvarez I (2009) Enhancement of the extraction of betanine from red beetroot by pulsed electric fields. J Food Eng 90(1):60–66. doi:http://dx.doi.org/10.1016/j.jfoodeng.2008.06.002
34. Yongguang Y, Yuzhu H, Yong H (2006) Pulsed electric field extraction of polysaccharide from Rana temporaria chensinensis David. Int J Pharm 312(1–2):33–36. doi:http://dx.doi.org/10.1016/j.ijpharm.2005.12.021
35. Puértolas E, Hernández-Orte P, Sladaña G, Álvarez I, Raso J (2010) Improvement of winemaking process using pulsed electric fields at pilot-plant scale. Evolution of chromatic parameters and phenolic content of Cabernet Sauvignon red wines. Food Res Int 43(3):761–766. doi:http://dx.doi.org/10.1016/j.foodres.2009.11.005
36. Puértolas E, Luengo E, Álvarez I, Raso J (2012) Improving mass transfer to soften tissues by pulsed electric fields: fundamentals and applications. Annu Rev Food Sci Technol 3(1):263–282
37. Loginova KV, Lebovka NI, Vorobiev E (2011) Pulsed electric field assisted aqueous extraction of colorants from red beet. J Food Eng 106(2):127–133. doi:http://dx.doi.org/10.1016/j.jfoodeng.2011.04.019
38. Chalermchat Y, Fincan M, Dejmek P (2004) Pulsed electric field treatment for solid–liquid extraction of red beetroot pigment: mathematical modelling of mass transfer. J Food Eng 64(2):229–236. doi:http://dx.doi.org/10.1016/j.jfoodeng.2003.10.002
39. Puértolas E, Cregenzán O, Luengo E, Álvarez I, Raso J (2013) Pulsed-electric-field-assisted extraction of anthocyanins from purple-fleshed potato. Food Chem 136(3–4):1330–1336. doi:http://dx.doi.org/10.1016/j.foodchem.2012.09.080
40. Abdullah SH, Zhao S, Mittal GS, Baik O-D (2012) Extraction of podophyllotoxin from Podophyllum peltatum using pulsed electric field treatment. Sep Purif Technol 93:92–97. doi:http://dx.doi.org/10.1016/j.seppur.2012.04.002
41. Goettel M, Eing C, Gusbeth C, Straessner R, Frey W (2013) Pulsed electric field assisted extraction of intracellular valuables from microalgae. Algal Res 2:401–408. doi:http://dx.doi.org/10.1016/j.algal.2013.07.004
42. Moubarik A, El-Belghiti K, Vorobiev E (2011) Kinetic model of solute aqueous extraction from Fennel (Foeniculum vulgare) treated by pulsed electric field, electrical discharges and ultrasonic irradiations. Food Bioprod Process 89(4):356–361. doi:http://dx.doi.org/10.1016/j.fbp.2010.09.002
43. Loginova KV, Vorobiev E, Bals O, Lebovka NI (2011) Pilot study of countercurrent cold and mild heat extraction of sugar from sugar beets, assisted by pulsed electric fields. J Food Eng 102(4):340–347. doi:http://dx.doi.org/10.1016/j.jfoodeng.2010.09.010
44. López N, Puértolas E, Hernández-Orte P, Álvarez I, Raso J (2009) Effect of a pulsed electric field treatment on the anthocyanins composition and other quality parameters of Cabernet Sauvignon freshly fermented model wines obtained after different maceration times. LWT Food Sci Technol 42(7):1225–1231. doi:http://dx.doi.org/10.1016/j.lwt.2009.03.009
45. Puértolas E, Saldaña G, Condón S, Álvarez I, Raso J (2010) Evolution of polyphenolic compounds in red wine from Cabernet Sauvignon grapes processed by pulsed electric fields during aging in bottle. Food Chem 119(3):1063–1070. doi:http://dx.doi.org/10.1016/j.foodchem.2009.08.018
46. Puértolas E, López N, Saldaña G, Álvarez I, Raso J (2010) Evaluation of phenolic extraction during fermentation of red grapes treated by a continuous pulsed electric fields process at pilot-plant scale. J Food Eng 98(1):120–125. doi:http://dx.doi.org/10.1016/j.jfoodeng.2009.12.017
47. Puértolas E, López N, Condón S, Álvarez I, Raso J (2010) Potential applications of PEF to improve red wine quality. Trends Food Sci Technol 21(5):247–255. doi:http://dx.doi.org/10.1016/j.tifs.2010.02.002
48. Duan M-H, Luo M, Zhao C-J, Wang W, Zu Y-G, Zhang D-Y, Fu Y-J, Yao X-H (2013) Ionic liquid-based negative pressure cavitation-assisted extraction of three main flavonoids from the pigeonpea roots and its pilot-scale application. Sep Purif Technol 107:26–36. doi:http://dx.doi.org/10.1016/j.seppur.2013.01.003

49. Zhang D-Y, Zu Y-G, Fu Y-J, Luo M, Wang W, Gu C-B, Zhao C-J, Jiao J, Efferth T (2012) Enzyme pretreatment and negative pressure cavitation extraction of genistein and apigenin from the roots of pigeon pea [Cajanus cajan (L.) Millsp.] and the evaluation of antioxidant activity. Ind Crop Prod 37(1):311–320. doi:http://dx.doi.org/10.1016/j.indcrop.2011.12.026
50. Zhang D-Y, Zhang S, Zu Y-G, Fu Y-J, Kong Y, Gao Y, Zhao J-T, Efferth T (2010) Negative pressure cavitation extraction and antioxidant activity of genistein and genistin from the roots of pigeon pea [Cajanus cajan (L.) Millsp.]. Sep Purif Technol 74(2):261–270. doi:http://dx.doi.org/10.1016/j.seppur.2010.06.015
51. Liu W, Fu Y, Zu Y, Kong Y, Zhang L, Zu B, Efferth T (2009) Negative-pressure cavitation extraction for the determination of flavonoids in pigeon pea leaves by liquid chromatography–tandem mass spectrometry. J Chromatogr A 1216(18):3841–3850. doi:http://dx.doi.org/10.1016/j.chroma.2009.02.073
52. Zhang D-Y, Zu Y-G, Fu Y-J, Luo M, Gu C-B, Wang W, Yao X-H (2011) Negative pressure cavitation extraction and antioxidant activity of biochanin A and genistein from the leaves of Dalbergia odorifera T. Chen. Sep Purif Technol 83:91–99. doi:http://dx.doi.org/10.1016/j.seppur.2011.09.017
53. Deng C, Yao N, Wang A, Zhang X (2005) Determination of essential oil in a traditional Chinese medicine, Fructus amomi by pressurized hot water extraction followed by liquid-phase microextraction and gas chromatography–mass spectrometry. Anal Chim Acta 536(1–2):237–244. doi:http://dx.doi.org/10.1016/j.aca.2004.12.044
54. Ndlela SC, Moura JMLN, Olson NK, Johnson LA (2012) Aqueous extraction of oil and protein from soybeans with subcritical water. J Am Oil Chem Soc 89(6):1145–1153. doi:10.1007/s11746-011-1993-7
55. Benito-Román Ó, Alonso E, Cocero MJ (2013) Pressurized hot water extraction of β-glucans from waxy barley. J Supercrit Fluids 73:120–125. doi:http://dx.doi.org/10.1016/j.supflu.2012.09.014
56. Chao Z, Ri-fu Y, Tai-qiu Q (2013) Ultrasound-enhanced subcritical water extraction of polysaccharides from Lycium barbarum L. Sep Purif Technol 120:141–147. doi:http://dx.doi.org/10.1016/j.seppur.2013.09.044
57. Pongnaravane B, Goto M, Sasaki M, Anekpankul T, Pavasant P, Shotipruk A (2006) Extraction of anthraquinones from roots of Morinda citrifolia by pressurized hot water: antioxidant activity of extracts. J Supercrit Fluids 37(3):390–396. doi:http://dx.doi.org/10.1016/j.supflu.2005.12.013
58. Cacace JE, Mazza G (2006) Pressurized low polarity water extraction of lignans from whole flaxseed. J Food Eng 77(4):1087–1095. doi:http://dx.doi.org/10.1016/j.jfoodeng.2005.08.039
59. Yang Y, Kayan B, Bozer N, Pate B, Baker C, Gizir AM (2007) Terpene degradation and extraction from basil and oregano leaves using subcritical water. J Chromatogr A 1152 (1–2):262–267. doi:http://dx.doi.org/10.1016/j.chroma.2006.11.037
60. Ko MJ, Cheigh CI, Chung MS (2014) Relationship analysis between flavonoids structure and subcritical water extraction (SWE). Food Chem 143:147–155. doi:10.1016/j.foodchem.2013.07.104
61. Ju ZY, Howard LR (2005) Subcritical water and sulfured water extraction of anthocyanins and other phenolics from dried red grape skin. J Food Sci 70(4):S270–S276
62. Tran Nguyen PL, Go AW, Huynh LH, Ju Y-H (2013) A study on the mechanism of subcritical water treatment to maximize extractable cellular lipids. Biomass Bioenergy 59:532–539. doi:http://dx.doi.org/10.1016/j.biombioe.2013.08.031
63. Schantz M (2006) Pressurized liquid extraction in environmental analysis. Anal Bioanal Chem 386(4):1043–1047. doi:10.1007/s00216-006-0648-2
64. Teo CC, Tan SN, Yong JWH, Hew CS, Ong ES (2008) Evaluation of the extraction efficiency of thermally labile bioactive compounds in Gastrodia elata Blume by pressurized hot water extraction and microwave-assisted extraction. J Chromatogr A 1182(1):34–40. doi:http://dx.doi.org/10.1016/j.chroma.2008.01.011

65. Heng MY, Tan SN, Yong JWH, Ong ES (2013) Emerging green technologies for the chemical standardization of botanicals and herbal preparations. TrAC Trends Anal Chem 50:1–10. doi:http://dx.doi.org/10.1016/j.trac.2013.03.012
66. Gil-Ramírez A, Mendiola JA, Arranz E, Ruíz-Rodríguez A, Reglero G, Ibáñez E, Marín FR (2012) Highly isoxanthohumol enriched hop extract obtained by pressurized hot water extraction (PHWE). Chemical and functional characterization. Innov Food Sci Emerg Technol 16:54–60. doi:http://dx.doi.org/10.1016/j.ifset.2012.04.006
67. Kronholm J, Hartonen K, Riekkola M-L (2007) Analytical extractions with water at elevated temperatures and pressures. TrAC Trends Anal Chem 26(5):396–412. doi:http://dx.doi.org/10.1016/j.trac.2007.03.004
68. Teo CC, Tan SN, Yong JWH, Hew CS, Ong ES (2010) Pressurized hot water extraction (PHWE). J Chromatogr A 1217(16):2484–2494. doi:http://dx.doi.org/10.1016/j.chroma.2009.12.050
69. Smith RM (2002) Extractions with superheated water. J Chromatogr A 975(1):31–46. doi:http://dx.doi.org/10.1016/S0021-9673(02)01225-6
70. Rovio S, Hartonen K, Holm Y, Hiltunen R, Riekkola ML (1999) Extraction of clove using pressurized hot water. Flavour Frag J 14(6):399–404. doi:10.1002/(sici)1099-1026(199911/12)14:6<399::aid-ffj851>3.0.co;2-a
71. Eng ATW, Heng MY, Ong ES (2007) Evaluation of surfactant assisted pressurized liquid extraction for the determination of glycyrrhizin and ephedrine in medicinal plants. Anal Chim Acta 583(2):289–295. doi:http://dx.doi.org/10.1016/j.aca.2006.09.019
72. Pronyk C, Mazza G (2009) Design and scale-up of pressurized fluid extractors for food and bioproducts. J Food Eng 95(2):215–226. doi:http://dx.doi.org/10.1016/j.jfoodeng.2009.06.002
73. Mustafa A, Turner C (2011) Pressurized liquid extraction as a green approach in food and herbal plants extraction: a review. Anal Chim Acta 703(1):8–18. doi:http://dx.doi.org/10.1016/j.aca.2011.07.018
74. Ong ES, Cheong JSH, Goh D (2006) Pressurized hot water extraction of bioactive or marker compounds in botanicals and medicinal plant materials. J Chromatogr A 1112(1–2):92–102. doi:http://dx.doi.org/10.1016/j.chroma.2005.12.052
75. Lagadec AJM, Miller DJ, Lilke AV, Hawthorne SB (2000) Pilot-scale subcritical water remediation of polycyclic aromatic hydrocarbon- and pesticide-contaminated soil. Environ Sci Technol 34(8):1542–1548. doi:10.1021/es990722u
76. Rodríguez-Meizoso I, Castro-Puyana M, Börjesson P, Mendiola JA, Turner C, Ibáñez E (2012) Life cycle assessment of green pilot-scale extraction processes to obtain potent antioxidants from rosemary leaves. J Supercrit Fluids 72:205–212. doi:http://dx.doi.org/10.1016/j.supflu.2012.09.005
77. Wang LK, Chen JP, Hung Y-T, Shammas NK (2011) Membrane and desalination technologies. Handbook of environmental engineering, vol 13. Humana Press Inc. NewYork.
78. Baker R (2004) Microfiltration. Membrane technology and applications, 2nd edn. Wiley, Weinheim.
79. Tamime AY (2013) Membrane processing: dairy and beverage applications. Willey-Blackwell, Weinheim.
80. Boerlage SFE (2001) Scaling and particulate fouling in membrane filtration systems. Swets and Zeitlinger Publishers, Lisse
81. Baker R (2004) Ultrafiltration. Membrane technology and applications, 2nd edn. Wiley, Weinheim.
82. Jitsuhara I, Kimura S (1983) Structure and properties of charged ultrafiltration membranes made of sulfonated polysulfone. J Chem Eng Jpn 16(5):389–393
83. Schäfer AI, Fane AG, Waite TD (2004) Nanofiltration: principles and applications. Elsevier, Oxford
84. Scott K, Hughes R (1996) Industrial membrane separation technology, revised edn. Chapman and Hall, Glasgow
85. Mulder M (1996) Basic principles of membrane technology, 2 edn. Kluwer Academic Publishers, Dordrecht.

86. Meng L, Lozano Y, Bombarda I, Gaydou EM, Li B (2009) Polyphenol extraction from eight Perilla frutescens cultivars. C R Chimie 12(5):602–611. doi:http://dx.doi.org/10.1016/j.crci.2008.04.011
87. Meng L, Lozano YF, Gaydou EM, Li B (2009) Antioxidant activities of polyphenols extracted from Perilla frutescens varieties. Molecules 14(1):133–140
88. Meng L, Lozano Y, Bombarda I, Gaydou E, Li B (2006) Anthocyanin and flavonoid production from Perilla frutescens: pilot plant scale processing including cross-flow microfiltration and reverse osmosis. J Agric Food Chem 54(12):4297–4303
89. Adjé FA, Lozano YF, Le Gernevé C, Lozano PR, Meudec E, Adima AA, Gaydou EM (2012) Phenolic acid and flavonol water extracts of Delonix regia red flowers. Ind Crop Prod 37(1):303–310. doi:http://dx.doi.org/10.1016/j.indcrop.2011.12.008
90. Koffi EN, Le Guernevé C, Lozano PR, Meudec E, Adjé FA, Bekro Y-A, Lozano YF (2013) Polyphenol extraction and characterization of Justicia secunda Vahl leaves for traditional medicinal uses. Ind Crop Prod 49:682–689. doi:http://dx.doi.org/10.1016/j.indcrop.2013.06.001
91. Adjé F, Lozano YF, Lozano P, Adima A, Chemat F, Gaydou EM (2010) Optimization of anthocyanin, flavonol and phenolic acid extractions from Delonix regia tree flowers using ultrasound-assisted water extraction. Ind Crop Prod 32(3):439–444. doi:http://dx.doi.org/10.1016/j.indcrop.2010.06.011

The manufacturer's authorised representative in the EU is Springer Nature Customer Service Centre GmbH, Europaplatz 3, 69115 Heidelberg, Germany. If you have any concerns regarding our products, please contact ProductSafety@springernature.com

Printed and bound by CPI Group (UK) Ltd, Croydon, CR0 4YY

23/03/2026

02076658-0003